Fundamentals of Biomechanics

Equilibrium, Motion, and Deformation

Second Edition

Springer
New York
Berlin
Heidelberg
Barcelona
Hong Kong
London
Milan
Paris
Singapore
Tokyo

Fundamentals of Biomechanics

Equilibrium, Motion, and Deformation

Second Edition

Nihat Özkaya   Margareta Nordin

Project Editor: Dawn L. Leger

With Forewords by Victor H. Frankel and Richard Skalak

With 500 Illustrations

Springer

Nihat Özkaya, Ph.D.
(deceased)
Occupational and Industrial Orthopaedic Center
Hospital for Joint Diseases Orthopaedic Institute
New York University Medical Center
63 Downing Street
New York, NY 10014-4331

Margareta Nordin, Dr. Sci.
Occupational and Industrial Orthopaedic Center
Hospital for Joint Diseases Orthopaedic Institute
New York University Medical Center
63 Downing Street
New York, NY 10014-4331

Project Editor:
Dawn L. Leger, Ph.D.
Occupational and Industrial Orthopaedic Center
Hospital for Joint Diseases Orthopaedic Institute
New York University Medical Center
63 Downing Street
New York, NY 10014-4331

Library of Congress Cataloging-in-Publication Data
Özkaya, Nihat, 1956–
    Fundamentals of biomechanics : equilibrium, motion, and
  deformation / Nihat Özkaya, Margareta Nordin. – 2nd ed.
      p.   cm.
  Includes bibliographical references and index.
  ISBN 0-387-98283-3 (hbk. : alk. paper)
  1. Biomechanics. I. Nordin, Margareta. II. Title .
QP303.O95   1998
612.7'6 – dc21   98-30564

Printed on acid-free paper.

Production managed by Terry Kornak; manufacturing supervised by Jacqui Ashri.
Typeset by TechBooks, Fairfax, VA.
Printed and bound by Maple-Vail Book Manufacturing Group, York, PA.
Printed in the United States of America.

9 8 7 6 5 4 3 2 1

ISBN 0-387-98283-3 Springer-Verlag New York Berlin Heidelberg SPIN 10632760

It was Dr. Özkaya's intention to dedicate this book to Professor Richard Skalak. We honor his choice, and remember both men as teachers, scholars, friends.

Richard Skalak *Nihat Özkaya*
1923–1997   1956–1998

Gone to the big classroom in the sky.

Foreword to the Second Edition

Biomechanics is a discipline utilized by different groups of professionals. It is a required basic science for orthopaedic surgeons, physiatrists, rheumatologists, physical and occupational therapists, and athletic trainers. These medical and paramedical specialists usually do not have a strong mathematics and physics background. Biomechanics must be presented to them in a rather nonmathematical way so that they may learn the concepts of mechanics without a rigorous mathematical approach.

On the other hand, many engineers work in fields in which biomechanics plays a significant role. Human factors engineering, ergonomics, biomechanics research, and prosthetic research and development all require that the engineers working in the field have a strong knowledge of biomechanics. They are equipped to learn biomechanics through a rigorous mathematical approach. Classical textbooks in the engineering fields do not approach the biological side of biomechanics.

The *Fundamentals of Biomechanics* by Drs. Nihat Özkaya and Margareta Nordin approaches biomechanics through a rigorous mathematical standpoint while emphasizing the biological side. This book will be very useful for engineers studying biomechanics and for medical specialists enrolled in courses who desire a more intensive study of biomechanics and are equipped through previous study of mathematics to develop a deeper comprehension of engineering as it applies to the human body.

This work was prepared in a combined clinical setting at the Hospital for Joint Diseases Orthopaedic Institute and teaching setting within the Program of Ergonomics and Biomechanics at New York University. The authors of this volume have the unique experience of teaching biomechanics in a clinical setting to professionals with diverse backgrounds. This work reflects their many years of classroom teaching, rehabilitation treatment, and practical and research experience.

Victor H. Frankel, M.D., Ph.D., K.N.O.
President
Hospital for Joint Diseases Orthopaedic Institute
New York University School of Medicine

Foreword to the First Edition

Bioengineering is a relatively recent field that has developed worldwide from the recognition that the theories and methods developed in conventional engineering are often useful for understanding and solving problems in physiology and medicine as well. Biomechanics is an important part of bioengineering, particularly for the field of orthopaedics. Biomechanics considers the applications of classical mechanics, including statics, dynamics, solid mechanics, and fluid mechanics to biological problems.

In the theoretical and practical advances that have been made to date, there has usually been a cooperation of a biological specialist or physician with an engineer or physicist who is thoroughly grounded in mathematics and physical sciences. Such a dialogue requires a certain amount of common vocabulary. The engineer must learn some anatomy and physiology to be useful. The medical personnel need to learn the basic concepts and vocabulary of the physical science and mathematics involved. This is often a difficult task and an impediment ot bonafide collaboration because most textbooks are writen for their own specialists and are not easily accessible to others.

This is where the present book will be useful. It provides a sound introduction to applied mechanics concepts that are useful in biomechanics. The basics given here will allow understanding of the results of more advanced theories than are presented here. It wil also serve as an introduction for engineers who know mechanics, but wish to explore other aspects of interest in biomechanics.

On a personal level, it is a pleasure to see this book come to fruition because it represents a second and successful stage in the development of bioengineering. Dr. Nihat Özkaya was a doctoral student at the Columbia University School of Engineering and Applied Sciences while I was a Professor there, and I can vouch for the rigor, accuracy, and care he has brought to all his work, including this book. It is a second stage in that most books and papers in bioengineering tend to be written by bioengineers for engineers or by biologists for biologists. But here is a genuine effort by an engineer to explain the foundations of mechanics in a manner suitable for study by biologists and medical personnel. This dialogue is enhanced by Dr. Margareta Nordin as a coauthor which also insures the accuracy of the medical content.

Finally, it is a pleasure to see a recent student publish a useful book. It makes the effort of being involved in education seem worthwhile and lends assurance that we are collectively on a useful track.

Richard Skalak, Ph.D., M.D. (Hon.)
Professor of Bioengineering
University of California, San Diego

Preface to the Second Edition

In the seven years since the publication of the first edition of *Fundamentals of Biomechanics*, the authors have used the book and refined its contents in accordance with feedback from students and other professors. There are substantial modifications to the new edition. More example related to biomechanics have been included, and exercise problems have been added.

Dr. Özkaya was devoted to teaching, and his dedication to his students is reflected in the refinements made in this new edition. Shortly before completing the manuscript, Dr. Özkaya passed away. As co-author, Dr. Nordin worked with Dr. Özkaya's editor and wife, Dr. Dawn Leger, to complete the book. The book could not have been completed without the assistance of Aydin Tözeren, Ph.D., of the Catholic University of America, and David Goldsheyder, MA, CIE, of the Hospital for Joint Diseases and New York University.

This text is divided into three parts. The first part (Chapters 1 through 5, and Appendices A and B) introduces the basic concepts of mechanics, provides the mathematical tools necessary to explain these concepts, outlines the procedure for analyzing systems in "equilibrium," and applies this procedure to relatively simple mechanical systems and to the human musculoskeletal system. The second part of the text (Chapters 6 through 9) provides the techniques for analyzing the "deformation" characteristics of materials with applications to orthopaedic biomechanics. The last part of the text is devoted to the analyses of "moving" systems with applications to human motion analyses and sports mechanics (Chapters 10 through 15, and Appendix C).

While preparing this text we paid particular attention to the applications of the concepts introduced, methods explained, and procedures outlined by providing many solved example problems. Most of these examples are constructed to illustrate the relevance of engineering knowledge to human physiology. To provide proper visual aids, special attention is given to the quality and quantity of illustrations. To avoid overwhelming the reader with extensive lists of references, we direct the interested reader to sources that contain more complete literature surveys.

It was our purpose to illustrate how biological phenomena can be described in terms of mechanical concepts. We believe that the knowledge of the biomechanical aspects and structural behavior of the human musculoskeletal system is an essential prerequisite for any experimental, theoretical, or analytical approach to analyze its physiological function in the body. By preparing this text, we hope to contribute to an improved dialogue between those professionals who are primarily interested in the biological and physiological aspects of the human body and those who are interested in the structural behavior of the human body through an engineering approach.

Margareta Nordin, Dr.Sci.
Dawn L. Leger, Ph.D.

Acknowledgments

The first thing, and always the focus, is the students. We would like to thank all the students who contributed to the development and refinement of this textbook by their participation in the biomechanics courses offered by the Graduate Program of Ergonomics and Biomechanics at New York University. Their questions and comments enhance the teaching and learning process. There is nothing more exciting to a teacher than the spark of recognition when a student suddenly grasps a difficult concept. That is the priceless reward that brings enrichment to the lives of all who labor in academia.

Special thanks to Dr. Robert Garber, formerly with Springer-Verlag New York, for initiating this project; and Dr. Robin Smith, Senior Editor for Life Sciences at Springer, for helping us to bring it to fruition. We are greatly indebted to Prof. Aydin Tözeren from the Catholic University of America, for encouragement and suggestions over many years of friendship; and to David Goldsheyder for rising to the challenge of taking over the Biomechanics classes at NYU and for assisting us with corrections to the manuscripts. Special thanks to Dr. Victor H. Frankel for his support and encouragement; and to all the staff at the Occupational and Industrial Orthopaedic Center, the Hospital for Joint Diseases, and New York University, for providing the bricks and mortar that support our efforts.

Contents

Appendix C   Calculus   359

Chapter 1

Introduction

1.1 Mechanics

Mechanics is a branch of physics that is concerned with the motion and deformation of bodies that are acted on by mechanical disturbances called forces. Mechanics is the oldest of all physical sciences, dating back to the times of Archimedes (287–212 B.C.). Galileo (1564–1642) and Newton (1642–1727) were the most prominent contributors to this field. Galileo made the first fundamental analyses and experiments in dynamics, and Newton formulated the laws of motion and gravity.

Engineering mechanics or *applied mechanics* is the science of applying the principles of mechanics. Applied mechanics is concerned with both the analysis and design of mechanical systems. The broad field of applied mechanics can be divided into three main parts, as illustrated in Table 1.1.

Table 1.1 *Classification of applied mechanics.*

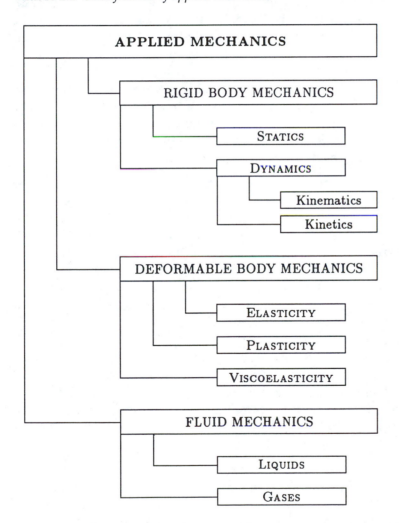

In general, a material can be categorized as either a solid or fluid. Solid materials can be rigid or deformable. *A rigid body* is one that cannot be deformed. In reality, every object or material does undergo deformation to some extent when acted upon by external forces. In some cases the amount of deformation is so small that it does not affect the desired analysis. In such cases, it is preferable to consider the body as rigid and carry out the analysis with relatively simple computations.

Statics is the study of forces on rigid bodies at rest or moving with a constant velocity. *Dynamics* deals with bodies in motion. *Kinematics* is a branch of dynamics that deals with the geometry and time-dependent aspects of motion without considering the forces causing the motion. *Kinetics* is based on kinematics, and it includes the effects of forces and masses in the analysis.

Statics and dynamics are devoted primarily to the study of the external effects of forces on rigid bodies, bodies for which the deformation (change in shape) can be neglected. On the other hand, the *mechanics of deformable bodies* deals with the relations between externally applied loads and their internal effects on bodies. This field of applied mechanics does not assume that the bodies of interest are rigid, but considers the true nature of their material properties. The mechanics of deformable bodies has strong ties with the field of material science that deals with the atomic and molecular structure of materials. The principles of deformable body mechanics have important applications in the design of structures and machine elements. In general, analyses in deformable body mechanics are more complex as compared to the analyses required in rigid body mechanics.

The mechanics of deformable bodies is the field that is concerned with the deformability of objects. An *elastic* body is defined as one in which all deformations are recoverable upon removal of external forces. This feature of some materials can easily be visualized by observing a spring or a rubber band. If you gently stretch (deform) a spring and then release it (remove the applied force), it will resume its original (undeformed) size and shape. A *plastic* body, on the other hand, undergoes permanent (unrecoverable) deformations. One can observe this behavior again by using a spring. Apply a large force on a spring so as to stretch the spring extensively, and then release it. The spring will bounce back, but there may be an increase in its length. This increase illustrates the extent of plastic deformation in the spring. Note that depending on the extent and duration of applied forces, a material may exhibit elastic or elastoplastic behavior as in the case of the spring.

To explain viscoelasticity, we must first define what is known as a *fluid*. In general, materials are classified as either solid or fluid. When an external force is applied to a solid body, the body will deform to a certain extent. The continuous application of the same force will not necessarily deform the solid body continuously. On the other hand, a continuously applied force on a fluid body will cause a continuous deformation (flow). *Viscosity* is a fluid property that is a quantitative measure of resistance to flow. In nature there are some materials that have both fluid and solid properties. The term *viscoelastic* is used to refer to the mechanical properties of such materials. Many biological materials exhibit viscoelastic properties.

Note that the distinctions between the various areas of applied mechanics are not sharp. For example, viscoelasticity simultaneously utilizes the principles of fluid and solid mechanics.

1.2 Biomechanics

Biomechanics combines the field of engineering mechanics with the fields of biology and physiology. Biomechanics is concerned with the human body. In biomechanics, the principles of mechanics are applied to the conception, design, development, and analysis of equipment and systems in biology and medicine. Although biomechanics is a relatively young and dynamic field, its history can be traced back to the fifteenth century, when Leonardo da Vinci (1452–1519) noted the significance of mechanics in his biological studies. As a result of contributions of researchers in the fields of biology, medicine, basic sciences, and engineering, the interdisciplinary field of biomechanics has been growing steadily in the last two decades.

The development of the field of biomechanics has improved our understanding of many things, including normal and pathological situations, mechanics of neuromuscular control, mechanics of blood flow in the microcirculation, mechanics of air flow in the lung, and mechanics of growth and form. It has contributed to the development of medical diagnostic and treatment procedures. It has provided the means for designing and manufacturing medical instruments, devices for the handicapped, artificial replacements, and implants. It has suggested the means for improving human performance in the workplace and in athletic competition.

Different aspects of biomechanics utilize different components of applied mechanics. For example, the principles of statics are applied to determine the magnitude and nature of forces involved in various joints and muscles of the musculoskeletal system. The principles of dynamics are utilized for motion description and have many applications in sports mechanics. The principles of the mechanics of deformable bodies provide the necessary tools for developing the field and constitutive equations for biological materials and systems, which in turn are used to evaluate their functional behavior under different conditions. The principles of fluid mechanics are used to investigate the blood flow in the human circulatory system and air flow in the lung.

It is the aim of this textbook to expose the reader to the principles and applications of biomechanics. For this purpose, the basic tools and principles will first be introduced. Next, systematic and comprehensive applications of these principles will be carried out with many solved example problems. Attention will be focused on the applications of statics, dynamics, and the mechanics of deformable bodies (i.e., solid mechanics). A limited study of fluid mechanics and its applications in biomechanics will be provided.

1.3 Basic Concepts

Engineering mechanics is based on Newtonian mechanics in which the basic concepts are length, time, and mass. These are absolute concepts because they are independent of each other. *Length* is a concept for describing size quantitatively. *Time* is a concept for ordering the flow of events. *Mass* is the property of all matter and is the quantitative measure of inertia. *Inertia* is the resistance to the change in motion of matter.

Other important concepts in mechanics are not absolute but derived from the basic concepts. These include force, moment or torque, velocity, acceleration, work, energy, power, impulse, momentum, stress, and strain. *Force* can be defined in many ways, such as mechanical disturbance or load. Force is the action of one body on another. It is the force applied on a body that causes the body to move and/or deform. *Moment* or *torque* is the quantitative measure of the rotational, bending, or twisting action of a force applied on a body. *Velocity* is defined as the time rate of change of position. The time rate of increase of velocity, on the other hand, is termed *acceleration*. Detailed descriptions of these and other concepts will be provided in subsequent chapters.

1.4 Newton's Laws

The entire field of mechanics rests on a few basic laws. Among these, the laws of mechanics introduced by Sir Isaac Newton form the basis for analyses in statics and dynamics.

Newton's first law states that a body that is originally at rest will remain at rest, or a body in motion will move in a straight line with constant velocity, if the net force acting upon it is zero. In analyzing this law, we must pay extra attention to a number of key words. The term "rest" implies no motion. For example, a book lying on a desk is said to be at rest. To be able to explain the concept of "net force" fully, we need to introduce vector algebra (see Chapter 2). The net force simply refers to the combined effect of all forces acting on a body. If the net force acting on a body is zero, it does not necessarily mean that there are no forces acting on the body. For example, there may be two equal and opposite forces applied on a body so that the combined effect of the two forces on the body is zero, assuming that the body is rigid. Note that if a body is either at rest or moving in a straight line with a constant velocity, then it is said to be in *equilibrium*.

Therefore, the first law states that if the net force acting on a body is zero, then the body is in equilibrium.

Newton's second law states that a body with a net force acting on it will accelerate in the direction of that force, and that the magnitude of the acceleration will be proportional to the magnitude of the net force. The important terms in the statement of the second law are "magnitude" and "direction," and they will be explained in detail in Chapter 2, within the context of vector algebra.

Newton's third law states that to every action there is always an equal reaction, and that the forces of action and reaction between interacting bodies are equal in magnitude, opposite in direction, and have the same line of action. This law can be simplified by saying that if you push a body, the body will push you back. This law has important applications in constructing *free body diagrams* of components constituting large systems. The free-body diagram of a component of a structure is one in which the surrounding parts of the structure are replaced by equivalent forces. It is an effective aid to study the forces involved in the structure.

Newton's laws will be explained in detail in subsequent chapters, and they will be utilized extensively throughout this text.

1.5 Dimensional Analysis

The term "dimension" has several uses in mechanics. It is used to describe space, as, for example, while referring to one-dimensional, two-dimensional, or three-dimensional situations. Dimension is also used to denote the nature of quantities. Every measurable quantity has *a dimension* and a *unit* associated with it. Dimension is a general description of a quantity, whereas unit is associated with a system of units (see Section 1.6). Whether a distance is measured in meters or feet, it is a distance. We say that its dimension is "length." There are two sets of dimensions. *Primary*, or *basic*, *dimensions* are those associated with the basic concepts of mechanics. In this text, we shall use L, T, and M to specify the primary dimensions length, time, and mass, respectively. We shall use square brackets to denote the dimensions of a physical quantity. The basic dimensions are:

$$[\text{LENGTH}] = L$$
$$[\text{TIME}] = T$$
$$[\text{MASS}] = M$$

Secondary dimensions are associated with dependent concepts, which are derived from basic concepts. For example, the area of a rectangle can be calculated by multiplying its width and length, both of which have the dimension of length. Therefore, the dimension of area is:

$$[\text{AREA}] = [\text{LENGTH}][\text{LENGTH}] = \text{LL} = \text{L}^2$$

The secondary dimensional quantities are established as a consequence of certain natural laws. If we know the definition of a physical quantity, we can easily determine the dimension of that quantity in terms of basic dimensions. If the dimension of a physical quantity is known, then the units of that quantity in different systems of units can easily be determined. Furthermore, the validity of an equation relating a number of physical quantities can be verified by analyzing the dimensions of terms forming the equation or formula. In this regard, the *law of dimensional homogeneity* imposes restrictions on the formulation of such relations. To explain this law, consider the following arbitrary equation:

$$Z = a\,X + b\,Y + c$$

For this equation to be dimensionally homogeneous, every grouping in the equation must have the same dimensional representation. In other words, if Z refers to a quantity whose dimension is length, then products $a\,X$ and $b\,Y$ and quantity c must all have the dimension of length. The numerical equality between both sides of the equation must also be maintained for all systems of units.

1.6 Systems of Units

There have been a number of different systems of units adopted in different parts of the world. For example, there is the British gravitational or foot-pound-second system, the Gaussian (metric absolute) or centimeter-gram-second (cgs) system, and the metric gravitational or meter-kilogram-second (mks) system. The lack of a universal standard in units of measure often causes confusion.

In 1960, an International Conference on Weights and Measures was held to bring order to the confusion surrounding the units of measure. Based on the metric system, this conference adopted a system called *Le Système International d'Unités* in French, which is abbreviated as SI. In English it is known as the International System of Units. Today, nearly the entire world is either using this modernized metric system or committed to its adoption. In the International System of Units, the units of length, time, and mass are meter (m), second (s), and kilogram (kg), respectively. The units of measure of these fundamental concepts in three

different systems of units are listed in Table 1.2. Throughout this
text, we shall use the International System of Units. Other units
will be defined for informational purposes.

Table 1.2 *Units of fundamental quantities of mechanics.*

SYSTEM	LENGTH	MASS	TIME
SI	meter (m)	kilogram (kg)	second (s)
c-g-s	centimeter (cm)	gram (gm)	second (s)
British	foot (ft)	slug (slug)	second (s)

Once the units of measure for the primary concepts are agreed
upon, the units of measure for the derived concepts can easily be
determined provided that the dimensional relationship between
the basic and derived quantities is known. All that is required
is replacing the dimensional representation of length, mass, and
time with their appropriate units. For example, the dimension of
force of ML/T^2. Therefore, according to the International System
of Units, force has the unit of $kg\text{-}m/s^2$, which is also known as
Newton (N). Similarly, the unit of force is $lb\text{-}ft/s^2$ in the British
system of units, and is $gm\text{-}cm/s^2$ or dyne (dyn) in the metric
absolute or cgs system. Table 1.3 lists the dimensional representations of some of the derived quantities and their units in the
International System of Units.

Table 1.3 *Dimensions and units of selected quantities in SI.*

QUANTITY	DIMENSION	SI UNIT	SPECIAL NAME
Area	2	m^2	
Volume	3	m^3	
Velocity		m/s	
Acceleration	2	m/s^2	
Force	2	$kg\text{-}m/s^2$	Newton (N)
Pressure & Stress	2	N/m^2	Pascal (Pa)
Moment (Torque)	22	N-m	
Work & Energy	22	N-m	Joule (J)
Power	23	J/s	Watt (W)

Note that "kilogram" is the unit of mass in SI. For example,
consider a 60 kg object. The weight of the same object in SI is

$(60 \, \text{kg}) \times (9.8 \, \text{m}/\text{s}^2) = 588 \, \text{N}$, the factor $9.8 \, \text{m}/\text{s}^2$ being the magnitude of the gravitational acceleration.

In addition to the primary and secondary units that are associated with the basic and derived concepts in mechanics, there are *supplementary units* such as plane angle and temperature. The common measure of an angle is degree (°). Three hundred sixty degrees is equal to one revolution (rev) or 2π radians (rad), where $\pi = 3.1416$. The SI unit of temperature is Kelvin (°K). However, degree Celsius (°C) is more commonly used. The British unit of temperature is degree Fahrenheit (°F).

It should be noted that in most cases, a number has a meaning only if the correct unit is displayed with it. In performing calculations, the ideal method is to show the correct units with each number throughout the solution of equations. This approach helps in detecting conceptual errors and eliminates the need for determining the unit of the calculated quantity separately. Another important aspect of using units is consistency. One must not use the units of one system for some quantities and the units of another system for other quantities while carrying out calculations.

1.7 Conversion of Units

The International System of Units is a revised version of the metric system, which is based on the decimal system. Table 1.4 lists the SI multiplication factors and corresponding prefixes. Table 1.5 lists factors needed to convert quantities expressed in British and metric systems to corresponding units in SI.

Table 1.4 *SI multiplication factors and prefixes.*

Multiplication Factor	SI Prefix	SI Symbol
$1\,000\,000\,000 = 10^9$	giga	G
$1\,000\,000 = 10^6$	mega	M
$1\,000 = 10^3$	kilo	k
$100 = 10^2$	hecto	h
$10 = 10^1$	deka	da
$.1 = 10^{-1}$	deci	d
$.01 = 10^{-2}$	centi	c
$.001 = 10^{-3}$	milli	m
$.000\,001 = 10^{-6}$	micro	μ
$.000\,000\,001 = 10^{-9}$	nano	n
$.000\,000\,000\,001 = 10^{-12}$	pico	p

Table 1.5 *Conversion of units.*

LENGTH	
	1 centimeter (cm) = 0.01 meter (m)
	1 inch (in) = 0.0254 m
	1 foot (ft) = 0.3048 m
	1 yard (yd) = 0.9144 m
	1 mile = 1609 m
	1 angstrom (Å) = 10^{-10} m
TIME	
	1 minute (min) = 60 second (s)
	1 hour (h) = 3600 s
	1 day (d) = 86400 s
MASS	
	1 pound mass (lbm) = 0.4536 kilogram (kg)
	1 slug = 14.59 kg
FORCE	
	1 kilogram force (kgf) = 9.807 Newton (N)
	1 pound force (lbf) = 4.448 N
	1 dyne (dyn) = 10^{-5} N
PRESSURE & STRESS	
	1 kg/m-s^2 = 1 N/m^2 = 1 Pascal (Pa)
	1 lbf/in^2 (psi) = 6896 Pa
	1 lbf/ft^2 (psf) = 992966 Pa
	1 dyn/cm^2 = 0.1 Pa
MOMENT (TORQUE)	
	1 dyn-cm = 10^{-7} N-m
	1 lbf-ft = 1.356 N-m
WORK & ENERGY	
	1 kg-m^2/s^2 = 1 N-m = 1 Joule (J)
	1 dyn-cm = 1 erg = 10^{-7} J
	1 lbf-ft = 1.356 J
POWER	
	1 kg-m^2/s^3 = 1 J/s = 1 Watt (W)
	1 horsepower (hp) = 550 lbf-ft/s = 746 W
PLANE ANGLE	
	1 degree (°) = $\pi/180$ radian (rad)
	1 revolution (rev) = 360°
	1 rev = 2π rad = 6.283 rad
TEMPERATURE	
	°C = °K − 273.2
	°C = 5(°F − 32)/9

1.8 Mathematics

The applications of biomechanics require a certain knowledge of mathematics. This knowledge includes simple geometry, properties of the right-triangle, basic algebra, differentiation, and integration. The appendices that follow the last chapter contain a summary of the mathematical techniques needed to carry out the calculations in this book. The reader may find it useful to examine them now, and review them later when those concepts are needed. In subsequent chapters throughout the text, the mathematics required will be reviewed and the corresponding appendix will be indicated.

During the formulation of the problems, we shall use Greek letters as well as the letters of the Latin alphabet. Greek letters will be used for example to refer to angles. The Greek alphabet is provided in Table 1.6 for quick reference.

Table 1.6 *Greek alphabet.*

alpha	A	α	iota	I	ι	rho	P	ρ
beta	B	β	kappa	K	κ	sigma	Σ	σ
gamma	Γ	γ	lambda	Λ	λ	tau	T	τ
delta	Δ	δ	mu	M	μ	upsilon	Υ	υ
epsilon	E	ϵ	nu	N	ν	phi	Φ	ϕ
zeta	Z	ζ	xi	Ξ	ξ	chi	X	χ
eta	H	η	omicron	O	o	psi	Ψ	ψ
theta	Θ	θ	pi	Π	π	omega	Ω	ω

1.9 Scalars and Vectors

In mechanics, two kinds of quantities are distinguished. A *scalar* quantity, such as mass, temperature, work, and energy, has magnitude. A *vector* quantity, such as force, velocity, and acceleration, has both a magnitude and a direction. Unlike scalars, vector quantities add according to a rule called the *parallelogram law*. Vector algebra will be covered in detail in Chapter 2.

1.10 Modeling and Approximations

One needs to make certain assumptions to simplify complex systems and problems so as to achieve analytical solutions. The complete model is the one that includes the effects of all parts constituting a system. However, the more detailed the model, the more difficult the formulation and solution of the problem. It is not always possible and in some cases it may not be necessary to include every detail in the analysis. For example, during most human activities, there is more than one muscle group activated at a time. If the task is to analyze the forces involved in the joints and muscles during a particular human activity, the best approach is to predict which muscle group is the most active and set up a model that neglects all other muscle groups. As we shall see in the following chapters, bone is a deformable body. If the forces involved are relatively small, then the bone can be treated as a rigid body. This approach may reduce the complexity of the problem under consideration.

In general, it is always best to begin with a simple basic model that represents the system. Gradually, the model can be expanded on the basis of experience gained and the results obtained from simpler models. The guiding principle is to make simplifications that are consistent with the required accuracy of the results. In this way, the researcher can set up a model that is simple enough to analyze and exhibit satisfactorily the phenomena under consideration. The more we learn, the more detailed our analysis can become.

1.11 Generalized Procedure

The general method of solving problems in biomechanics may be outlined as follows:

1. Select the system of interest.
2. Postulate the characteristics of the system.
3. Simplify the system by making proper approximations. Explicitly state important assumptions.
4. Form an analogy between the human body parts and basic mechanical elements.
5. Construct a mechanical model of the system.
6. Apply principles of mechanics to formulate the problem.
7. Solve the problem for the unknowns.
8. Compare the results with the behavior of the actual system. This may involve tests and experiments.
9. If satisfactory agreement is not achieved, steps 3 through 7 must be repeated by considering different assumptions and a new model of the system.

1.12 Scope of the Text

Courses in biomechanics are taught within a wide variety of academic programs to students with quite different backgrounds. This text is prepared to provide a teaching and learning tool to health care professionals who are seeking a graduate degree in biomechanics but have limited backgrounds in calculus, physics, and engineering mechanics. This text can also be a useful reference for undergraduate biomedical, biomechanical, or bioengineering programs.

This text is divided into three parts. The first part (Chapters 1 through 5, and Appendices A and B) will introduce the basic concepts of mechanics including force and moment vectors, provide the mathematical tools (geometry, algebra, and vector algebra) so that complete definitions of these concepts can be given, explain the procedure for analyzing the systems at "static equilibrium," and apply this procedure to analyze simple mechanical systems and the forces involved at various muscles and joints of the human musculoskeletal system. It should be noted here that the topics covered in the first part of this text are prerequisites for both parts two and three. The second part of the text (Chapters 6 through 9) provides the techniques for analyzing the "deformation" characteristics of materials under different load conditions. For this purpose, the concepts of stress and strain are defined. Classifications of materials based on their stress–strain diagrams are given. The concepts of elasticity, plasticity, and viscoelasticity are also introduced and explained. Topics such as torsion, bending, fatigue, endurance, and factors affecting the strength of materials are provided. The emphasis is placed upon applications to orthopedic biomechanics. The last section of the text (Chapters 10 through 15, and Appendix C) is devoted to "dynamic" analyses. The concepts introduced in the second part are position, velocity and acceleration vectors, work, energy, power, impulse, and momentum. Also provided in the second part are the techniques for kinetic and kinematic analyses of systems undergoing translational and rotational motions. These techniques are applied for human motion analyses of various sports activities.

1.13 Notation

While preparing this text, special attention was given to the consistent use of notation. Important terms are italicized where they are defined or described (such as, *force* is defined as load or mechanical disturbance). Symbols for quantities are also italicized (for example, m for mass). Units are not italicized (for example, kg for kilogram). Underlined letters are used to refer to vector quantities (for example, force vector $\underline{F}$). Sections and subsections marked with a star ($*$) are considered optional. In other words, the reader can omit a section or subsection marked with a star without losing the continuity of the topics covered in the text.

1.14 References and Suggested Reading

We believe that this text is a self-sufficient teaching and learning tool. While preparing it, we utilized the information provided in a number of sources, some of which are listed below. Note, however, that it is not our intention to promote these titles, or to suggest that these are the only texts available on the subject matter. The field of biomechanics has been growing very rapidly. There are many other sources of information available, including scientific journals presenting peer-reviewed research articles in biomechanics.

Black, J. 1988. *Orthopaedic Biomaterials in Research and Practice*. New York: Churchill Livingstone.

Chaffin, D.B., and Andersson, G.B.J. 1991. *Occupational Biomechanics*. 2nd ed. New York: John Wiley & Sons.

Burnstein, A.H., and Wright, T.M. 1994. *Fundamentals of Orthopaedic Biomechanics*, Philadelphia: Williams & Wilkins.

Hay, J. 1978. *The Biomechanics of Sports Techniques*, Englewood Cliffs, NJ: Prentice-Hall.

Hay, J.G., and Reid, J. G. 1988. *Anatomy, Mechanics and Human Motion*, 2nd ed. Englewood Cliffs, NJ: Prentice-Hall.

Kelly, D.L. 1971. *Kinesiology: Fundamentals of Motion Description*. Englewood Cliffs, NJ: Prentice-Hall.

Mow, V.C., and Hayes, W.C. 1997. *Basic Orthopaedic Biomechanics*. 2nd ed. Philadelphia: Lippincott-Raven.

Nahum, A.M., and Melvin, J. (eds.) 1985. *The Biomechanics of Trauma*, Norwalk, CT: Appleton-Century-Crofts.

Nordin, M., Andersson, G.B.J., and Pope, M.H. 1997. *Musculoskeletal Disorders in the Workplace: Principles and Practice*. Philadelphia: Mosby-Year Book.

Nordin, M., and Frankel, V.H. 1989. *Basic Biomechanics of the Musculoskeletal System*. 2nd ed. Philadelphia: Lea & Febiger.

Thompson, C.W. 1989. *Manual of Structural Kinesiology*. 11th ed. St. Louis, MO: Times Mirror/Mosby.

Williams, M., and Lissner, H.R. 1977. *Biomechanics of Human Motion*. 2nd ed. (B. LeVeau, ed.) Philadelphia: Saunders.

Winter, D.A. 1990. *Biomechanics and Motor Control of Human Behavior*, 2nd ed. New York: John Wiley & Sons.

The following sources contain advanced topics and up-to-date information in the fields of biomechanics and bioengineering.

Bronzino, J.D. (Ed.) 1995. *The Biomedical Engineering Handbook*. Boca Raton, FL: CRC Press.

Fung, Y.C. 1981. *Biomechanics: Mechanical Properties of Living Tissues*. New York: Springer-Verlag.

Fung, Y.C. 1990. *Biomechanics: Motion, Flow, Stress, and Growth*. New York: Springer-Verlag.

Mow, V.C., Ratcliff, A., and Woo, S.L.-Y. (eds.) 1990. *Biomechanics of Diarthrodial Joints*. New York: Springer-Verlag.

Schmid-Schönbein, G.W., Woo, S.L.-Y., and Zweifach, B.W. (eds.) 1985. *Frontiers in Biomechanics*. New York: Springer-Verlag.

Skalak, R., and Chien, S. (eds.) 1987. *Handbook of Bioengineering*. New York: McGraw-Hill.

Winters, J.M., and Woo, S.L.-Y. (eds.) 1990. *Multiple Muscle Systems*. New York: Springer-Verlag.

In addition to the biomechanics related books listed above, there are many physics and engineering mechanics books dedicated to undergraduate engineering education. These books may be useful for reviewing the basic concepts of mechanics. A few examples of such titles are listed below.

Sandor, B.I. 1983. *Engineering Mechanics: Statics and Dynamics*. 2nd ed. Englewood Cliffs, NJ: Prentice-Hall.

Servay, R.A. 1982. *Physics: For Scientists and Engineers*. Philadelphia: Saunders.

Shames, I.H. 1980. *Engineering Mechanics: Statics and Dynamics*. 3rd ed. Englewood Cliffs, NJ: Prentice-Hall.

The second part of this text will introduce the principles of mechanics of deformable bodies and use these principles to explain certain aspects of biomechanics. The following books can be reviewed to gain more detailed information on these principles.

Crandall, S.H., Dahl, N.C., and Lardner, T.J. 1978. *An Introduction to the Mechanics of Solids*. 2nd ed. New York: McGraw-Hill.

Popov, E.P. 1978. *Mechanics of Materials*. 2nd ed. Englewood Cliffs, NJ: Prentice-Hall.

Pytel, A., and Singer, F.L. 1987. *Strength of Materials*. 4th ed. New York: Harper & Row.

Chapter 2

Force Vector

2.1 Definition of Force

Force may be defined as mechanical disturbance or load. When you pull or push an object, you apply a force to it. You also exert a force when you throw or kick a ball. In all of these cases, the force is associated with the result of muscular activity. A force acting on an object can deform the object, change its state of motion, or both. Although forces cause motion, it does not necessarily follow that force is always associated with motion. For example, a person sitting on a chair applies his/her weight on the chair, and yet the chair remains stationary. There are relatively few basic laws that govern the relationship between force and motion. These laws will be discussed in detail in later chapters.

2.2 Properties of Force as a Vector Quantity

Forces are vector quantities and the principles of vector algebra (see Appendix B) must be applied to analyze problems involving forces. To describe a force fully, its magnitude and direction must be specified. As illustrated in Figure 2.1, a force vector can be illustrated graphically with an arrow such that the orientation of the arrow indicates the line of action of the force vector, the arrowhead identifies the direction and sense along which the force is acting, and the base of the arrow represents the point of application of the force vector. If there is a need for showing more than one force vector in a single drawing, then the length of each arrow must be proportional to the magnitude of the force vector it is representing.

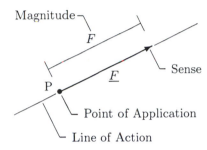

Figure 2.1 *Graphical representation of the force vector.*

Like other vector quantities, forces may be added by utilizing graphical and trigonometric methods. For example, consider the partial knee illustrated in Figure 2.2. Forces applied by the quadriceps ($\underline{F}_Q$) and patellar tendon ($\underline{F}_P$) on the patella are shown. The resultant force $\underline{F}_R$ on the patella due to the forces applied by the quadriceps and patellar tendon can be determined by considering the vector sum of these forces:

$$\underline{F}_R = \underline{F}_Q + \underline{F}_P \tag{2.1}$$

If the magnitude of the resultant force needs to be calculated, then the Pythagorean theorem can be utilized:

$$F_R = \sqrt{F_Q^2 + F_P^2} \tag{2.2}$$

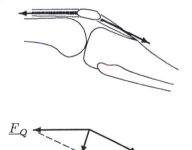

Figure 2.2 *Resultant force.*

2.3 Dimension and Units of Force

By definition, force is equal to mass times acceleration, acceleration is the time rate of change of velocity, and velocity is the time rate of change of relative position. The change in position is measured in terms of length units. Therefore, velocity has a dimension of length divided by time, acceleration has a dimension of velocity divided by time, and force has a dimension of mass times acceleration:

$$[\text{VELOCITY}] = \frac{[\text{POSITION}]}{[\text{TIME}]} = \frac{L}{T}$$

$$[\text{ACCELERATION}] = \frac{[\text{VELOCITY}]}{[\text{TIME}]} = \frac{L/T}{T} = \frac{L}{T^2}$$

$$[\text{FORCE}] = [\text{MASS}]\,[\text{ACCELERATION}] = \frac{M\,L}{T^2}$$

Units of force in different unit systems are provided in Table 2.1.

Table 2.1 *Units of force* (1 N = 10^5 dyn, 1 N = 0.225 lb).

System	Units of Force	Special Name
SI	kilogram-meter/second^2	Newton (N)
c-g-s	gram-centimeter/second^2	dyne (dyn)
British	slug-foot/second^2	pound (lb)

2.4 Force Systems

Any two or more forces acting on a single body form a *force system*. Forces constituting a force system may be classified in various ways. Forces may be classified according to their effect on the bodies upon which they are applied or according to their orientation as compared to one another.

2.5 External and Internal Forces

A force may be broadly classified as *external* or *internal*. Almost all commonly known forces are external forces. For example, when you push a cart, hammer a nail, sit on a chair, kick a football, or shoot a basketball, you apply an external force on the cart, nail, chair, football, or basketball. Internal forces, on the other hand, are the ones that hold a body together when the body is under the effect of externally applied forces. For example, a piece of string does not necessarily break when it is pulled from both ends. When a rubber band is stretched, the band elongates to a certain extent. What holds any material together under externally applied forces is the internal forces generated within that material. If we consider the human body as a whole, then the forces generated by muscle contractions are also internal forces. The significance and details of internal forces will be studied by introducing the concept of "stress" in later chapters.

2.6 Normal and Tangential Forces

In mechanics, the word "normal" implies perpendicular. If a force acting on a surface is applied in a direction perpendicular to that surface, then the force is called a *normal force* (Figure 2.3). For example, a book resting on a flat horizontal desk applies a normal force on the desk, the magnitude of which is equal to the weight of the book.

A *tangential force* is that applied on a surface in the direction parallel to the surface. A good example of a tangential force is the frictional force. As illustrated in Figure 2.4, pushing or pulling a block will cause a frictional force to occur between the bottom surface of the block and the floor. The line of action of the frictional force is always tangential to the surfaces in contact.

2.7 Tensile and Compressive Forces

A *tensile force* applied on a body will tend to stretch or elongate the body, whereas a *compressive force* will tend to shrink the body in the direction of the applied force (Figure 2.5). For example, a tensile force applied on a rubber band will stretch the band. Poking into an inflated balloon will produce a compressive force on the balloon. It must be noted that there are certain materials upon which only tensile forces can be applied. For example, a rope, a cable, or a string cannot withstand compressive forces. The shapes of these materials will be completely distorted under compressive forces. Similarly, muscles contract to produce tensile forces that pull together the bones to which they are attached. Muscles can neither produce compressive forces nor exert a push.

2.8 Coplanar Forces

A system of forces is said to be *coplanar* if all the forces are acting on a two-dimensional (plane) surface. Forces forming a coplanar system have at most two non-zero components. Therefore, with respect to the Cartesian (rectangular) coordinate frame, it is sufficient to analyze coplanar force systems by considering the x and y components of the forces involved.

2.9 Collinear Forces

A system of forces is *collinear* if all the forces have a common line of action. For example, the forces applied on a rope in a rope-pulling contest form a collinear force system (Figure 2.6).

2.10 Concurrent Forces

A system of forces is *concurrent* if the lines of action of the forces have a common point of intersection. Examples of concurrent force systems can be seen in various traction devices, as

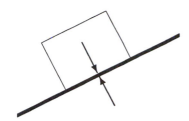

Figure 2.3 *Forces normal to the surfaces in contact.*

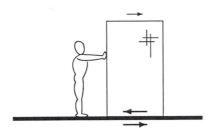

Figure 2.4 *Frictional forces are tangential forces.*

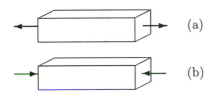

Figure 2.5 *(a) Tensile and (b) compressive forces.*

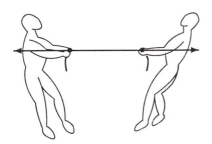

Figure 2.6 *Collinear forces.*

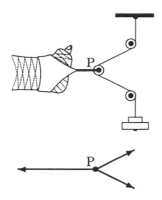

Figure 2.7 *Concurrent forces.*

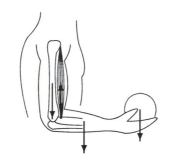

Figure 2.8 *Parallel forces.*

illustrated in Figure 2.7. Owing to the weight in the weight pan, the cables stretch and forces are applied on the pulleys and the leg. The force applied on the leg holds the leg in place.

2.11 Parallel Forces

A set of forces form a *parallel* force system if the lines of action of the forces are parallel to each other. An example of a parallel force system is illustrated in Figure 2.8 by a human arm flexed at a right angle and holding an object. The forces on the forearm are the weight of the object, the weight of the arm itself, the tension in the biceps muscle, and the joint reaction force at the elbow.

2.12 Gravitational Force or Weight

The force exerted by Earth on an object is called the *gravitational force* or *weight* of the object. The magnitude of weight of an object is equal to the mass of the object times the magnitude of gravitational acceleration. The magnitude of gravitational acceleration for different unit systems is listed in Table 2.2. These values are valid only on the surface of Earth. The magnitude of the gravitational acceleration can vary slightly with altitude. For our applications, we shall assume g to be a constant.

Table 2.2 *Gravitational acceleration on Earth.*

SYSTEM	GRAVITATIONAL ACCELERATION
SI	9.81 m/s^2
c-g-s	981 cm/s^2
British	32.2 ft/s^2

The terms mass and weight are often confused with one another. Mass is a property of a body. Weight is the force of gravity acting on the mass of the body. A body has the same mass on Earth and on the moon. However, the weight of a body is about six times as much on Earth as on the moon, because the magnitude of the gravitational acceleration on the moon is about one-sixth of what it is on Earth. Therefore, a 10 kg mass on Earth weighs about 98 N on Earth, while it weighs about 17 N on the moon.

Like force, acceleration is a vector quantity. The direction of gravitational acceleration and gravitational force vectors is always toward the center of Earth, or always vertically downward. The force of gravity acts on an object at all times. If we drop an object from a height, it is the force of gravity that will pull the object downward. When an object is at rest on the ground, the gravitational force does not disappear. An object at rest or in static equilibrium simply means that the net force acting on the object is zero (Figure 2.9).

2.13 Distributed Force Systems and Pressure

Consider a pile of sand lying on a flat horizontal surface, as illustrated in Figure 2.10a. The sand exerts force or load on the surface, which is distributed over the area under the sand. The load is not uniformly distributed over this area. The marginal regions under the pile are loaded less as compared to the central regions (Figure 2.10b). For practical purposes, the distributed load applied by the sand may be represented by a single force, called the *equivalent force* or *concentrated load*. The magnitude of the equivalent force would be equal to the total weight of the sand (Figure 2.10c). The line of action of this force would pass through a point called the *center of gravity*. For some applications, we can assume that the entire weight of the pile is concentrated at the center of gravity of the load. For uniformly distributed loads, such as the load applied by the rectangular block on the horizontal surface shown in Figure 2.11, the center of gravity coincides with the geometric center of the load. For nonuniformly distributed loads, the center of gravity can be determined by experimentation (see Chapter 4).

Center of gravity is associated with the gravitational force of Earth. There is another concept called *center of mass*, which is independent of gravitational effects. For a large object or a structure, such as the Empire State building in New York City, the center of gravity may be different than the center of mass because the magnitude of gravitational acceleration varies with altitude. For relatively small objects and for our applications, the difference between the two can be ignored.

Another important concept associated with distributed force systems is *pressure*, which is a measure of the intensity of distributed loads. By definition, *average pressure* is equal to total applied force divided by the area of the surface over which the force is applied in a direction perpendicular to the surface. It is also known as *load intensity*. For example, if the bottom surface area of the rectangular block in Figure 2.11 is A and the total weight of the block is W, then the magnitude p of the pressure exerted by the block on the horizontal surface can be calculated by:

$$p = \frac{W}{A} \tag{2.3}$$

It follows that the dimension of pressure has the dimension of force (ML/T^2) divided by the dimension of area (L^2):

$$[\text{PRESSURE}] = \frac{[\text{FORCE}]}{[\text{AREA}]} = \frac{M}{LT^2}$$

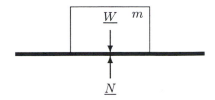

Figure 2.9 *The net force on an object at rest is zero.*

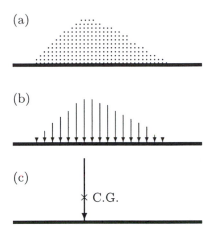

Figure 2.10 *A pile of sand (a), distributed load on the ground (b), and an equivalent force (c).*

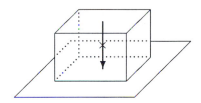

Figure 2.11 *Rectangular block.*

Units of pressure in different unit systems are listed in Table 2.3.

Table 2.3 *Units of pressure.*

System	Units of Pressure	Special Name
SI	$kg/m\text{-}s^2$ or N/m^2	Pascal (Pa)
c-g-s	$gm/cm\text{-}s^2$ or dyn/cm^2	
British	lb/ft^2 or lb/in^2	psf or psi

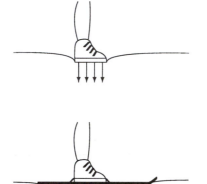

Figure 2.12 *Intensity of force (pressure) applied on the snow by a pair of boots is higher than that applied by a pair of skis.*

The principles behind the concept of pressure have many applications. Note that the larger the area over which a force is applied, the lower the magnitude of pressure. If we observe two people standing on soft snow, one wearing a pair of boots and the other wearing skis, we can easily notice that the person wearing boots stands deeper in the snow than the skier. This is simply because the weight of the person wearing boots is distributed over a smaller area on the snow, and therefore applies a larger force per unit area of snow (Figure 2.12). It is obvious that the sensation and pain induced by a sharp object is much more severe than that produced by a force that is applied by a dull object. A prosthesis that fits the amputated limb, or a set of dentures that fits the gum and the bony structure properly, would feel and function better than an improperly fitted implant or replacement device. The idea is to distribute the forces involved as uniformly as possible over a large area.

2.14 Frictional Forces

Frictional forces occur between two surfaces in contact when one surface slides or tends to slide over the other. When a body is in motion on a rough surface or when an object moves in a fluid (a viscous medium such as water), there is resistance to motion because of the interaction of the body with its surroundings. In some applications friction may be desirable, while in others it may have to be reduced to a minimum. For example, it would be impossible to start walking in the absence of frictional forces. Automobile, bicycle, and wheelchair brakes utilize the principles of friction. On the other hand, friction can cause heat to be generated between the surfaces in contact. Excess heat can cause early, unexpected failure of machine parts. Friction may also cause wear.

There are several factors that influence frictional forces. Friction depends on the nature of the two sliding surfaces. For example, if all other conditions are the same, the friction between two metal surfaces would be different than the friction between two wood surfaces in contact. Friction is larger for materials that strongly interact. Friction depends on the surface quality and surface finish. A good surface finish can reduce frictional effects. The frictional force does not depend on the total surface area of contact.

Consider the block resting on the floor, as shown in Figure 2.13. The block is applying its weight $\underline{W}$ on the floor. In return the floor is applying a normal force $\underline{N}$ on the block, such that the magnitudes of the two forces are equal ($N = W$). Now consider that a horizontal force $\underline{F}$ is applied on the block to move it toward the right. This will cause a frictional force f to develop between the block and the floor. As long as the block remains stationary (in static equilibrium), the magnitude f of the frictional force would be equal to the magnitude F of the applied force. This frictional force is called the *static friction*.

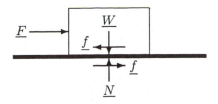

Figure 2.13 *Friction occurs on surfaces when one surface slides or tends to slide over the other.*

If the magnitude of the applied force is increased, the block will eventually slip or begin sliding over the floor. When the block is on the verge of sliding (the instant just before sliding occurs), the magnitude of the static friction is maximum (f_{max}). When the magnitude of the applied force exceeds f_{max}, the block moves toward the right. When the block is in motion, the resistance to motion at the surfaces of contact is called the *kinetic* or *dynamic friction*, $\underline{f}_k$. In general, the magnitude of the force of kinetic friction is lower than the maximum static friction ($f_k < f_{max}$) and the magnitude of the applied force ($f_k < F$). The difference between the magnitudes of the applied force and kinetic friction causes the block to accelerate toward the right.

It has been determined experimentally that the magnitudes of both static and kinetic friction are directly proportional to the normal force (N in Figure 2.13) acting on the surfaces in contact. The constant of proportionality is commonly referred to with μ (mu) and is called the *coefficient of friction*, which depends on such factors as the material properties, the quality of the surface finish, and the conditions of the surfaces in contact. The coefficient of friction also varies depending on whether the bodies in contact are stationary or sliding over each other. To be able to distinguish the frictional forces involved at static and dynamic conditions, two different friction coefficients are defined. The *coefficient of static friction* (μ_s) is associated with static friction, and the *coefficient of kinetic friction* (μ_k) is associated with kinetic or dynamic friction. The magnitude of the static frictional force is such that $f_s = \mu_s N = f_{max}$ when the block is on the verge of sliding, and $f_s < \mu_s N$ when the magnitude of the applied force is less than the maximum frictional force, in which case the magnitude of the force of static friction is equal in magnitude to the applied force ($f_s = F$). The formula relating the kinetic friction and the normal force is:

$$f_k = \mu_k N \qquad (2.4)$$

The variations of friction coefficients with respect to the force

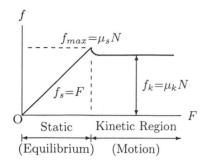

Figure 2.14 *The variation of frictional force as a function of applied force.*

applied in a direction parallel (tangential) to the surfaces in contact are shown in Figure 2.14. For any given pair of materials, the coefficient of kinetic friction is usually lower than the coefficient of static friction. The coefficient of kinetic friction is approximately constant at moderate sliding speeds. At higher speeds, μ_k may decrease because of the heat generated by friction. Sample coefficients of friction are listed in Table 2.4. Note that the figures provided in Table 2.4 are some average ranges and do not distinguish between static and kinetic friction coefficients.

Table 2.4 *Coefficients of friction.*

SURFACES IN CONTACT	FRICTION COEFFICIENT
Wood on wood	$0.25 - 0.50$
Metal on metal	$0.30 - 0.80$
Plastic on plastic	$0.10 - 0.30$
Metal on plastic	$0.10 - 0.20$
Rubber on concrete	$0.60 - 0.70$
Rubber on tile	$0.20 - 0.40$
Rubber on wood	$0.70 - 0.75$
Bone on metal	$0.10 - 0.20$
Cartilage on cartilage	$0.001 - 0.002$

Frictional forces always act in a direction tangent to the surfaces in contact. If one of the two bodies in contact is moving, then the frictional force acting on that body has a direction opposite to the direction of motion. For example, under the action of applied force, the block in Figure 2.13 tends to move toward the right. The direction of the frictional force on the block is toward the left, trying to stop the motion of the block. The frictional forces always occur in pairs because there are always two surfaces in contact for friction to occur. Therefore, in Figure 2.13, a frictional force is also acting on the floor. The magnitude of the frictional force on the floor is equal to that of the frictional force acting on the block. However, the direction of the frictional force on the floor is toward the left.

The effects of friction and wear may be reduced by introducing additional materials between the sliding surfaces. These materials may be solids or fluids, and are called *lubricants*. Lubricants placed between the moving parts reduce frictional effects and wear by reducing direct contact between the moving parts. In the case of the human body, the diarthrodial joints (such as the elbow, hip, and knee joints) are lubricated by the *synovial fluid*. The synovial fluid is a viscous material that reduces frictional effects, reduces wear and tear of articulating surfaces by limiting direct contact between them, and nourishes the articular cartilage lining the joint surfaces. Although diarthrodial joints are subjected to very large loading conditions, the cartilage surfaces undergo little wear under normal, daily conditions. It is important to note that introducing a fluid as a lubricant between two solid surfaces undergoing relative motion changes the discussion of how to assess the frictional effects. For example, frictional force with a viscous medium present is not only a function of the normal forces (pressure) involved, but also depends on the relative velocity of the moving parts. A number of lubrication modes have been defined to account for frictional effects at diarthrodial joints under different loading and motion conditions. These modes include hydrodynamic, boundary, elastohydrodynamic, squeeze-film, weeping, and boosted lubrication.

2.15 Exercise Problems

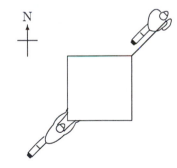

Problem 2.1 As illustrated in Figure 2.15, consider two workers who are trying to move a block. Assume that both workers are applying equal magnitude forces of 200 N. One of the workers is pushing the block toward the north and the other worker is pushing it toward the east. Determine the magnitude and direction of the net force applied by the workers on the block.

Figure 2.15 *Problem 2.1.*

Answer: 283 N, northeast.

Problem 2.2 As illustrated in Figure 2.16, consider two workers who are trying to move a block. Assume that both workers are applying equal magnitude forces of 200 N. One of the workers is pushing the block toward the northeast, while the other is pulling it in the same direction. Determine the magnitude and direction of the net force applied by the workers on the block.

Answer: 400 N, northeast.

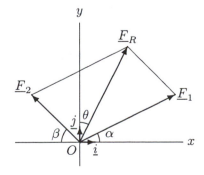

Figure 2.16 *Problem 2.2.*

Problem 2.3 Consider the two forces, $\underline{F}_1$ and $\underline{F}_2$, shown in Figure 2.17. Assume that these forces are applied on an object in the xy-plane. The first force has a magnitude $F_1 = 15$ N and is applied in a direction that makes an angle $\alpha = 30°$ with the positive x axis, and the second force has a magnitude $F_2 = 10$ N and is applied in a direction that makes an angle $\beta = 45°$ with the negative x axis.

(a) Calculate the scalar components of $\underline{F}_1$ and $\underline{F}_2$ along the x and y directions.
(b) Express $\underline{F}_1$ and $\underline{F}_2$ in terms of their components.

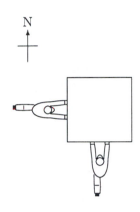

Figure 2.17 *Problem 2.3.*

(c) Determine an expression for the resultant force vector, $\underline{F}_R$.
(d) Calculate the magnitude of the resultant force vector.
(e) Calculate angle θ that $\underline{F}_R$ makes with the positive y axis.

Answers:

(a) $F_{1x} = 13.0$ N, $F_{1y} = 7.5$ N, $F_{2x} = 7.1$ N, $F_{2y} = 7.1$ N
(b) $\underline{F}_1 = 13.0\underline{i} + 7.5\underline{j}$ and $\underline{F}_2 = -7.1\underline{i} + 7.1\underline{j}$
(c) $\underline{F}_R = 5.9\underline{i} + 14.6\underline{j}$
(d) $F_R = 15.7$ N
(e) $\theta = 22°$

Figure 2.18 *Problem 2.4.*

Problem 2.4 As illustrated in Figure 2.18, consider a 2 kg, 20 cm × 30 cm book resting on a table. Calculate the average pressure applied by the book on the table top.

Answer: 327 Pa.

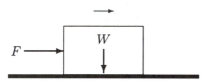

Figure 2.19 *Problem 2.5.*

Problem 2.5 As illustrated in Figure 2.19, consider a block that weighs 400 N and is resting on a horizontal surface. Assume that the coefficient of static friction between the block and the horizontal surface is 0.3. What is the minimum horizontal force required to move the block toward the right?

Answer: Slightly greater than 300 N.

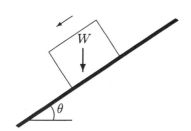

Figure 2.20 *Problem 2.6.*

Problem 2.6 As shown in Figure 2.20, consider a block that weighs W. Owing to the effect of gravity, the block is sliding down a slope that makes an angle θ with the horizontal. The coefficient of kinetic friction between the block and the slope is μ_k.

Show that the magnitude of the frictional force generated between the block and the slope is $f = \mu_k W \cos\theta$.

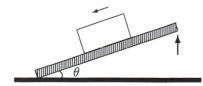

Figure 2.21 *An experimental method to determine the coefficient of friction.*

Problem 2.7 Figure 2.21 shows a simple experimental method to determine the coefficient of static friction between surfaces in contact. This method is applied by placing a block on a horizontal plate, tilting the plate slowly until the block starts sliding over the plate, and recording the angle the plate makes with the horizontal at the instant when the sliding occurs. This critical angle (θ_c) is called the *angle of repose*. At the instant just before the sliding occurs, the block is in static equilibrium which will be discussed in later chapters.

Through force equilibrium, show that the coefficient of static friction just before motion starts is $\mu = \tan\theta_c$.

Chapter 3

Moment and Torque

3.1 Definitions of Moment and Torque Vectors

A force applied to an object can translate, rotate, and/or deform the object. The effect of a force on the object to which it is applied depends on how the force is applied and how the object is supported. For example, when pulled, an open door will swing about the edge along which it is hinged to the door frame (Figure 3.1). What causes the door to swing is the torque generated by the applied force about an axis that passes through the hinges of the door. If one stands on the free-end of a diving board, the board will bend (Figure 3.2). What bends the board is the moment of the body weight about the fixed end of the board. In general, *torque* is associated with the rotational and twisting actions of applied forces, while *moment* is related to their bending effect. However, the mathematical definition of moment and torque is the same. Therefore, it is sufficient to use moment to discuss the common properties of moment and torque vectors.

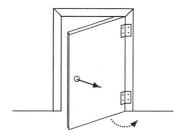

Figure 3.1 *Rotational effect.*

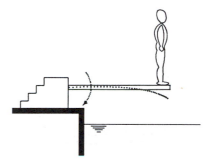

Figure 3.2 *Bending effect.*

3.2 Magnitude of Moment

The magnitude of the moment of a force about a point is equal to the magnitude of the force times the length of the shortest distance between the point and the line of action of the force, which is known as the *lever* or *moment arm*. Consider a person on an exercise apparatus who is holding a handle that is attached to a cable (Figure 3.3). The cable is wrapped around a pulley and attached to a weight pan. The weight in the weight pan stretches the cable and produces a tensile force $\underline{F}$ in the cable. This force is transmitted to the person's hand through the handle. Assume that the magnitude of the moment of force $\underline{F}$ about point O at the elbow joint is to be determined. To determine the shortest distance between O and the line of action of the force, extend the line of action of $\underline{F}$ and drop a line from O that cuts the line of action of $\underline{F}$ at right angles. If the point of intersection of the two lines is R, then the distance d between O and R is the lever arm, and the magnitude of the moment $\underline{M}$ of force $\underline{F}$ about point O is:

$$M = d\,F \qquad (3.1)$$

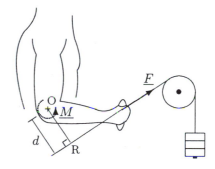

Figure 3.3 *Moment of a force about a point.*

3.3 Direction of Moment

The moment of a force about a point acts in a direction perpendicular to the plane upon which the point and the force lie. For example, in Figure 3.4, point O and the line of action of force $\underline{F}$ lie on plane A. The line of action of moment $\underline{M}$ of force $\underline{F}$ about point O is perpendicular to plane A. The direction and sense of the moment vector along its line of action can be determined using the

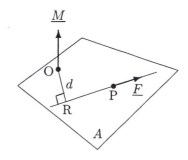

Figure 3.4 *Direction of the moment vector.*

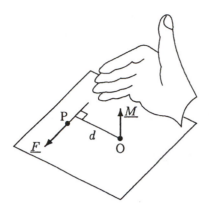

Figure 3.5 *The right-hand-rule.*

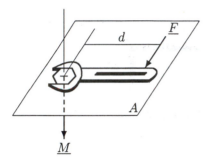

Figure 3.6 *Wrench and bolt.*

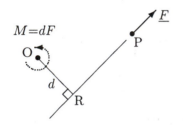

Figure 3.7 *The magnitude and direction of moment of F about O.*

right-hand-rule. As illustrated in Figure 3.5, when the fingers of the right hand curl in the direction that the applied force tends to rotate the body about point O, the right hand thumb points in the direction of the moment vector. More specifically, extend the right hand with the thumb at a right angle to the rest of the fingers, position the finger tips in the direction of the applied force, and position the hand so that the point about which the moment is to be determined faces toward the palm. The tip of the thumb points in the direction of the moment vector.

The line of action and direction of the moment vector can also be explained using a wrench and a right-treaded bolt (Figure 3.6). When a force is applied on the handle of the wrench, a torque is generated that rotates the wrench. The line of action of this torque coincides with the centerline of the bolt. Owing to the torque, the bolt will either advance into or retract from the board depending on how the force is applied. As in Figure 3.6, if the force causes a clockwise rotation, then the direction of torque is "into" the board and the bolt will advance into the board. If the force causes a counterclockwise rotation, then the direction of torque is "out of" the board and the bolt will retract from the board.

In Figure 3.7, if point O and force F lie on the surface of the page, then the line of action of moment M is perpendicular to the page. If you pin the otherwise unencumbered page at point O, F will rotate the page in the counterclockwise direction. Using the right-hand-rule, that corresponds to the direction away from the page. To refer to the direction of the moments of coplanar force systems, it may be sufficient to say that a particular moment is either clockwise (cw) or counterclockwise (ccw).

3.4 Dimension and Units of Moment

By definition, moment is equal to the product of applied force and the length of the moment arm. Therefore, the dimension of moment is equal to the dimension of force (ML/T^2) times the dimension of length (L):

$$[\text{MOMENT}] = [\text{FORCE}]\,[\text{MOMENT ARM}] = \frac{ML}{T^2}\,L = \frac{ML^2}{T^2}$$

The units of moment in different systems are listed in Table 3.1.

Table 3.1 *Units of moment and torque (1 lb-ft = 1.3573 N-m).*

System	Units of Moment and Torque
SI	Newton-meter (N-m)
c-g-s	dyne-centimeter (dyn-cm)
British	pound-foot (lb-ft)

3.5 Some Fine Points About the Moment Vector

• The moment of a force is invariant under the operation of sliding the force vector along its line of action, which is illustrated in Figure 3.8. For all cases illustrated, the moment of force about point O is:

$$M = d\,F \qquad (\text{ccw})$$

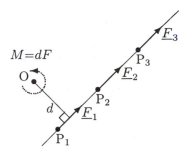

Figure 3.8 *Moment is invariant under the operation of sliding the force along its line of action.*

where length d is always the shortest distance between point O and the line of action of $\underline{F}$. Again for all three cases shown in Figure 3.8, the forces generate a counterclockwise moment.

• Let $\underline{F}_1$ and $\underline{F}_2$ shown in Figure 3.9 be two forces with equal magnitude ($F_1 = F_2 = F$) and the same line of action, but acting in opposite directions. The moment $\underline{M}_1$ of force $\underline{F}_1$ and the moment $\underline{M}_2$ of force $\underline{F}_2$ about point O have an equal magnitude ($M_1 = M_2 = M = d\,F$), but opposite directions ($\underline{M}_1 = -\underline{M}_2$).

• The magnitude of the moment of an applied force increases with an increase in the length of the moment arm. That is, the greater the distance the point about which the moment is to be calculated from the line of action of the force vector, the higher the magnitude of the corresponding moment vector.

• The moment of a force about a point that lies on the line of action of the force is zero, because the length of the moment arm is zero (Figure 3.10).

• A force applied to a body may tend to rotate or bend the body in one direction with respect to one point and in the opposite direction with respect to another point in the same plane.

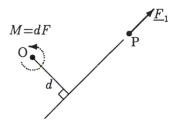
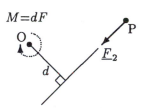

Figure 3.9 *Opposite moments.*

• The principles of resolution of forces into their components along appropriate directions can be utilized to simplify the calculation of moments. For example, in Figure 3.11, the applied force $\underline{F}$ is resolved into its components $\underline{F}_x$ and $\underline{F}_y$ along the x and y directions, such that:

$$F_x = F\cos\theta$$
$$F_y = F\sin\theta$$

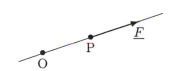

Figure 3.10 *The moment of $\underline{F}$ about point O is zero.*

Since point O lies on the line of action of $\underline{F}_x$, the moment arm of $\underline{F}_x$ relative to point O is zero. Therefore, the moment of $\underline{F}_x$ about point O is zero. On the other hand, r is the length of the moment arm for force $\underline{F}$ relative to point O. Therefore, the moment of force $\underline{F}_y$ about point O is:

$$M = r\,F_y = r\,F\sin\theta \qquad (\text{cw})$$

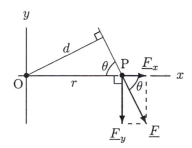

Figure 3.11 *Component of $\underline{F}$.*

Note that this is also the moment $d\,F$ generated by the resultant force vector $\underline{F}$ about point O, because $d = r\sin\theta$.

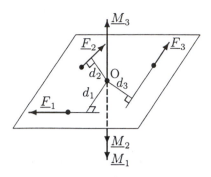

Figure 3.12 *The net moment.*

3.6 The Net or Resultant Moment

When there is more than one force applied on a body, the net or resultant moment can be calculated by considering the vector sum of the moments of all forces. For example, consider the coplanar three-force system shown in Figure 3.12. Let d_1, d_2, and d_3 be the moment arms of $\underline{F}_1$, $\underline{F}_2$, and $\underline{F}_3$ relative to point O. These forces produce moments $\underline{M}_1$, $\underline{M}_2$, and $\underline{M}_3$ about point O, which can be calculated as follows:

$$M_1 = d_1\,F_1 \qquad \text{(cw)}$$
$$M_2 = d_2\,F_2 \qquad \text{(cw)}$$
$$M_3 = d_3\,F_3 \qquad \text{(ccw)}$$

The net moment $\underline{M}_{net}$ generated on the body due to forces $\underline{F}_1$, $\underline{F}_2$, and $\underline{F}_3$ about point O is equal to the vector sum of the moments of all forces about the same point:

$$\underline{M}_{net} = \underline{M}_1 + \underline{M}_2 + \underline{M}_3 \qquad (3.2)$$

A practical way of determining the magnitude and direction of the net moment for coplanar force systems will be discussed next. Note that for the case illustrated in Figure 3.12, the individual moments are either clockwise or counterclockwise. Therefore, the resultant moment must be either clockwise or counterclockwise. Choose or guess the direction of the resultant moment. For example, if we assume that the resultant moment vector is clockwise, then the clockwise moments $\underline{M}_1$ and $\underline{M}_2$ are positive and the counterclockwise moment $\underline{M}_3$ is negative. The magnitude of the net moment can now be determined by simply adding the magnitudes of the positive moments and subtracting the negatives:

$$M_{net} = M_1 + M_2 - M_3 \qquad (3.3)$$

Depending on the numerical values of M_1, M_2, and M_3, this equation will give a positive, negative, or zero value for the net moment. If the computed value is positive, then it is actually the magnitude of the net moment and the chosen direction for the net moment was correct (in this case, clockwise). If the value calculated is negative, then the chosen direction for the net moment was wrong, but can readily be corrected. If the chosen direction was clockwise, a negative value will indicate that the correct direction for the net moment is counterclockwise. (Note that magnitudes of vector quantities are scalar quantities that are always positive.) Once the correct direction for the net moment is indicated, the negative sign in front of the value calculated can be eliminated. The third possibility is that the value calculated from Eq. (3.3) may be zero. If the net moment is equal to zero, then the body is said to be in *rotational equilibrium*. This case will be discussed in detail in the following chapter.

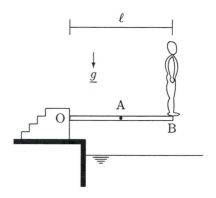

Example 3.1 Figure 3.13 illustrates a person preparing to dive into a pool. The horizontal diving board has a uniform thickness, mounted to the ground at O, has a mass of 120 kg, and is $\ell = 4$ m in length. The person has a mass of 90 kg and stands at B which is the free-end of the board. Point A indicates the location of the center of gravity of the board. Point A is equidistant from points O and B.

Determine the moments generated about point O by the weights of the person and the board. Calculate the net moment about point O.

Figure 3.13 *A person is preparing to dive.*

Solution: Let m_1 and m_2 be the masses and W_1 and W_2 be the weights of the person and diving board, respectively. W_1 and W_2 can be calculated because the masses of the person and the board are given:

$$W_1 = m_1\, g = (90 \text{ kg})(9.8 \text{ m/s}^2) = 882 \text{ N}$$

$$W_2 = m_2\, g = (120 \text{ kg})(9.8 \text{ m/s}^2) = 1176 \text{ N}$$

The person is standing at B. Therefore, the weight $\underline{W}_1$ of the person is applied on the board at B. The mass of the diving board produces a force system distributed over the entire length of the board. The resultant of this distributed force system is equal to the weight $\underline{W}_2$ of the board. For practical purposes and since the board has a uniform thickness, we can assume that the weight of the board is a concentrated force acting at A, which is the center of gravity of the board.

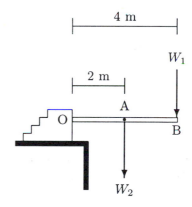

As shown in Figure 3.14, weights $\underline{W}_1$ and $\underline{W}_2$ act vertically downward or in the direction of gravitational acceleration. The diving board is horizontal. Therefore, the distance between O and B (ℓ) is the length of the moment arm for $\underline{W}_1$ and the distance between O and A ($\ell/2$) is the length of the moment arm for $\underline{W}_2$. Therefore, moments $\underline{M}_1$ and $\underline{M}_2$ due to $\underline{W}_1$ and $\underline{W}_2$ about point O are:

Figure 3.14 *Forces acting on the diving board.*

$$M_1 = \ell\, W_1 = (4 \text{ m})(882 \text{ N}) = 3528 \text{ N-m} \quad \text{(cw)}$$

$$M_2 = \frac{\ell}{2}\, W_2 = (2 \text{ m})(1176 \text{ N}) = 2352 \text{ N-m} \quad \text{(cw)}$$

Since both moments have a clockwise direction, the net moment must have a clockwise direction as well. The magnitude of the net moment about point O is:

$$M_{net} = M_1 + M_2 = 5880 \text{ N-m} \quad \text{(cw)}$$

(a)

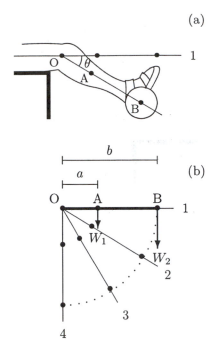

(b)

Figure 3.15 *Example 3.2.*

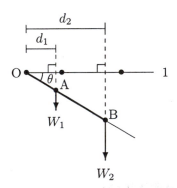

Figure 3.16 *Forces and moment arms when the lower leg makes an angle θ with the horizontal.*

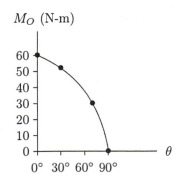

Figure 3.17 *Variation of moment with angle θ.*

Example 3.2 As illustrated in Figure 3.15a, consider an athlete wearing a weight boot, and from a sitting position, doing lower leg flexion/extension exercises to strengthen quadricep muscles. The weight of the athlete's lower leg is $W_1 = 50\,N$ and the weight of the boot is $W_2 = 100\,N$. As measured from the knee joint at O, the center of gravity (A) of the lower leg is located at a distance $a = 20$ cm and the center of gravity (B) of the weight boot is located at a distance $b = 50$ cm.

Determine the net moment generated about the knee joint when the lower leg is extended horizontally (position 1), and when the lower leg makes an angle of 30° (position 2), 60° (position 3), and 90° (position 4) with the horizontal (Figure 3.15b).

Solution: At position 1, the lower leg is extended horizontally and the long axis of the leg is perpendicular to the lines of action of W_1 and W_2. Therefore, a and b are the lengths of the moment arms for W_1 and W_2, respectively. Both W_1 and W_2 apply clockwise moments about the knee joint. The net moment M_O about the knee joint when the lower leg is at position 1 is:

$$M_O = a\,W_1 + b\,W_2$$
$$= (0.20)(50) + (0.50)(100)$$
$$= 60\,\text{N-m} \qquad \text{(cw)}$$

Figure 3.16 illustrates the external forces acting on the lower leg and their moment arms (d_1 and d_2) when the lower leg makes an angle θ with the horizontal. From the geometry of the problem:

$$d_1 = a\,\cos\,\theta$$
$$d_2 = b\,\cos\,\theta$$

Therefore, the net moment about O is:

$$M_O = d_1\,W_1 + d_2\,W_2$$
$$= a\,\cos\,\theta\,W_1 + b\,\cos\,\theta\,W_2$$
$$= (a\,W_1 + b\,W_2)\cos\,\theta$$

The term in the parentheses has already been calculated as 60 N-m. Therefore, we can write:

$$M_O = 60\,\cos\,\theta$$

For position 1: $\theta = 0°$ $M_O = 60\,\text{N-m}$ (cw)
For position 2: $\theta = 30°$ $M_O = 52\,\text{N-m}$ (cw)
For position 3: $\theta = 60°$ $M_O = 30\,\text{N-m}$ (cw)
For position 4: $\theta = 90°$ $M_O = 0$ (cw)

In Figure 3.17, the moment generated about the knee joint is plotted as a function of angle θ.

Example 3.3 Figure 3.18a illustrates an athlete doing shoulder muscle strengthening exercises by lowering and raising a barbell with straight arms. The position of the arms when they make an angle θ with the vertical is simplified in Figure 3.18b. O represents the shoulder joint, A is the center of gravity of one arm, and B is a point of intersection of the centerline of the barbell and the extension of line OA. The distance between O and A is $a = 24$ cm and the distance between O and B is $b = 60$ cm. Each arm arm weighs $W_1 = 50$ N and the total weight of the barbell is $W_2 = 300$ N.

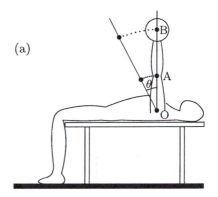
(a)

Determine the net moment due to $\underline{W}_1$ and $\underline{W}_2$ about the shoulder joint as a function of θ, which is the angle the arm makes with the vertical. Calculate the moments for $\theta = 0°$, $15°$, $30°$, $45°$, and $60°$.

Solution: To calculate the moments generated about the shoulder joint by $\underline{W}_1$ and $\underline{W}_2$, we need to determine the moment arms d_1 and d_2 of forces $\underline{W}_1$ and $\underline{W}_2$ relative to O. From the geometry of the problem (Figure 3.18b), the lengths of the moment arms are:

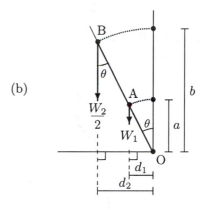
(b)

$$d_1 = a \, \sin\theta$$

$$d_2 = b \, \sin\theta$$

Figure 3.18 *An exercise to strengthen the shoulder muscles, and a simple model of the arm.*

Since the athlete is using both arms, the total weight of the barbell is assumed to be shared equally by each arm. Also note that relative to the shoulder joint, both the weight of the arm and the weight of the barbell are trying to rotate the arm in the counterclockwise direction. Moments $\underline{M}_1$ and $\underline{M}_2$ due to $\underline{W}_1$ and $\underline{W}_2$ about point O are:

$$M_1 = d_1 W_1 = a \, W_1 \sin\theta = (0.24)(50) \sin\theta = 12 \, \sin\theta$$

$$M_2 = d_2\frac{W_2}{2} = b\frac{W_2}{2}\sin\theta = (0.60)\left(\frac{300}{2}\right)\sin\theta = 90 \, \sin\theta$$

Since both moments are counterclockwise, the net moment must be counterclockwise as well. Therefore, the net moment $\underline{M}_O$ generated about the shoulder joint is:

$$M_O = M_1 + M_2 = 12 \, \sin\theta + 90 \, \sin\theta = 102 \, \sin\theta \quad \text{N-m} \quad \text{(ccw)}$$

To determine the magnitude of the moment about O, for $\theta = 0°$, $15°$, $30°$, $45°$, and $60°$, all we need to do is evaluate the sines and carry out the multiplications. The results are provided in Table 3.2.

Table 3.2 *Moment about the shoulder joint (Example 3.3).*

θ	$\sin\theta$	M_O (N-m)
0°	0.000	0.0
15°	0.259	26.4
30°	0.500	51.0
45°	0.707	72.1
60°	0.866	88.3

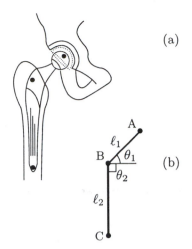

Figure 3.19 *Total hip joint prosthesis.*

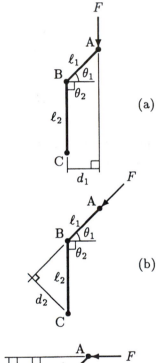

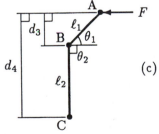

Figure 3.20 *Example 3.4.*

Example 3.4 Consider the total hip joint prosthesis shown in Figure 3.19. The geometric parameters of the prosthesis are such that $\ell_1 = 50$ mm, $\ell_2 = 100$ mm, $\theta_1 = 45°$, and $\theta_2 = 90°$. Assume that, when standing symmetrically on both feet, a joint reaction force of $F = 400$ N is acting at the femoral head due to the body weight of the patient. For the sake of illustration, consider three different lines of action for the applied force, which are shown in Figure 3.20.

Determine the moments generated about points B and C on the prosthesis for all cases shown.

Solution: For each case shown in Figure 3.20, the line of action of the joint reaction force is different, and therefore the lengths of the moment arms are different.

From the geometry of the problem in Figure 3.20a, we can see that the moment arm of force F about points B and C are the same:

$$d_1 = \ell_1 \cos\theta_1 = (50)(\cos 45°) = 35 \text{ mm}$$

Therefore, the moments generated about points B and C are:

$$M_B = M_C = d_1 F = (0.035)(400) = 14 \text{ N-m} \qquad (\text{cw})$$

For the case shown in Figure 3.20b, point B lies on the line of action of the joint reaction force. Therefore, the length of the moment arm for point B is zero, and:

$$M_B = 0$$

For the same case, the length of the moment arm and the moment about point C are:

$$d_2 = \ell_2 \cos\theta_1 = (100)(\cos 45°) = 71 \text{ mm}$$
$$M_C = d_2 F = (0.071)(400) = 28 \text{ N-m} \qquad (\text{ccw})$$

For the case shown in Figure 3.20c, the moment arms relative to B and C are:

$$d_3 = \ell_1 \sin\theta_1 = (50)(\sin 45°) = 35 \text{ mm}$$
$$d_4 = d_3 + \ell_2 = (35) + (100) = 135 \text{ mm}$$

Therefore, the moments generated about points B and C are:

$$M_B = d_3 F = (0.035)(400) = 14 \text{ N-m} \qquad (\text{ccw})$$
$$M_C = d_4 F = (0.135)(400) = 54 \text{ N-m} \qquad (\text{ccw})$$

3.7 The Couple and Couple-Moment

A special arrangement of forces that is of importance is called *couple*, which is formed by two parallel forces with equal magnitude and opposite directions. On a rigid body, the couple has a pure rotational effect. The rotational effect of a couple is quantified with *couple-moment*.

Consider the forces shown in Figure 3.21, which are applied at A and B. Note that the net moment about point A is $M = dF$ (cw), which is due to the force applied at B. The net moment about point B is also $M = dF$ (cw), which is due to the force applied at A. Consider point C. The distance between C and B is b, and therefore, the distance between C and A is $d - b$. The net moment about C is equal to the sum of the clockwise moments of forces applied at A and B with moment arms $d - b$ and b. Therefore:

$$M = (d - b) F + b F = d F \quad \text{(cw)}$$

It can be concluded without further proof that the couple has the same moment about every point in space. If F is the magnitude of the forces forming the couple and d is the perpendicular distance between the lines of actions of the forces, then the magnitude of the couple-moment is:

$$M = d F \quad (3.4)$$

The direction of the couple-moment can be determined by the right-hand-rule.

3.8 Translation of Forces

The overall effect of a pair of forces applied on a rigid body is zero if the forces have an equal magnitude and the same line of action, but are acting in opposite directions. Keeping this in mind, consider the force with magnitude F applied at point P_1 in Figure 3.22a. As illustrated in Figure 3.22b, this force may be translated to point P_2 by placing a pair of forces at P_2 with equal magnitude (F), having the same line of action, but acting in opposite directions. Note that the original force at P_1 and the force at P_2 that is acting in a direction opposite to that of the original force form a couple. This couple produces a counterclockwise moment with magnitude $M = dF$, where d is the shortest distance between the lines of action of forces at P_1 and P_2. Therefore, as illustrated in Figure 3.22c, the couple can be replaced by the couple-moment.

Provided that the original force was applied to a rigid body, the one-force system in Figure 3.22a, the three-force system in Figure 3.22b, and the one-force and one couple-moment system in Figure 3.22c are mechanically equivalent.

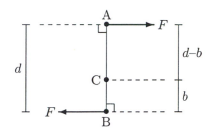

Figure 3.21 *A couple.*

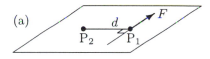

(a)

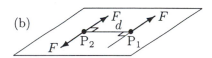

(b)

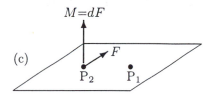

(c)

Figure 3.22 *Translation of a force from point P_1 to P_2.*

3.9 Moment as a Vector Product

We have been applying the scalar method of determining the moment of a force about a point. The scalar method is satisfactory to analyze relatively simple coplanar force systems and systems in which the perpendicular distance between the point and the line of action of the applied force are easy to calculate. The analysis of more complex problems can be simplified by utilizing additional mathematical tools.

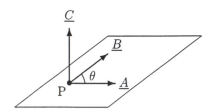

Figure 3.23 $\underline{C}$ *is the vector product of $\underline{A}$ and $\underline{B}$.*

The concept of the *vector (cross) product* of two vectors was introduced in Appendix B and will be reviewed here. Consider vectors $\underline{A}$ and $\underline{B}$, shown in Figure 3.23. The cross product of $\underline{A}$ and $\underline{B}$ is equal to a third vector, $\underline{C}$:

$$\underline{C} = \underline{A} \times \underline{B} \tag{3.5}$$

The product vector $\underline{C}$ has the following properties:

• The magnitude of $\underline{C}$ is equal to the product of the magnitude of $\underline{A}$, the magnitude of $\underline{B}$, and $\sin\theta$, where θ is the smaller angle between $\underline{A}$ and $\underline{B}$.

$$C = A B \sin\theta \tag{3.6}$$

• The line of action of $\underline{C}$ is perpendicular to the plane formed by vectors $\underline{A}$ and $\underline{B}$.

• The direction and sense of $\underline{C}$ obeys the right-hand-rule.

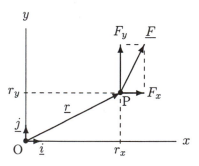

Figure 3.24 *The moment about O is $\underline{M} = \underline{r} \times \underline{F}$.*

The principle of vector or cross product can be applied to determine the moments of forces. The moment of a force about a point is defined as the vector product of the position and force vectors. The *position vector* of a point P with respect to another point O is defined by an arrow drawn from point O to point P. To help understand the definition of moment as a vector product, consider Figure 3.24. Force $\underline{F}$ acts in the xy-plane and has a point of application at P. Force $\underline{F}$ can be expressed in terms of its components F_x and F_y along the x and y directions:

$$\underline{F} = F_x \underline{i} + F_y \underline{j} \tag{3.7}$$

The position vector of point P with respect to point O is represented by vector $\underline{r}$, which can be written in terms of its components:

$$\underline{r} = r_x \underline{i} + r_y \underline{j} \tag{3.8}$$

The components r_x and r_y of the position vector are simply the x and y coordinates of point P as measured from point O.

The moment of force $\underline{F}$ about point O is equal to the vector product of the position vector $\underline{r}$ and force vector $\underline{F}$:

$$\underline{M} = \underline{r} \times \underline{F} \tag{3.9}$$

Using Eqs. (3.7) and (3.8), Eq. (3.9) can alternatively be written as:

$$\begin{aligned}\underline{M} &= (r_x\,\underline{i} + r_y\,\underline{j}) \times (F_x\,\underline{i} + F_y\,\underline{j})\\ &= r_x F_x\,(\underline{i} \times \underline{i}) + r_x F_y\,(\underline{i} \times \underline{j})\\ &\quad + r_y F_x\,(\underline{j} \times \underline{i}) + r_y F_y\,(\underline{j} \times \underline{j})\end{aligned} \qquad (3.10)$$

Recall that $\underline{i} \times \underline{i} = \underline{j} \times \underline{j} = 0$ since the angle that a unit vector makes with itself is zero, and $\sin 0° = 0$. $\underline{i} \times \underline{j} = \underline{k}$ because the angle between the positive x axis and the positive y axis is $90°$ ($\sin 90° = 1$). On the other hand, $\underline{j} \times \underline{i} = -\underline{k}$. For the last two cases, the product is either in the positive z (counterclockwise or out of the page) or negative z (clockwise or into the page) direction. z and unit vector $\underline{k}$ designate the direction perpendicular to the xy-plane (Figure 3.25). Now, Eq. (3.10) can be simplified as:

$$\underline{M} = (r_x F_y - r_y F_x)\,\underline{k} \qquad (3.11)$$

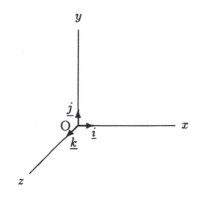

Figure 3.25 *The right-hand-rule applies to Cartesian coordinate directions as well.*

To show that the definition of moment as the vector product of the position and force vectors is consistent with the scalar method of finding the moment, consider the simple case illustrated in Figure 3.26. The force vector $\underline{F}$ is acting in the positive y direction and its line of action is d distance away from point O. Applying the scalar method, the moment about O is:

$$M = d\,F \qquad \text{(ccw)} \qquad (3.12)$$

The force vector is acting in the positive y direction. Therefore:

$$\underline{F} = F\,\underline{j} \qquad (3.13)$$

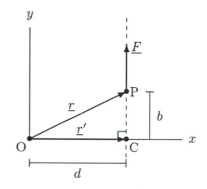

Figure 3.26 $M = d\,F$ *(ccw).*

If b is the y coordinate of the point of application of the force, then the position vector of point P is:

$$\underline{r} = d\,\underline{i} + b\,\underline{j} \qquad (3.14)$$

Therefore, the moment of $\underline{F}$ about point O is:

$$\begin{aligned}\underline{M} &= \underline{r} \times \underline{F}\\ &= (d\,\underline{i} + b\,\underline{j}) \times (F\,\underline{j})\\ &= d\,F\,(\underline{i} \times \underline{j}) + b\,F\,(\underline{j} \times \underline{j})\\ &= d\,F\,\underline{k}\end{aligned} \qquad (3.15)$$

Equations (3.12) and (3.15) carry exactly the same information, in two different ways. Furthermore, the y coordinate (b) of point P does not appear in the solution. This is consistent with the definition of the moment, which is the vector product of the position vector of any point on the line of action of the force and the force itself. In Figure 3.26, if C is the point of intersection of the line of action of force $\underline{F}$ and the x axis, then the position vector $\underline{r}'$ of point C with respect to O is:

$$\underline{r}' = d\,\underline{i} \qquad (3.16)$$

Therefore, the moment of $\underline{F}$ about point O can alternatively be determined as:

$$\begin{aligned}\underline{M} &= \underline{r}'\,\underline{F} \\ &= (d\,\underline{i}) \times (F\,\underline{j}) \\ &= d\,F\,\underline{k}\end{aligned} \tag{3.17}$$

For any two-dimensional problem composed of a system of coplanar forces in the xy-plane, the resultant moment vector has only one component. The resultant moment vector has a direction perpendicular to the xy-plane, acting in the positive or negative z direction.

The method we have outlined to study coplanar force systems using the concept of the vector (cross) product can easily be expanded to analyze three-dimensional situations. In general, the force vector $\underline{F}$ and the position vector $\underline{r}$ of a point on the line of action of $\underline{F}$ about a point O would have up to three components. With respect to the Cartesian coordinate frame:

$$\underline{F} = F_x\,\underline{i} + F_y\,\underline{j} + F_z\,\underline{k} \tag{3.18}$$
$$\underline{r} = r_x\,\underline{i} + r_y\,\underline{j} + r_z\,\underline{k} \tag{3.19}$$

The moment of $\underline{F}$ about point O can be determined as:

$$\begin{aligned}\underline{M} &= \underline{r} \times \underline{F} \\ &= (r_x\,\underline{i} + r_y\,\underline{j} + r_z\,\underline{k}) \times (F_x\,\underline{i} + F_y\,\underline{j} + F_z\,\underline{k}) \\ &= (r_yF_z - r_zF_y)\,\underline{i} + (r_zF_x - r_xF_z)\,\underline{j} \\ &\quad + (r_xF_y - r_yF_x)\,\underline{k}\end{aligned} \tag{3.20}$$

The moment vector can be expressed in terms of its components along the x, y, and z directions:

$$\underline{M} = M_x\,\underline{i} + M_y\,\underline{j} + M_z\,\underline{k} \tag{3.21}$$

By comparing Eqs. (3.20) and (3.21), we can conclude that:

$$\begin{aligned}M_x &= r_yF_z - r_zF_y \\ M_y &= r_zF_x - r_xF_z \\ M_z &= r_xF_y - r_yF_x\end{aligned} \tag{3.22}$$

The following example provides an application of the analysis outlined in the last three sections of this chapter.

Example 3.5 Figure 3.27a illustrates a person using an exercise machine. The "L" shaped beam shown in Figure 3.27b represents the left arm of the person. Points A and B correspond to the shoulder and elbow joints, respectively. Relative to the person, the upper arm (AB) is extended toward the left (x direction)

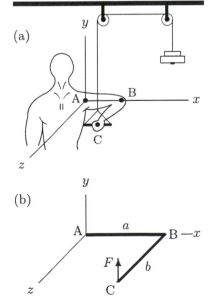

Figure 3.27 *Example 3.5.*

and the lower arm (BC) is extended forward (z direction). At this instant the person is holding a handle that is connected by a cable to a suspending weight. The weight applies an upward (in the y direction) force with magnitude F on the arm at point C. The lengths of the upper arm and lower arm are $a = 25$ cm and $b = 30$ cm, respectively, and the magnitude of the applied force is $F = 200$ N.

Explain how force F can be translated to the shoulder joint at A, and determine the magnitudes and directions of moments developed at the lower and upper arms by $\underline{F}$.

Solution 1:   Scalar method.
The scalar method of finding the moments generated on the lower arm (BC) and upper arm (AB) is illustrated in Figure 3.28, and it utilizes the concepts of couple and couple-moment.

The first step is placing a pair of forces at B with equal magnitude (F) and opposite directions, both being parallel to the original force at C (Figure 3.28a). The upward force at C and the downward force at B form a couple. Therefore, they can be replaced by a couple-moment (shown by a double-headed arrow in Figure 3.28b). The magnitude of the couple-moment is bF. Applying the right-hand-rule, we can see that the couple-moment acts in the negative x direction. If M_x refers to the magnitude of this couple-moment, then:

$$M_x = b\,F \qquad (-x \text{ direction})$$

The next step is to place another pair of forces at A where the shoulder joint is located (Figure 3.28c). This time, the upward force at B and the downward force at A form a couple, and again, they can be replaced by a couple-moment (Figure 3.28d). The magnitude of this couple-moment is aF, and it has a direction perpendicular to the xy-plane, or, it is acting in the positive z direction. Referring to the magnitude of this moment as M_z, then:

$$M_z = a\,F \qquad (+z \text{ direction})$$

To evaluate the magnitudes of the couple-moments, the numerical values of a, b, and F must be substituted in the above equations:

$$M_x = (0.30)(200) = 60 \text{ N-m}$$
$$M_z = (0.25)(200) = 50 \text{ N-m}$$

The effect of the force applied at C is such that at the elbow joint, the person feels an upward force with magnitude F and a moment with magnitude M_x that is trying to *rotate* the lower arm in the yz-plane. At the shoulder joint, the feeling is such that there is an upward force of F, a torque with magnitude M_x that

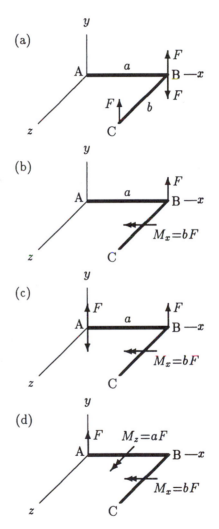

Figure 3.28 *Scalar method.*

is trying to twist the upper arm in the yz-plane, and a moment M_z that is trying to rotate or bend the upper arm in the xy-plane. If the person is able to hold the arm in this position, then he/she is producing sufficient muscle forces to counterbalance these applied forces and moments.

Solution 2:   Vector product method.

The definition of moment as the vector product of the position and force vectors is more straightforward to apply. The position vector of point C (where the force is applied) with respect to point A (where the shoulder joint is located) and the force vector shown in Figure 3.29 can be expressed as follows:

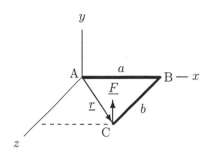

Figure 3.29 *Vector product method (Example 3.5).*

$$\underline{r} = a\,\underline{i} + b\,\underline{k}$$
$$\underline{F} = F\,\underline{j}$$

The cross product of $\underline{r}$ and $\underline{F}$ will yield the moment of force $\underline{F}$ about point A:

$$
\begin{aligned}
\underline{M} &= \underline{r} \times \underline{F} \\
&= (a\,\underline{i} + b\,\underline{k}) \times (F\,\underline{j}) \\
&= a\,F\,(\underline{i} \times \underline{j}) + b\,F\,(\underline{k} \times \underline{j}) \\
&= a\,F\,\underline{k} - b\,F\,\underline{i} \\
&= (0.25)(200)\,\underline{k} - (0.30)(200)\,\underline{i} \\
&= 50\,\underline{k} - 60\,\underline{i}
\end{aligned}
$$

The negative sign in front of $60\,\underline{i}$ indicates that the x component of the moment vector is acting in the negative x direction. Furthermore, there is no component associated with the unit vector $\underline{j}$, which implies that the y component of the moment vector is zero. Therefore, the moment about point A has the following components:

$$M_x = 60\ \text{N-m} \qquad (-x \text{ direction})$$
$$M_y = 0$$
$$M_z = 50\ \text{N-m} \qquad (+z \text{ direction})$$

These results are consistent with those obtained using the scalar method.

3.10 Exercise Problems

Problem 3.1 Consider a person using an exercise apparatus who is holding a handle that is attached to a cable (Figure 3.30). The cable is wrapped around a pulley and attached to a weight pan.

The weight in the weight pan stretches the cable and produces a tensile force F in the cable. This force is transmitted to the person's hand through the handle. The force makes an angle θ with the horizontal and applied to the hand at B. Point A represents the center of gravity of the person's lower arm and O is a point along the center of rotation of the elbow joint. Assume that points O, A, and B and force F all lie on a plane surface.

If the horizontal distance between O and A is $a = 15$ cm, distance between O and B is $b = 35$ cm, total weight of the lower arm is $W = 20$ N, magnitude of the applied force is $F = 50$ N, and angle $\theta = 30°$, determine the net moment generated about O by F and W.

Answer: $M_O = 5.75$ N-m (ccw)

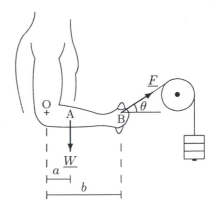

Figure 3.30 *Problem 3.1.*

Problem 3.2 Figure 3.31 illustrates a simplified version of a hamstring strength training system for rehabilitation and athlete training protocols. From a seated position, a patient or athlete flexes the lower leg against a set resistance provided through a cylindrical pad that is attached to a load. For the position illustrated, the lower leg makes an angle θ with the horizontal. Point O represents the knee joint, A is the center of gravity of the lower leg, W is the total weight of the lower leg, F is the magnitude of the force applied by the pad on the lower leg in a direction perpendicular to the long axis of the lower leg, a is the distance between O and A, and b is the distance between O and the line of action of F measured along the long axis of the lower leg.

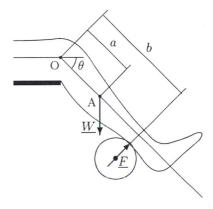

Figure 3.31 *Problem 3.2.*

(a) Determine an expression for the net moment about O due to W and F.
(b) If $a = 20$ cm, $b = 40$ cm, $\theta = 30°$, $W = 60$ N, and $F = 200$ N, calculate the net moment about O.

Answers: (a) $M_O = bF - aW\cos\theta$ (b) $M_O = 69.6$ N-m (ccw)

Problem 3.3 Figure 3.32 illustrates a bench experiment designed to test the strength of materials. In the case illustrated, an intertrochanteric nail that is commonly used to stabilize fractured femoral heads is firmly clamped to the bench such that the distal arm (BC) of the nail is aligned vertically. The proximal arm (AB) of the nail has a length a and makes an angle θ with BC.

As illustrated in Figure 3.32, the intertrochanteric nail is subjected to three experiments by applying forces F_1 (horizontal, toward the right), F_2 (aligned with AB, toward A), and F_3 (vertically downward). Determine expressions for the moment

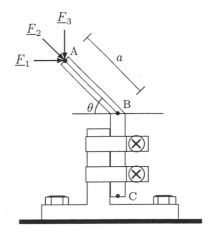

Figure 3.32 *A bench test.*

generated at B by the three forces in terms of force magnitudes and geometric parameters a and θ.

Answers: $M_1 = a\,F_1 \sin\theta$ (cw) $M_2 = 0$ $M_3 = a\,F_3 \cos\theta$ (ccw)

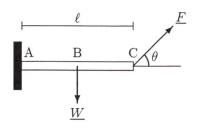

Figure 3.33 *A cantilever beam.*

Problem 3.4 The simple structure shown in Figure 3.33 is called a *cantilever beam* and is one of the fundamental mechanical elements in engineering. A cantilever beam is fixed at one end and free at the other. In Figure 3.33, the fixed and free ends of the beam are identified as A and C, respectively. Point B corresponds to the center of gravity of the beam.

Assume that the beam shown has a weight $W = 100$ N and a length $\ell = 1$ m. A force with magnitude $F = 150$ N is applied at the free-end of the beam in a direction that makes an angle $\theta = 45°$ with the horizontal.

Determine the magnitude and direction of the net moment developed at the fixed-end of the beam.

Answer: $M_A = 56$ N-m (ccw)

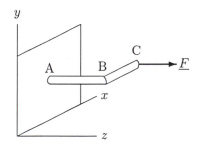

Figure 3.34 *Problem 3.5.*

Problem 3.5 Consider the L-shaped beam illustrated in Figure 3.34. The beam is mounted to the wall at A, the arm AB extends in the z direction, and the arm BC extends in the x direction. A force $\underline{F}$ is applied in the z direction at the free-end of the beam.

(a) If the lengths of arms AB and BC are a and b, respectively, and the magnitude of the applied force is F, observe that the position vector of point C relative to point A can be written as $\underline{r} = b\,\underline{i} + a\,\underline{k}$ and the force vector can be expressed as $\underline{F} = F\,\underline{k}$, where $\underline{i}$, $\underline{j}$, and $\underline{k}$ are unit vectors indicating positive x, y, and z directions, respectively.
(b) Using the cross product of position and force vectors, determine an expression for the moment generated by $\underline{F}$ about A in terms of a, b, and F.
(c) If $a = b = 30$ cm and $F = 20$ N, calculate the magnitude of the moment about A due to $\underline{F}$.

Answers: (b) $\underline{M}_A = -b\,F\,\underline{j}$ (c) $M_A = 6$ N-m

---

Chapter 4

Statics: Analyses of Systems
in Equilibrium

4.1 Overview

Statics is an area within the field of applied mechanics that is concerned with the analysis of rigid bodies in equilibrium. In mechanics, the term *equilibrium* implies that the body of concern is either at rest or moving with a constant velocity. A *rigid body* is one that is assumed to undergo no deformation under the effect of externally applied forces. In reality, there is no rigid material and the concept is an approximation. For some applications, the extent of deformations involved may be so small that the inclusion of the deformation characteristics of the material may not influence the desired analysis. In such cases, the material may be treated as a rigid body.

It should be noted that although the field of statics deals with the analyses of rigid bodies in equilibrium, analyses that deal with the deformation characteristics and strength of materials also start with static analysis. In this chapter, the principles of statics will be introduced and the applications of these principles to relatively simple systems will be provided. In Chapter 5, applications of the same principles to analyze forces involved at and around the major joints of the human body will be demonstrated.

4.2 Newton's Laws of Mechanics

The entire structure of mechanics is based on a few basic laws that were established by Sir Isaac Newton. *Newton's first law* states that a body that is originally at rest will remain at rest, or a body moving with a constant velocity in a straight line will maintain its motion unless an external non-zero resultant force acts on the body. Newton's first law must be considered in conjunction with his second law.

Newton's second law states that if the net or the resultant force acting on a body is not zero, then the body will accelerate in the direction of the resultant force. Furthermore, the magnitude of the acceleration of the body will be directly proportional to the magnitude of the net force acting on the body and inversely proportional to its mass. Newton's second law can be formulated as:

$$\underline{F} = m\,\underline{a} \tag{4.1}$$

This is also known as the *equation of motion*. In Eq. (4.1), $\underline{F}$ is the net or the resultant force (vector sum of all forces) acting on the body, m is the mass of the body, and $\underline{a}$ is its acceleration. Note that both force and acceleration are vector quantities while mass is a scalar quantity. Mass is a measure of the inertia of a body. The tendency of a body to maintain its state of rest or uniform motion along a straight line is called *inertia*. The more inertia a body has, the more difficult it is to start moving it from rest, to change its motion, or to change its direction of motion.

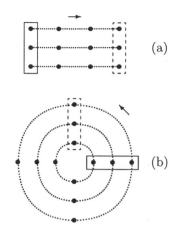

Figure 4.1 *(a) Linear and (b) angular movements.*

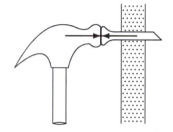

Figure 4.2 *Hammering.*

Figure 4.3 *The harder you push, the harder you will be pushed.*

Figure 4.4 *An ice skater.*

Equation (4.1) is valid for translational or linear motion analyses such as those along a straight path. As illustrated in Figure 4.1a, *linear motion* occurs if all parts of a body move the same distance at the same time and in the same direction. A typical example of linear motion is the vertical movement of an elevator in a shaft.

Equation (4.1) is only one way of formulating Newton's second law, which can alternatively be formulated for rotational or angular motion analysis as:

$$\underline{M} = I \, \underline{\alpha} \qquad (4.2)$$

In Eq. (4.2), $\underline{M}$ is the net moment or torque acting on the body, I is the *mass moment of inertia* of the object, and $\underline{\alpha}$ is its *angular acceleration*. As illustrated in Figure 4.1b, *angular motion* occurs when a body moves in a circular path such that all parts of the body move through the same angle at the same time. These concepts will be discussed in detail in later chapters that cover dynamic analysis.

Newton's third law is based on the observation that there are always two sides when it comes to forces. A force applied to an object is always applied by another object. When a worker pushes a cart, a child pulls a wagon, and a hammer hits a nail, force is applied by one body onto another. Newton's third law states that if two bodies are in contact and body 1 is exerting a force on body 2, then body 2 will apply a force on body 1 in such a way that the two forces will have an equal magnitude but opposite directions.

For example, consider the case case of a hammer pushing a nail (Figure 4.2). It is the force applied by the hammer on the nail that causes the nail to advance. However, it is the force of the nail applied back on the hammer that causes the hammer to stop after every impact upon the nail.

If you press your hand against the edge of a desk, you can see the shape of your hand change, and also feel the force exerted by the desk on your hand (Figure 4.3). The harder you press your hand against the desk, the harder the desk will push your hand back.

Perhaps the best example that can help us understand Newton's third law is a skater applying a force on a wall (Figure 4.4). By pushing against the wall, the skater can move backwards. It is the force exerted by the wall back on the skater that causes the motion of the skater.

Newton's third law can be summarized as "to every action there is an equal and opposite reaction." This law is particularly useful in analyzing complex problems in mechanics and biomechanics in which there are several interacting bodies.

4.3 Conditions for Equilibrium

According to Newton's second law as formulated in Eqs. (4.1) and (4.2), a body will have linear and angular accelerations if the net force and the net moment acting on it are not zero. If the net force and the net moment are zero, then the acceleration (linear and angular) of the body is zero, and, consequently, the velocity (linear and angular) of the body is either constant or zero. When the acceleration is zero, the body is said to be in *equilibrium*. If the velocity is zero as well, then the body is in *static equilibrium* or at *rest*.

Therefore, there are two conditions that need to be satisfied for equilibrium. The body is said to be in *translational equilibrium* if the net force (vector sum of all forces) acting on it is zero:

$$\sum \underline{F} = 0 \tag{4.3}$$

The body is in *rotational equilibrium* if the net moment (vector sum of all moments) acting on it is zero:

$$\sum \underline{M} = 0 \tag{4.4}$$

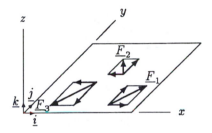

Figure 4.5 *A coplanar force system in the xy-plane.*

Note that there may be a number of forces acting on a body. For example, consider the coplanar force system in Figure 4.5. Assume that the three forces are acting on an object in the xy-plane. These forces can be expressed in terms of their components along the x and y directions:

$$\begin{aligned}
\underline{F}_1 &= F_{1x}\,\underline{i} + F_{1y}\,\underline{j} \\
\underline{F}_2 &= -F_{2x}\,\underline{i} + F_{2y}\,\underline{j} \\
\underline{F}_3 &= -F_{3x}\,\underline{i} - F_{3y}\,\underline{j}
\end{aligned} \tag{4.5}$$

In Eq. (4.5), F_{1x}, F_{2x}, and F_{3x} are the scalar components of $\underline{F}_1$, $\underline{F}_2$, and $\underline{F}_3$ in the x direction, F_{1y}, F_{2y}, and F_{3y} are their scalar components in the y direction, and $\underline{i}$ and $\underline{j}$ are the unit vectors indicating positive x and y directions. These equations can be substituted into Eq. (4.3), and the x and y components of all forces can be grouped together to write:

$$\sum \underline{F} = (F_{1x} - F_{2x} - F_{3x})\,\underline{i} + (F_{1y} + F_{2y} - F_{3y})\,\underline{j} = 0$$

For this equilibrium to hold, each group must individually be equal to zero. That is:

$$\begin{aligned}
F_{1x} - F_{2x} - F_{3x} &= 0 \\
F_{1y} + F_{2y} - F_{3y} &= 0
\end{aligned} \tag{4.6}$$

In other words, for a body to be in translational equilibrium, the net force acting in the x and y directions must be equal to zero. For three-dimensional force systems, the net force in the z direction must also be equal to zero. These results may be summarized as follows:

$$\sum F_x = 0$$
$$\sum F_y = 0 \qquad (4.7)$$
$$\sum F_z = 0$$

Caution. The equilibrium conditions given in Eq. (4.7) are in scalar form and must be handled properly. For example, while applying the translational equilibrium condition in the x direction, the forces acting in the positive x direction must be added together and the forces acting in the negative x direction must be subtracted.

As stated earlier, for a body to be in equilibrium, it has to be both in translational and rotational equilibrium. For a body to be in rotational equilibrium, the net moment about any point must be zero. Consider the coplanar system of forces illustrated in Figure 4.6. Assume that these forces are acting on an object in the xy-plane. Let $d_1, d_2,$ and d_3 be the moment arms of forces $F_1, F_2,$ and F_3 relative to point O. Therefore, the moments due to these forces about point O are:

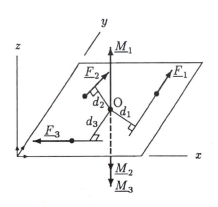

Figure 4.6 *Moments due to coplanar forces.*

$$\underline{M}_1 = d_1\, F_1\, \underline{k}$$
$$\underline{M}_2 = -d_2\, F_2\, \underline{k} \qquad (4.8)$$
$$\underline{M}_3 = -d_3\, F_3\, \underline{k}$$

In Eq. (4.8), $\underline{k}$ is the unit vector indicating the positive z direction. Moment $\underline{M}_1$ is acting in the positive z direction while $\underline{M}_2$ and $\underline{M}_3$ are in the negative z direction. Equation (4.8) can be substituted into the rotational equilibrium condition in Eq. (4.4) to obtain:

$$\sum \underline{M} = \underline{M}_1 + \underline{M}_2 + \underline{M}_3 = (d_1 F_1 - d_2 F_2 - d_3 F_3)\, \underline{k} = 0$$

For this equilibrium to hold:

$$d_1\, F_1 - d_2\, F_2 - d_3\, F_3 = 0 \qquad (4.9)$$

That is, in this case, the resultant moment in the z direction must be zero. Instead of a coplanar system, if we had a three-dimensional force system, then we could have moments in the x and y directions as well. The rotational equilibrium condition would also require us to have zero net moment in the x and y directions. For three-dimensional systems:

$$\sum M_x = 0$$
$$\sum M_y = 0 \qquad (4.10)$$
$$\sum M_z = 0$$

Caution. The rotational equilibrium conditions given in Eq. (4.10) are also in scalar form and must be handled properly. For example, while applying the rotational equilibrium condition in the z direction, moments acting in the positive z direction must be added together and the moments acting in the negative z direction must be subtracted.

Note that for the coplanar force system shown in Figure 4.6, the moments can alternatively be expressed as:

$$\begin{aligned} M_1 &= d_1\,F_1 \quad (\text{ccw}) \\ M_2 &= d_2\,F_2 \quad (\text{cw}) \\ M_3 &= d_3\,F_3 \quad (\text{cw}) \end{aligned} \qquad (4.11)$$

In other words, with respect to the xy-plane, moments are either clockwise or counterclockwise. Depending on the choice of the positive direction as either clockwise or counterclockwise, some of the moments are positive and others are negative. If we consider the counterclockwise moments to be positive, then the rotational equilibrium about point O would again yield Eq. (4.9):

$$d_1\,F_1 - d_2\,F_2 - d_3\,F_3 = 0$$

4.4 Free-Body Diagrams

Free-body diagrams are constructed to help identify the forces and moments acting on individual parts of a system and to ensure the correct use of the equations of statics. For this purpose, the parts constituting a system are isolated from their surroundings, and the effects of the surroundings are replaced by proper forces and moments. In a free-body diagram, all known and unknown forces and moments are shown. A force is unknown if its magnitude or direction is not known. For the known forces, we indicate the correct directions. If the direction of a force is not known, then we try to predict it. Our prediction may be correct or not. However, the correct direction of the force will be an outcome of the static analysis.

For example, consider a person trying to push a file cabinet to the right on a horizontal surface (Figure 4.7). There are three parts constituting this system: the person, the file cabinet, and the horizontal surface representing Earth. The free-body diagrams of these parts are shown in Figure 4.8. F is the magnitude of the horizontal force applied by the person on the file cabinet to move the file cabinet to the right. Since action and reaction must have equal magnitudes, F is also the magnitude of the force applied by the file cabinet on the person. Since action and reaction must have opposite directions, the force applied by the file cabinet on the person tends to push the person to the left. W_1 and W_2 are the weights of the file cabinet and the person, respectively, and are always directed vertically downward. N_1 and N_2 are the magnitudes of the normal forces on the horizontal surface

Figure 4.7 *A person pushing a file cabinet.*

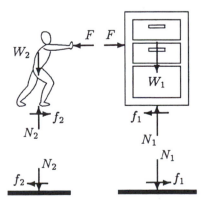

Figure 4.8 *Free-body diagrams.*

applied by the file cabinet and the person, respectively. They are also the magnitudes of the forces applied by the horizontal surface on the file cabinet and the person. f_1 and f_2 are the magnitudes of the frictional forces $\underline{f}_1$ and $\underline{f}_2$ between the file cabinet, the person, and the horizontal surface, both acting in the horizontal direction parallel to the surfaces of contact. Since the file cabinet tends to move toward the right, $\underline{f}_1$ on the file cabinet acts toward the left. Since the person tends to move toward the left, $\underline{f}_2$ on the person acts toward the right, and it is the driving force for the person to push the file cabinet.

For many applications, it is sufficient to consider the free-body diagrams of only a few of the individual parts forming a system. For the example illustrated in Figures 4.7 and 4.8, the free-body diagram of the file cabinet provides sufficient detail to proceed with further analysis.

4.5 Procedure to Analyze Systems in Equilibrium

The general procedure for analyzing the forces and moments acting on rigid bodies in equilibrium is outlined below.

- Draw a simple, neat diagram of the system to be analyzed.

- Draw the free-body diagrams of the parts constituting the system. Show all of the known and unknown forces and moments on the free-body diagrams. Indicate the correct directions of the known forces and moments. If the direction of a force or a moment is not known, try to predict the correct direction, indicate it on the free-body diagram, and do not change it in the middle of the analysis. The correct direction for an unknown force or moment will appear in the solutions. For example, a positive numerical value for an unknown force in the result will indicate that the correct direction was assumed. A negative numerical value, on the other hand, will indicate that the force has a direction opposite to that assumed. In this case, the force magnitude must be made positive, the proper unit of force must be indicated, and the correct direction must be identified in the solution.

- Adopt a proper coordinate system. If the rectangular coordinate system (x, y, z) will be used, one can orient the coordinate axes differently for different free-body diagrams within a single system. A good orientation of the coordinate axes may simplify the analyses. Resolve all forces and moments into their components, and express forces and moments in terms of their components.

- For each free-body diagram, apply the translational and rotational equilibrium conditions. For three-dimensional problems, the number of equations available are six (three translational and three rotational):

$$\sum F_x = 0 \qquad \sum M_x = 0$$
$$\sum F_y = 0 \qquad \sum M_y = 0$$
$$\sum F_z = 0 \qquad \sum M_z = 0$$

For two-dimensional force systems in the xy-plane, the number of equations available is three (two translational and one rotational, and the other three are automatically satisfied):

$$\sum F_x = 0$$
$$\sum F_y = 0$$
$$\sum M_z = 0$$

- Solve these equations simultaneously for the unknowns. Include the correct directions and the units of forces and moments in the solution.

4.6 Notes Concerning the Equilibrium Equations

- Remember that the equilibrium equations given above are in "scalar" form and they do not account for the directions of forces and moments involved. Therefore, for example, while applying the translational equilibrium condition in the x direction, let a force be positive if it is acting in the positive x direction and let it be a negative force if it is acting in the negative x direction. Apply this rule in every direction and also for moments.

- For coplanar force systems and while applying the rotational equilibrium condition, choose a proper point about which all moments are to be calculated. The choice of this point is arbitrary, but a good choice can simplify the calculations.

- For some problems, you may not need to use all of the equilibrium conditions available. For example, if there are no forces in the x direction, then the translational equilibrium condition in the x direction will be satisfied automatically.

- Sometimes, it may be more convenient to apply each rotational equilibrium condition more than once. For example, for a coplanar force system in the xy-plane, the rotational equilibrium condition in the z direction may be applied twice by considering the moments of forces about two different points. In such cases, the third independent equation that might be used is the translational equilibrium condition either in the x or y direction. The rotational equilibrium condition may also be applied three times in a single problem. However, the three points about which moments are calculated must not all lie along a straight line.

- *Two-force systems:* If there are only two forces acting on a body, then the body will be in equilibrium if and only if the forces are

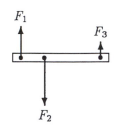

Figure 4.9 *A collinear, two-force system.*

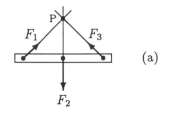

Figure 4.10 *A parallel, three-force system.*

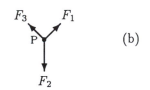

Figure 4.11 *A concurrent, three-force system.*

collinear, have an equal magnitude, and act in opposite directions (Figure 4.9).

• *Three-force systems:* If there are only three forces acting on a body and if the body is in equilibrium, then the forces must form either a parallel or a concurrent system of forces. Furthermore, for either case, the forces form a coplanar force system. A parallel three-force system is illustrated in Figure 4.10. If two of the three forces are known to be parallel to one another, then we can readily assume that the third force is parallel to the other two and establish the line of action of the unknown force.

A concurrent three-force system is illustrated in Figure 4.11. In this case, the lines of action of the forces have a common point of intersection. If the lines of action of two of three forces acting on a body are known and if they are not parallel, then the line of action of the third force can be determined by extending the lines of action of the first two forces until they meet and by drawing a straight line that connects the point of intersection and the point of application of the unknown force. The straight line thus obtained will represent the line of action of the unknown force. To analyze a concurrent force system, the forces forming the system can be translated to the point of intersection (Figure 4.11b). The forces can then be resolved into their components along two perpendicular directions and the translational equilibrium conditions can be employed to determine the unknowns. In such cases, the rotational equilibrium condition is satisfied automatically.

• *Statically determinate and statically indeterminate systems:* For two-dimensional problems in statics, the number of equilibrium equations available is three. Therefore, the maximum number of unknowns (forces and/or moments) that can be determined by applying the equations of equilibrium is limited to three. For two-dimensional problems, if there are three or less unknowns, then the problem is a *statically determinate* one. If the number of unknowns exceeds three, then the problem is *statically indeterminate*. This means that the information we have through statics is not sufficient to solve the problem fully and that we need additional information. Usually, the additional information comes from the material characteristics and properties of the parts constituting the system. The extension of this argument for three-dimensional problems is straightforward. Since there are six equilibrium equations, the maximum number of unknowns that can be determined is limited to six.

4.7   Constraints and Reactions

One way of classifying forces is by saying that they are either active or reactive. Active forces include externally applied loads and gravitational forces. The constraining forces and moments acting on a body are called *reactions*. Usually, reactions are unknowns and must be determined by applying the equations of equilibrium.

Reactive forces and moments are those exerted by the ground, supporting elements such as rollers, wedges, and knife-edges, and connecting members such as cables, pivots, and hinges. Knowing the common characteristics of these members and elements, which are summarized in Table 4.1, can facilitate drawing free-body diagrams.

Table 4.1 *Commonly encountered supports, connections, and connecting elements, and some of their characteristics.*

TYPE OF SUPPORT OR JOINT	SAMPLE GRAPHICAL REPRESENTATIONS	REPRESENTATIONS FOR USE IN FREE-BODY DIAG.	BIOMECHANICS EXAMPLES	UNKNOWNS
Flexible members (cables, ropes)		T	Muscles, ligaments	Magnitude of the tension in the cable or muscle
Two-force members		F F	Muscles	Force magnitude
Rollers and other simple supports (no friction)		F	Bone-to-bone contact	Force magnitude
Hinge or pin connection		F_x F_y	Elbow	Magnitude and direction of force (force components)
Ball-and-socket		F_y F_x F_z	Hip	Magnitude and direction of force
Fixed, welded, or built-in		F_y M_x F_x F_z M_z M_y	Skull	Magnitude and direction of force and moment

4.8 Simply Supported Structures

Mechanical systems are composed of a number of elements linked together in various ways. "Beams" constitute the fundamental elements that form the building blocks in mechanics. A beam connected to other structural elements or to the ground by means of rollers, knife-edges, hinges, pin connections, pivots, or cables forms a mechanical system. We shall first analyze relatively simple examples of such cases to demonstrate the use of the equations of equilibrium.

Example 4.1 As illustrated in Figure 4.12, consider a person standing on a uniform, horizontal beam that is resting on frictionless knife-edge (wedge) and roller supports. Let A and B be where the knife-edge and roller supports contact the beam, C be the center of gravity of the beam, and D be a point on the beam directly under the center of gravity of the person. Assume that the length of the beam (the distance between A and B) is $\ell = 5$ m, the distance between A and D is $d = 3$ m, the weight of the beam is $W_1 = 900$ N, and the mass of the person is $m = 60$ kg.

Calculate the reactions on the beam at A and B.

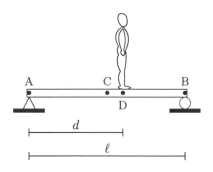

Figure 4.12 *Example 4.1.*

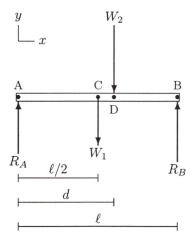

Figure 4.13 *Free-body diagram of the beam.*

Solution: The free-body diagram of the beam is shown in Figure 4.13. The forces acting on the beam include the weights $\underline{W}_1$ and $\underline{W}_2$ of the beam and the person, respectively. Note that the weight of the beam is assumed to be a concentrated load applied at the center of gravity of the beam, which is equidistant from A and B. The magnitude of the weight of the beam is given as $W_1 = 900$ N. The mass of the person is given as $m = 60$ kg. Therefore, the weight of the person is $W_2 = mg = 60 \times 9.8 = 588$ N.

Also acting on the beam are reaction forces $\underline{R}_A$ and $\underline{R}_B$ due to the knife-edge and roller supports, respectively. To a horizontal beam, frictionless knife-edge and roller supports can only apply forces in the vertical. Furthermore, the direction of these forces is toward the beam, not away from the beam. Therefore, we know the directions of the reaction forces, but we do not know their magnitudes. Note that the forces acting on the beam form a parallel force system.

There are two unknowns: R_A and R_B. For the solution of the problem, we can either apply the vertical equilibrium condition along with the rotational equilibrium condition about point A or B, or consider the rotational equilibrium of the beam about points A and B. Here, we shall use the latter approach, and check the results by considering the equilibrium in the vertical direction. Note that the horizontal equilibrium condition is automatically satisfied because there are no forces in that direction.

Consider the rotational equilibrium of the beam about point A. Relative to point A, the length of the moment arm of R_A is zero, the moment arm of W_1 is $\ell/2$, the moment arm of W_2 is d, and the moment arm of R_B is ℓ. Assuming that clockwise moments are positive, then the moments due to W_1 and W_2 are positive, while the moment due to R_B is negative. Therefore, for the rotational equilibrium of the beam about point A:

$$\sum M_A = 0: \qquad \frac{\ell}{2} W_1 + d\, W_2 - \ell\, R_B = 0$$

$$R_B = \frac{1}{\ell}\left(\frac{\ell}{2} W_1 + d\, W_2\right)$$

$$R_B = \frac{1}{5}\left(\frac{5}{2} 900 + 3 \times 588\right) = 802.8 \text{ N} \ (\uparrow)$$

Now, consider the rotational equilibrium of the beam about point B. Relative to point B, the length of the moment arm of R_B is zero, the moment arm of W_2 is $\ell - d$, the moment arm of W_1 is $\ell/2$, and the moment arm of R_A is ℓ. Assuming that clockwise moments are positive, then the moments due to W_1 and W_2 are negative while the moment due to R_A is positive. Therefore, for the rotational equilibrium of the beam about point B:

$$\sum M_B = 0: \qquad \ell\, R_A - \frac{\ell}{2} W_1 - (\ell - d)\, W_2 = 0$$

$$R_A = \frac{1}{\ell}\left(\frac{\ell}{2} W_1 + (\ell - d)\, W_2\right)$$

$$R_A = \frac{1}{5}\left(\frac{5}{2} 900 + (5 - 3)\, 588\right) = 685.2 \text{ N} \ (\uparrow)$$

Remarks:

• The arrows in parentheses in the solutions indicate the correct directions of the forces.

• We can check these results by considering the equilibrium of the beam in the vertical direction:

$$\sum F_y = 0: \qquad R_A - W_1 - W_2 + R_B \overset{?}{=} 0$$

$$685.2 - 900 - 588 + 802.8 \overset{\checkmark}{=} 0$$

• Because of the type of supports used, the beam analyzed in this example has a limited use. A frictionless knife-edge, point support, fulcrum, or roller can provide a support only in the direction perpendicular (normal) to the surfaces of contact. Such supports cannot provide reaction forces in the direction tangent to the contact surfaces. Therefore, if there were active forces on the beam applied along the long axis (x) of the beam, these supports could not provide the necessary reaction forces to maintain the horizontal equilibrium of the beam.

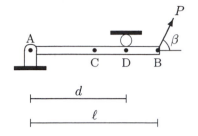

Figure 4.14 *Example 4.2.*

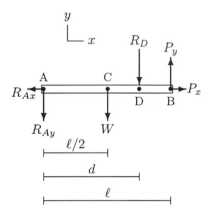

Figure 4.15 *Free-body diagram of the beam.*

Example 4.2 The uniform, horizontal beam shown in Figure 4.14 is hinged to the ground at A. A frictionless roller is placed between the beam and the ceiling at D to constrain the counterclockwise rotation of the beam about the hinge joint. A force that makes an angle $\beta = 60°$ with the horizontal is applied at B. The magnitude of the applied force is $P = 1000$ N. Point C represents the center of gravity of the beam. The distance between A and B is $\ell = 4$ m and the distance between A and D is $d = 3$ m. The beam weighs $W = 800$ N.

Calculate the reactions on the beam at A and D.

Solution: Figure 4.15 illustrates the free-body diagram of the beam under consideration. The horizontal and vertical directions are identified by the x and y axes, respectively. The hinge joint at A constrains the translational movement of the beam both in the x and y directions. Therefore, there exists a reaction force $\underline{R}_A$ at A. We know neither the magnitude nor the direction of this force (two unknowns). As illustrated in Figure 4.15, instead of a single resultant force with two unknowns (magnitude and direction), the reaction force at A can be represented in terms of its components R_{Ax} and R_{Ay} (still two unknowns).

The frictionless roller at D functions as a "stop." It prevents the counterclockwise rotation of the beam. Under the effect of the applied force $\underline{P}$, the beam compresses the roller, and as a reaction, the roller applies a force $\underline{R}_D$ back on the beam. This force is applied in a direction perpendicular to the length of the beam, or vertically downward. The magnitude R_D of this force is not known either. The weight of the beam is given as $W = 800$ N and is assumed to be acting at the center of gravity of the beam. P_x and P_y are the scalar components of the applied force along the x and y directions such that:

$$P_x = P \cos \beta = (1000)(\cos 60) = 500 \text{ N}$$
$$P_y = P \sin \beta = (1000)(\sin 60) = 866 \text{ N}$$

There are three unknowns (R_{Ax}, R_{Ay}, R_B) and we need three equations to solve this problem. Consider the equilibrium condition in the x direction:

$$\sum F_x = 0 : \quad -R_{Ax} + P_x = 0$$
$$R_{Ax} = P_x = 500 \text{ N } (\leftarrow)$$

Next, consider the rotational equilibrium of the beam about point A. Relative to point A, lengths of moment arms for R_{Ax} and R_{Ay} are zero. The length of the moment arm for P_x is zero as well because its line of action passes through A. On the other hand, $\ell/2$ is the length of the moment arm for $\underline{W}$, d is the moment arm for R_D, and ℓ is the moment arm for P_y. Assuming that clockwise moments are positive:

$$\sum M_A = 0 : \qquad \frac{\ell}{2} W + d\, R_D - \ell\, P_y = 0$$

$$R_D = \frac{1}{d}\left(\ell\, P_y - \frac{\ell}{2}\, W \right)$$

$$R_D = \frac{1}{3}\left(4 \times 866 - \frac{4}{2}\, 800 \right) = 621\ \text{N}\ (\downarrow)$$

Now, consider the translational equilibrium of the beam in the y direction:

$$\sum F_y = 0 : \qquad -R_{Ay} - W - R_D + P_y = 0$$

$$R_{Ay} = P_y - W - R_D$$

$$R_{Ay} = 866 - 800 - 621 = -555$$

R_{Ay} is determined to have a negative value. This is not permitted because R_{Ay} corresponds to a scalar quantity and scalar quantities cannot take negative values. The negative sign implies that while drawing the free-body diagram of the beam, we assumed the wrong direction (downward) for the vertical component of the reaction force at A. We can now correct it by writing:

$$R_{Ay} = 555\ \text{N}\ (\uparrow)$$

The corrected free-body diagram of the beam is shown in Figure 4.16. Since we already calculated the components of the reaction force at A, we can also determine the magnitude and direction of the resultant reaction force $\underline{R}_A$ at A. The magnitude of $\underline{R}_A$ is:

$$R_A = \sqrt{(R_{Ax})^2 + (R_{Ay})^2} = 747\ \text{N}$$

If α is the angle $\underline{R}_A$ makes with the horizontal, then:

$$\alpha = \tan^{-1}\left(\frac{R_{Ay}}{R_{Ax}}\right) = \tan^{-1}\left(\frac{555}{500}\right) = 50°$$

The modified free-body diagram of the beam showing the force resultants is illustrated in Figure 4.17.

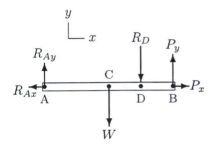

Figure 4.16 *Corrected free-body diagram of the beam.*

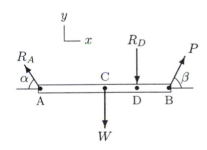

Figure 4.17 *Resultant forces acting on the beam.*

Example 4.3 The uniform, horizontal beam shown in Figure 4.18 is hinged to the wall at A and supported by a cable attached to the beam at B. At the other end, the cable is attached to the wall such that it makes an angle $\beta = 53°$ with the horizontal. Point C represents the center of gravity of the beam, which is equidistant from A and B. A load that weighs $W_2 = 400\ \text{N}$ is placed on the beam such that its center of gravity is directly above C. If the length of the beam is $\ell = 4\ \text{m}$ and the weight of the beam is $W_1 = 600\ \text{N}$, calculate the tension T in the cable and the reaction force on the beam at A.

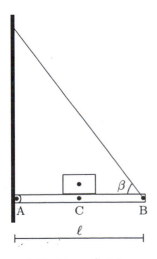

Figure 4.18 *Example 4.3.*

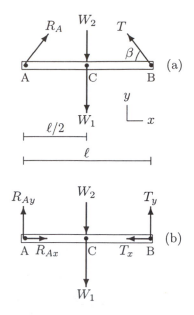

Figure 4.19 *Forces acting on the beam.*

Solution: Figure 4.19 illustrates the free-body diagram of the beam under consideration. The horizontal and vertical directions are identified by the x and y axes, respectively. The hinge joint at A constrains the translational movement of the beam both in the x and y directions. In Figure 4.19b, the reaction force $\underline{R}_A$ at A is represented in terms of its components R_{Ax} and R_{Ay} (two unknowns). Because of the weight of the beam and the load, the cable is stretched, or a tensile force $\underline{T}$ is developed in the cable. This force is applied by the cable on the beam in a direction along the length of the cable that makes an angle β with the horizontal. In other words, we know the direction of $\underline{T}$, but we do not know its magnitude (another unknown). In Figure 4.19b, the force applied by the cable on the beam is expressed in terms of its scalar components T_x and T_y.

We have three unknowns: R_{Ax}, R_{Ay}, and T. First, consider the rotational equilibrium of the beam about A. Relative to A, there are three moment-producing forces: W_1, W_2, and T_y with moment arms $\ell/2, \ell/2,$ and ℓ, respectively. Assuming that clockwise moments are positive:

$$\sum M_A = 0: \quad \frac{\ell}{2} W_1 + \frac{\ell}{2} W_2 - \ell\, T_y = 0$$

$$T_y = \frac{1}{2}(W_1 + W_2)$$

$$T_y = \frac{1}{2}(600 + 400) = 500\ \text{N}\ (\uparrow)$$

T_y is the vertical component of $\underline{T}$ that makes an angle $\beta = 53°$ with the horizontal. Therefore:

$$T = \frac{T_y}{\sin \beta} = \frac{500}{\sin 53°} = 626.1\ \text{N}$$

The x component of the tensile force applied by the cable on the beam is:

$$T_x = T\, \cos \beta = 626.1\, \cos 53° = 376.8\ \text{N}\ (\leftarrow)$$

Next, consider the translational equilibrium of the beam in the x direction:

$$\sum F_x = 0: \quad R_{Ax} - T_x = 0$$
$$R_{Ax} = T_x = 376.8\ \text{N}\ (\rightarrow)$$

Finally, consider the translational equilibrium of the beam in the y direction:

$$\sum F_y = 0: \quad R_{Ay} - W_1 - W_2 + T_y = 0$$
$$R_{Ay} = W_1 + W_2 - T_y$$
$$R_{Ay} = 600 + 400 - 500 = 500\ \text{N}\ (\uparrow)$$

Now that we determined the scalar components of the reaction force at A, we can also calculate its magnitude:

$$R_A = \sqrt{(R_{Ax})^2 + (R_{Ay})^2} = 626.1 \text{ N}$$

If α is the angle $\underline{R}_A$ makes with the horizontal, then:

$$\alpha = \tan^{-1}\left(\frac{R_{Ay}}{R_{Ax}}\right) = \tan^{-1}\left(\frac{500}{376.8}\right) = 53°$$

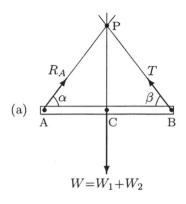

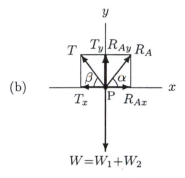

Remarks:

- The weights of the beam and load on the beam have a common line of action. As illustrated in Figure 4.20a, their effects can be combined and expressed by a single weight $W = W_1 + W_2 = 1000$ N that acts at the center of gravity of the beam. In addition to $\underline{W}$, we also have $\underline{R}_A$ and $\underline{T}$ applied on the beam. In other words, we have a three-force system. Since these forces do not form a parallel force system, they have to be concurrent. Therefore, the lines of action of the forces must meet at a single point, say P. As illustrated in Figure 4.20b, if we slide the forces to point P and express them in terms of their components, we can observe equilibrium in the x and y directions such that:

In the x direction: $R_{Ax} = T_x$

In the y direction: $R_{Ay} + T_y = W$

Figure 4.20 *A concurrent force system.*

- The fact that $\underline{R}_A$ and $\underline{T}$ have an equal magnitude and they both make a 53° angle with the horizontal is due to the symmetry of the problem with respect to a plane perpendicular to the xy-plane that passes through the center of gravity of the beam.

4.9 Cable-Pulley Systems and Traction Devices

Cable-pulley arrangements are commonly used to elevate weights and have applications in the design of traction devices used in patient rehabilitation. For example, consider the simple arrangement in Figure 4.21 where a person is trying to lift a load through the use of a cable-pulley system. Assume that the person lifted the load from the floor and is holding it in equilibrium. The cable is wrapped around the pulley, which is housed in a case that is attached to the ceiling. Figure 4.22 shows free-body diagrams of the pulley and the load. r is the radius of the pulley and O represents a point along the centerline (axle or shaft) of the pulley. When the person pulls the cable to lift the load, a force is applied on the pulley that is transmitted to the ceiling via the case housing the pulley. As a reaction, the ceiling applies a force back on the pulley through the shaft that connects the pulley and the case housing the pulley. In other words, there

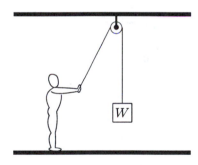

Figure 4.21 *A cable-pulley arrangement.*

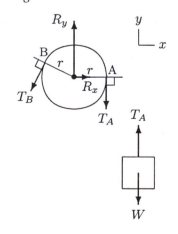

Figure 4.22 *Pulley and load.*

exists a reaction force $\underline{R}$ on the pulley. In Figure 4.22, the reaction force at O is represented by its scalar components R_x and R_y.

The cable is wrapped around the pulley between A and B. If we ignore the frictional effects between the cable and pulley, then the magnitude T of the tension in the cable is constant everywhere in the cable. To prove this point, let's assume that the magnitude of the tensile force generated in the cable is not constant. Let T_A and T_B be the magnitudes of the tensile forces applied by the cable on the pulley at A and B, respectively. Regardless of the way the cable is wrapped around the pulley, the cable is tangent to the circumference of the pulley at the initial and final contact points. This implies that straight lines drawn from O toward A and B will cut the cable at right angles. Therefore, as measured from O, the moment arms of the forces applied by the cable on the pulley are always equal to the radius of the pulley. Now, consider the rotational equilibrium of the pulley about O:

$$\sum M_O = 0 : \quad r\,T_B - r\,T_A = 0$$
$$T_B = T_A$$

Therefore, the tension T in the cable is the same on the two sides of the pulley. For the case illustrated in Figures 4.21 and 4.22, the vertical equilibrium of the load requires that the tension in the cable is equal to the weight W of the load to be lifted.

Note that R_x and R_y applied by the ground (through the ceiling) on the pulley do not produce any moment about O. If needed, these forces could be determined by considering the horizontal and vertical equilibrium of the pulley.

Figures 4.23 and 4.24 illustrate examples of simple traction devices. Such devices are designed to maintain parts of the human body in particular positions for healing purposes. For such devices to be effective, they must be designed to transmit forces properly to the body part in terms of force direction and magnitude. Different arrangements of cables and pulleys can transmit different magnitudes of forces and in different directions. For example, the traction in Figure 4.23 applies a horizontal force to the leg with magnitude equal to the weight in the weight pan. On the other hand, the traction in Figure 4.24 applies a horizontal force to the leg with magnitude twice as great as the weight in the weight pan.

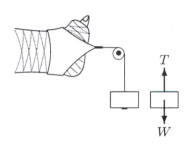

Figure 4.23 *One-pulley traction.*

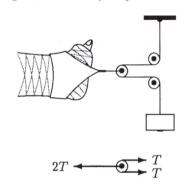

Figure 4.24 *3-pulley traction.*

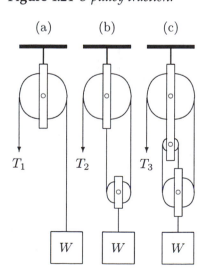

Figure 4.25 *Example 4.4.*

Example 4.4 Using three different cable-pulley arrangements shown in Figure 4.25, a block of weight W is elevated to a certain height. For each system, determine how much force is applied to the person holding the cable.

Solutions: The necessary free-body diagrams to analyze each system in Figure 4.25 are shown in Figure 4.26. For the analysis of the system in Figure 4.25a, all we need is the free-body diagram of the block (Figure 4.26a). For the vertical equilibrium of the block, tension in the cable must be $T_1 = W$. Therefore, a force with magnitude equal to the weight of the block is applied to the person holding the cable.

For the analysis of the system in Figure 4.25b, we need to examine the free-body diagram of the pulley closest to the block (Figure 4.26b). For the vertical equilibrium of the pulley, the tension in the cable must be $T_2 = W/2$. T_2 is the tension in the cable that is wrapped around the two pulleys and is held by the person. Therefore, a force with magnitude equal to half of the weight of the block is applied to the person holding the cable.

For the analysis of the system in Figure 4.25c, we again need to examine the free-body diagram of the pulley closest to the block (Figure 4.26c). For the vertical equilibrium of the pulley, the tension in the cable must be $T_3 = W/3$. T_3 is the tension in the cable that is wrapped around the three pulleys and is held by the person. Therefore, a force with a magnitude equal to one-third of the weight of the block is applied to the person holding the cable.

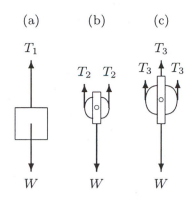

Figure 4.26 *Free-body diagrams.*

4.10 Built-in Structures

The beam shown in Figure 4.27 is built-in or welded to a wall and is called a *cantilever beam*. Cantilever beams can withstand externally applied moments as well as forces. The beam in Figure 4.27 is welded to the wall at A and a downward force with magnitude P is applied at the free end B. The length of the beam is ℓ. For the sake of simplicity, ignore the weight of the beam. The free-body diagram of the beam is shown in Figure 4.28. Instead of welding, if the beam were hinged to the wall at A, then under the effect of the applied force, the beam would undergo a clockwise rotation. The fact that the beam is not rotating indicates that there is rotational equilibrium. This rotational equilibrium is due to a reactive moment generated at the welded end of the beam. The counterclockwise moment with magnitude M in Figure 4.28 represents this reactive moment at A. Furthermore, the applied force tends to translate the beam downward. However, the beam is not translating. This indicates the presence of an upward reaction force at A that balances the effect of the applied force. Force R_A represents the reactive force at A. Since the beam is in equilibrium, we can utilize the conditions of equilibrium to determine the reactions. For example, consider the translational equilibrium of the beam in the y direction:

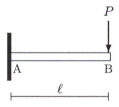

Figure 4.27 *Cantilever beam.*

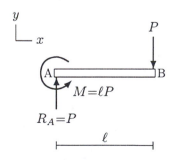

Figure 4.28 *Free-body diagram of the cantilever beam.*

$$\sum F_y = 0: \qquad R_A - P = 0$$
$$R_A = P$$

Now, consider the rotational equilibrium of the beam about A and assume that counterclockwise moments are positive:

$$\sum M_A = 0: \qquad M - \ell P = 0$$
$$M = \ell P$$

Note that the reactive moment with magnitude M is a "free vector" that acts everywhere along the beam.

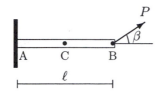

Figure 4.29 Example 4.5.

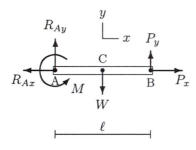

Figure 4.30 *Free-body diagram of the beam.*

Example 4.5 Consider the uniform, horizontal beam shown in Figure 4.29. The beam is fixed at A and a force that makes an angle $\beta = 30°$ with the horizontal is applied at B. The magnitude of the applied force is $P = 100$ N. C is the center of gravity of the beam. The beam weighs $W = 50$ N and has a length $\ell = 2$ m.

Determine the reactions generated at the fixed end of the beam.

Solution: The free-body diagram of the beam is shown in Figure 4.30. The horizontal and vertical directions are indicated by the x and y axes, respectively. P_x and P_y are the scalar components of the applied force $\underline{P}$. Since we know the magnitude and direction of $\underline{P}$, we can readily calculate P_x and P_y:

$$P_x = P \, \cos\beta = (100)(\cos 60) = 50.0 \text{ N } (+x)$$
$$P_y = P \, \sin\beta = (100)(\sin 60) = 86.6 \text{ N } (+y)$$

In Figure 4.30, the reactive force $\underline{R}_A$ at A is represented in terms of its scalar components R_{Ax} and R_{Ay}. We know neither the magnitude nor the direction of $\underline{R}_A$ (two unknowns). We also have a reactive moment at A with magnitude M acting in a direction perpendicular to the xy-plane. While drawing the free-body diagram, we assumed that this moment is counterclockwise with respect to the xy-plane.

In this problem, we have three unknowns: R_{Ax}, R_{Ay}, and M. To solve the problem, we need three conditions. First, consider the translational equilibrium of the beam in the x direction:

$$\sum F_x = 0: \qquad -R_{Ax} + P_x = 0$$
$$R_{Ax} = P_x = 86.6 \text{ N } (-x)$$

Next, consider the translational equilibrium of the beam in the y direction:

$$\sum F_y = 0: \qquad R_{Ay} - W + P_y = 0$$
$$R_{Ay} = W - P_y$$
$$R_{Ay} = 50 - 50 = 0$$

That is, the reactive force at A has no vertical component. Therefore, the resultant reactive force $\underline{R}_A$ at A is equal to $\underline{R}_{Ax}$. Finally, consider the rotational equilibrium of the beam about A.

Assuming that counterclockwise moments are positive:

$$\sum M_A = 0: \qquad M - \frac{\ell}{2} W + \ell \, P_y = 0$$

$$M = \frac{\ell}{2} W - \ell \, P_y$$

$$M = \frac{2}{2} 50 - (2)(50) = -50$$

Note that we calculated a negative value for M. This is not allowed because M is the magnitude of a moment vector and it corresponds to a scalar quantity. Scalar quantities cannot take negative values. The negative sign implies that while drawing the free-body diagram of the beam, we assumed the wrong direction (counterclockwise) for the reactive moment vector. We can now correct it by writing:

$$M = 50 \text{ N-m} \qquad \text{(cw)}$$

The corrected free-body diagram of the beam is shown in Figure 4.31.

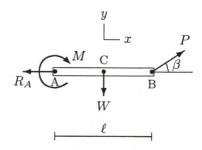

Figure 4.31 *Forces acting on the beam.*

Example 4.6 Consider the L-shaped beam illustrated in Figure 4.32. The beam is welded to the wall at A, the arm AB extends in the positive z direction, and the arm BC extends in the negative x direction. A force $\underline{P}$ is applied in the negative y direction at B. The lengths of arms AB and BC are $a = 20$ cm and $b = 30$ cm, respectively, and the magnitude of the applied force is $P = 120$ N.

Assuming that the weight of the beam is negligibly small as compared to the magnitude of the applied force, determine the reactions at the fixed end of the beam.

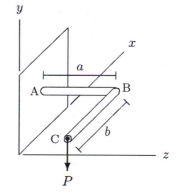

Figure 4.32 *Example 4.6.*

Solution A: Scalar method

This is a three-dimensional problem and we shall employ two methods to analyze it. The first method will utilize the concepts of couple and couple-moment introduced in the previous chapter, and the second method will utilize the vector properties of forces and moments.

The scalar method of analyzing the problem is described in Figure 4.33 and it involves the translation of the force applied on the beam from C to A. First, to translate $\underline{P}$ from C to B, place a pair of forces at B that are equal in magnitude (P) and acting in opposite directions such that the common line of action of the new forces is parallel to the line of action of the original force at C (Figure 4.33a). The downward force at C and the upward force at B form a couple. Therefore, they can be replaced by a couple-moment illustrated by a double-headed arrow in Figure 4.33b. The magnitude of the couple-moment is $M_1 = bP$. Applying the right-hand-rule, we can see that the couple-moment acts in

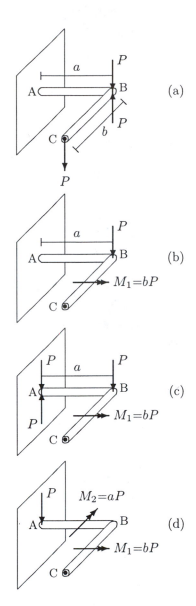

Figure 4.33 *Scalar method.*

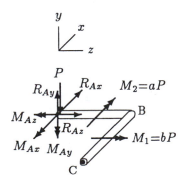

Figure 4.34 *Free-body diagram of the beam.*

the positive z direction. Therefore:

$$M_1 = b\,P = (0.30)(120) = 36 \text{ N-m} \ (+z)$$

As illustrated in Figure 4.33c, to translate the force from B to A, place another pair of forces at A with equal magnitude and opposite directions. This time, the downward force at B and the upward force at A form a couple, and again, they can be replaced by a couple-moment (Figure 4.33d). The magnitude of this couple-moment is $M_2 = a\,P$ and it acts in the positive x direction. Therefore:

$$M_2 = a\,P = (0.20)(120) = 24 \text{ N-m} \ (+x)$$

Figure 4.34 shows the free-body diagram of the beam. P is the magnitude of the force applied at C which is translated to A, and M_1 and M_2 are the magnitudes of the couple-moments. R_{Ax}, R_{Ay}, and R_{Az} are the scalar components of the reactive force at A, and M_{Ax}, M_{Ay}, and M_{Az} are the scalar components of the reactive moment at A. Consider the translational equilibrium of the beam in the x direction:

$$\sum F_x = 0: \qquad R_{Ax} = 0$$

The translational equilibrium of the beam in the y direction requires that:

$$\sum F_y = 0: \qquad R_{Ay} = P = 120 \text{ N} \ (+y)$$

For the translational equilibrium of the beam in the z direction:

$$\sum F_z = 0: \qquad R_{Az} = 0$$

Therefore, there is only one non-zero component of the reactive force on the beam at A and it acts in the positive y direction. Now, consider the rotational equilibrium of the beam in the x direction:

$$\sum M_x = 0: \qquad M_{Ax} = M_2 = 24 \text{ N-m} \ (-x)$$

For the rotational equilibrium of the beam in the y direction:

$$\sum M_y = 0: \qquad M_{Ay} = 0$$

Finally, the rotational equilibrium of the beam in the z direction requires that:

$$\sum M_z = 0: \qquad M_{Az} = M_1 = 36 \text{ N-m} \ (-z)$$

Therefore, the reactive moment at A has two non-zero components in the x and y directions. Now that we determined the components of the reactive force and moment at A, we can also calculate the magnitudes of the resultant force and moment at A:

$$R_A = \sqrt{R_{Ax}^2 + R_{Ay}^2 + R_{Az}^2} = R_{Ay} = 120 \text{ N}$$

$$M_A = \sqrt{M_{Ax}^2 + M_{Ay}^2 + M_{Az}^2} = 43.3 \text{ N-m}$$

Solution B: Vector method

The second method of analyzing the same problem utilizes the vector properties of the parameters involved. For example, the force applied at C and the position vector of point C relative to A can be expressed as (Figure 4.35):

$$\underline{P} = -P \, \underline{j} = -120 \, \underline{j}$$
$$\underline{r} = -b \, \underline{i} + a \, \underline{k} = -0.30 \, \underline{i} + 0.20 \, \underline{k}$$

Here, $\underline{i}$, $\underline{j}$, and $\underline{k}$ are unit vectors indicating positive x, y, and z directions, respectively. The free-body diagram of the beam is shown in Figure 4.36 where the reactive forces and moments are represented by their scalar components such that:

$$\underline{R}_A = R_{Ax} \, \underline{i} + R_{Ay} \, \underline{j} + R_{Az} \, \underline{k}$$
$$\underline{M}_A = M_{Ax} \, \underline{i} + M_{Ay} \, \underline{j} + M_{Az} \, \underline{k}$$

First, consider the translational equilibrium of the beam:

$$\sum \underline{F} = 0: \quad \underline{R}_A + \underline{P} = 0$$
$$(R_{Ax} \, \underline{i} + R_{Ay} \, \underline{j} + R_{Az} \, \underline{k}) + (-120 \, \underline{j}) = 0$$
$$R_{Ax} \, \underline{i} + (R_{Ay} - 120) \, \underline{j} + R_{Az} \, \underline{k} = 0$$

For this equilibrium to hold:

$$R_{Ax} = 0$$
$$R_{Ay} = 120 \text{ N} \ (+y)$$
$$R_{Az} = 0$$

As discussed in the previous chapter, by definition, moment is the cross (vector) product of the position and force vectors. Therefore, the moment $\underline{M}_C$ relative to A due to force $\underline{P}$ applied at C is:

$$\underline{M}_C = \underline{r} \times \underline{P}$$
$$= (-0.30 \, \underline{i} + 0.20 \, \underline{k}) \times (-120 \, \underline{j})$$
$$= (-0.30)(-120)(\underline{i} \times \underline{j}) + (0.20)(-120)(\underline{k} \times \underline{j})$$
$$= 36 \, \underline{k} + 24 \, \underline{i}$$

Now consider the rotational equilibrium of the beam about A:

$$\sum \underline{M} = 0: \quad \underline{M}_A + \underline{M}_C = 0$$
$$(M_{Ax} \, \underline{i} + M_{Ay} \, \underline{j} + M_{Az} \, \underline{k}) + (36 \, \underline{k} + 24 \, \underline{i}) = 0$$
$$(M_{Ax} + 24) \, \underline{i} + M_{Ay} \, \underline{j} + (M_{Az} + 36) \, \underline{k} = 0$$

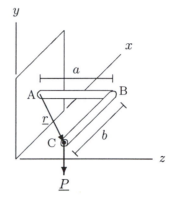

Figure 4.35 *Vector method.*

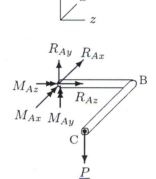

Figure 4.36 *Free-body diagram of the beam.*

For this equilibrium to hold:

$$M_{Ax} = 24 \text{ N-m } (-x)$$
$$M_{Ay} = 0$$
$$M_{Az} = 36 \text{ N-m } (-z)$$

All of these results are consistent with those obtained using the scalar method of analysis.

Remarks:

• We analyzed this problem in two ways. It is clear that the scalar method of analysis requires less rigorous mathematical manipulations. However, it has its limitations. For instance, in this example, the force applied at C had only one non-zero component. What would happen if we had a force at C with non-zero components in the x and z directions as well as the y direction? The scalar method could still be applied in a step-by-step manner by considering one component of the applied force at a time, solving the problem for that component alone, repeating this for all three components, and then superimposing the three solutions to obtain the final solution. Obviously, this would be quite time consuming. On the other hand, the extension of the vector method to analyze such problems is very straightforward. All one needs to do is to redefine the applied force vector as $\underline{P} = P_x\underline{i} + P_y\underline{j} + P_z\underline{k}$, and carry out exactly the same procedure outlined above.

• In this example, it is stated that the weight of the beam was negligibly small as compared to the force applied at C. What would happen if the weight of the beam was not negligible? As illustrated in Figure 4.37, assume that we know the weights W_1 and W_2 of the arms AB and BC of the beam along with their centers of gravity (points D and E). Also, assume that D is located in the middle of arm AB and E is located in the middle of arm BC. We have already discussed the vector representation of $\underline{R}_A$ and $\underline{P}$. The vector representations of $\underline{W}_1$ and $\underline{W}_2$ are:

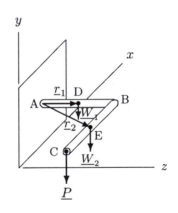

Figure 4.37 *Including the weights of the arms.*

$$\underline{W}_1 = -W_1\,\underline{j}$$
$$\underline{W}_2 = -W_2\,\underline{j}$$

The positions vectors of points D and E relative to A are:

$$\underline{r}_1 = \frac{a}{2}\,\underline{k}$$
$$\underline{r}_2 = -\frac{b}{2}\,\underline{i} + a\,\underline{k}$$

Therefore, the moments due to $\underline{W}_1$ and $\underline{W}_2$ relative to A are:

$$\underline{M}_D = \underline{r}_1 \times \underline{W}_1 = \frac{a\,W_1}{2}\,\underline{i}$$
$$\underline{M}_E = \underline{r}_2 \times \underline{W}_2 = a\,W_2\,\underline{i} - \frac{b\,W_2}{2}\,\underline{k}$$

The translational equilibrium of the beam would yield:

$$R_{Ax} = 0$$
$$R_{Ay} = W_1 + W_2 + P$$
$$R_{Az} = 0$$

The rotational equilibrium of the beam would yield:

$$M_{Ax} = -\frac{a\ W_1}{2} - a\ W_2 + a\,P$$
$$M_{Ay} = 0$$
$$M_{Az} = \frac{b\ W_2}{2} - b\,P$$

• Note the similarities between this and the shoulder example in the previous chapter (Example 3.5).

4.11   Systems Involving Friction

Frictional forces were discussed in detail in Chapter 2. Here, we shall analyze a problem in which frictional forces play an important role.

Example 4.7 Figure 4.38 illustrates a person trying to push a block up an inclined surface by applying a force parallel to the incline. The weight of the block is W, the coefficient of maximum friction between the block and the incline is μ, and the incline makes an angle θ with the horizontal.

Determine the magnitude P of the minimum force the person must apply in order to overcome the frictional and gravitational effects to start moving the block in terms of W, μ, and θ.

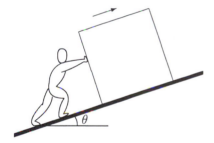

Figure 4.38 *Example 4.7.*

Solution: Note that if the person pushes the block by applying a force closer to the top of the block, the block may tilt (rotate in the clockwise direction) about its bottom right edge. Here, we shall assume that there is no such effect and that the bottom surface of the block remains in full contact with the ground.

The free-body diagram of the block is shown in Figure 4.39. x and y correspond to the directions parallel and perpendicular to the incline, respectively. P is the magnitude of the force applied by the person on the block in the x direction, f is the frictional force applied by the ground on the block in the negative x direction, N is the normal force applied by the ground on the block in the positive y direction, and W is the weight of the block acting vertically downward. The weight of the block has components in the x and y directions that can be determined from the geometry of the problem (see Figure A.3 in

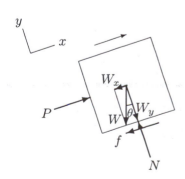

Figure 4.39 *Free-body diagram of the block.*

Appendix A):

$$W_x = W \sin\theta$$
$$W_y = W \cos\theta$$

Since the intended direction of motion of the block is in the positive x direction, the frictional force on the block is acting in the negative x direction trying to stop the block from moving. The magnitude f of the frictional force is directly proportional to the magnitude N of the normal force applied by the inclined surface on the block and the coefficient of friction, μ, is the constant of proportionality. Therefore, f, N, and μ are related through:

$$f = \mu N \qquad (i)$$

The unknowns in this example are P and N. Since we have an expression relating f and N, if N is known, so is f. For the solution of the problem, first consider the equilibrium of the block in the y direction:

$$\sum F_y = 0: \qquad N - W_y = 0$$
$$N = W_y = W \cos\theta \qquad (ii)$$

We are asked to determine the minimum force the person must apply to overcome frictional and gravitational effects to start moving the block. There is equilibrium at the instant just before the movement starts. Therefore, we can apply the equilibrium condition in the x direction:

$$\sum F_x = 0: \qquad P - f - W_x = 0$$
$$P = f + W_x \qquad (iii)$$

Substitute Eqs. (i) and (ii) into Eq. (iii) along with $W_x = W \sin\theta$:

$$P = \mu W \cos\theta + W \sin\theta$$

This is a general solution for P in terms of W, μ, and θ. This solution is valid for any θ less than $90°$, including $\theta = 0°$, which represents a flat horizontal surface (Figure 4.40). For $\theta = 0°$, $\sin 0 = 0$ and $\cos 0 = 1$. Therefore, the force required to start moving the same block on a horizontal surface is:

$$P = f = \mu W$$

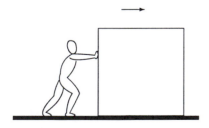

Figure 4.40 *Pushing a block on a horizontal surface.*

To have a numerical example, assume that $W = 1000\,\text{N}$, $\mu = 0.3$, and $\theta = 15°$. Then:

$$P = (0.3)(1000)(\cos 15) + (1000)(\sin 15) = 548.6\,\text{N}$$

Therefore, to be able to start moving a 1000 N block up the $15°$ incline which has a surface friction coefficient of 0.3, the person must apply a force slightly greater than 548.6 N in a direction parallel to the incline.

To start moving the same block on a horizontal surface with the same friction coefficient, the person must apply a horizontal force of:

$$P = (0.3)(1000) = 300 \text{ N}$$

As compared to a horizontal surface, the person must apply about 83% more force on the block to start moving the block on the 15° incline, which is calculated as:

$$\frac{548.6 - 300}{300} \times 100 = 82.9$$

4.12 Center of Gravity Determinations

As discussed in Chapter 2, every object may be considered to consist of an infinite number of particles that are acted upon by the force of gravity, thus forming a distributed force system. The resultant of these forces or individual weights of particles is equal to the total weight of the object, acting as a concentrated load at the *center of gravity* of the object. A concept related to the center of gravity is that of the *center of mass*, which is a point at which the entire mass of an object is assumed to be concentrated. In general, there is a difference between the centers of mass and gravity of an object. This may be worth considering if the object is large enough for the magnitude of the gravitational acceleration to vary at different parts of the object. For our applications, the centers of mass and gravity of an object refer to the same point. There is also the concept of *gravity line* that is used to refer to the vertical line that passes through the center of gravity.

The center of gravity of an object is such that if the object is cut into two parts by any vertical plane that goes through the center of gravity, then the weight of each part would be equal. Therefore, the object can be balanced on a knife-edge that is located directly under its center of gravity or along its gravity line. If an object has a symmetric, well-defined geometry and uniform composition, then its center of gravity is located at the geometric center of the object. There are several methods of finding the centers of gravity of irregularly shaped objects. One method is by "suspending" the object. For the sake of illustration, consider the piece of paper shown in Figure 4.41. The center of gravity of the paper is located at C. Let O and Q be two points on the paper. If the paper is pinned to the wall at O, there will be a non-zero net clockwise moment acting on the paper about point O because the center of gravity of the paper is located to the

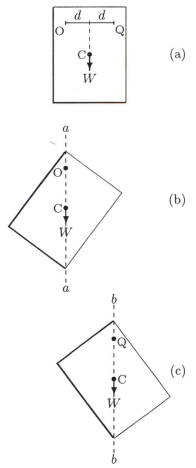

Figure 4.41 *Center of gravity of the paper is located at the intersection of lines aa and bb (suspension method).*

right of O. If the paper is released, the moment about O due to the weight W of the paper will cause the paper to swing in the clockwise direction first, oscillate, and finally come to rest at a position in which C lies along a vertical line aa (gravity line) passing through point O (Figure 4.41b). At this position, the net moment about O is zero because the length of the moment arm of $\underline{W}$ relative to O is zero. If the paper is then pinned at point Q, the paper will swing in the counterclockwise direction and soon come to rest at a position in which C is located directly under Q, or along the vertical line bb passing through Q (Figure 4.41c). The point at which lines aa and bb intersect will indicate the center of gravity of the paper, which is point C.

Note that a piece of paper is a plane object with negligible thickness and that suspending the paper at two points is sufficient to locate its center of gravity. For a three-dimensional object, the object must be suspended at three points in two different planes.

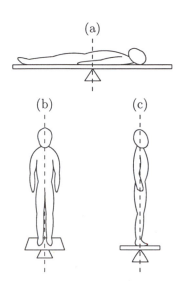

Another method of finding the center of gravity is by "balancing" the object on a knife-edge. As illustrated in Figure 4.42, to determine the center of gravity of a person, first balance a board on the knife-edge and then place the person supine on the board. Adjust the position of the person on the board until the board is again balanced (Figure 4.42a). The horizontal distance between the feet of the person and the point of contact of the knife-edge and the board is the height of the center of gravity of the person. Consider a plane that passes through the point of contact of the knife-edge with the board that cuts the person into upper and lower portions. The center of gravity of the person lies somewhere on this plane. Note that the center of gravity of a three-dimensional body, such as a human being, has three coordinates. Therefore, the same method must be repeated in two other planes to establish the exact center of gravity. For this purpose, consider the anteroposterior balance of the person that will yield two additional planes as illustrated in (Figures 4.42b and 4.42c). The intersection of these planes will correspond to the center of gravity of the person.

Figure 4.42 *Balance method.*

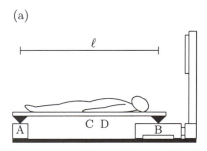

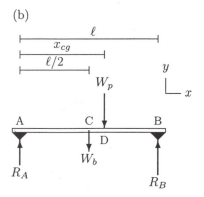

Figure 4.43 *Reaction board method.*

The third method of finding the center of gravity of a body involves the use of a "reaction board" with two knife-edges fixed to its undersurface (Figure 4.43). Assume that the weight W_b of the board and the distance ℓ between the knife-edges are known. One of the two edges (A) rests on a platform and the other edge (B) rests on a scale, such that the board is horizontal. The location of the center of gravity of a person can be determined by placing the person on the board and recording the weight indicated on the scale, which is essentially the magnitude R_B of the reaction force on the board at B. Figure 4.43b illustrates the free-body diagram of the board. R_A is the magnitude of the reaction force at A, W_p is the known weight of the person. The weight W_b of the board is acting at the geometric center (C) of the board, which is equidistant from A and B. D is a point on the board directly under the center of gravity of the person. The unknown distance between A and D is designated by x_{cg} which can be determined by considering the rotational equilibrium of the board about point A. Assuming that clockwise moments are positive:

$$\sum M_A = 0: \qquad \frac{\ell}{2} W_b + x_{cg} W_p - \ell R_B = 0$$

Solving this equation for x_{cg} will yield:

$$x_{cg} = \frac{\ell}{W_p}\left(R_B - \frac{W_b}{2}\right)$$

If the person is placed on the board so that the feet are directly above the knife-edge at A, x_{cg} will designate the height of the person's center of gravity as measured from the floor level.

Sometimes, we must deal with a system made up of parts with known centers of gravity where the task is to determine the center of gravity of the system as a whole. This can be achieved simply by utilizing the definition of the center of gravity. Consider the system shown in Figure 4.44, which is composed of three spheres with weights W_1, W_2, and W_3 connected to one another through rods. Assume that the weights of the rods are negligible. Let x_1, x_2, and x_3 be the x coordinates of the centers of gravity of each sphere. The net moment about point O due to the individual weights of the spheres is:

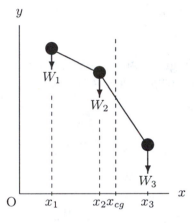

Figure 4.44 x_{cg} is the x component of the center of gravity.

$$M_O = x_1\, W_1 + x_2\, W_2 + x_3\, W_3$$

The total weight of the system is $W_1 + W_2 + W_3$, which is assumed to be acting at the center of gravity of the system as a whole. If x_{cg} is the x coordinate of the center of gravity of the entire system, then the moment of the total weight of the system about point O is:

$$M_O = x_{cg}\,(W_1 + W_2 + W_3)$$

The last two equations can be combined together so as to eliminate M_O, which will yield:

$$x_{cg} = \frac{x_1\, W_1 + x_2\, W_2 + x_3\, W_3}{W_1 + W_2 + W_3}$$

This result can be generalized for any system composed of n parts:

$$x_{cg} = \frac{\sum_{i=1}^{n} x_i\, W_i}{\sum_{i=1}^{n} W_i} \qquad (4.12)$$

Equation (4.12) provides only the x coordinate of the center of gravity of the system. To determine the exact center, the y coordinate of the center of gravity must also be determined. For this purpose, the entire system must be rotated by an angle, preferably 90°, as illustrated in Figure 4.45. If y_1, y_2, and y_3

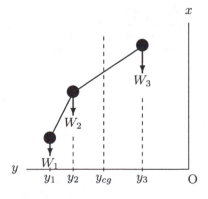

Figure 4.45 y_{cg} is the y component of the center of gravity.

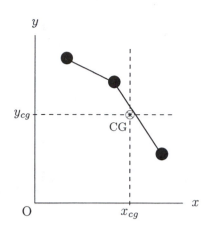

Figure 4.46 ⊗ *is the center of gravity of the system.*

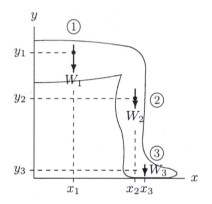

Figure 4.47 *Locating the center of gravity of a flexed leg.*

Table 4.2 *Example 4.8.*

PART	x (cm)	y (cm)	%W
①	17.3	51.3	10.6
②	42.5	32.8	4.6
③	45.0	3.3	1.7

correspond to the y coordinates of the centers of gravity of the spheres, then the y coordinate of the center of gravity of the system as a whole is:

$$y_{cg} = \frac{y_1\, W_1 + y_2\, W_2 + y_3\, W_3}{W_1 + W_2 + W_3}$$

For any system composed of n parts:

$$y_{cg} = \frac{\sum_{i=1}^{n} y_i\, W_i}{\sum_{i=1}^{n} W_i} \tag{4.13}$$

In Figure 4.46, the center of gravity of the entire system is located at the point of intersection of the perpendicular lines passing through x_{cg} and y_{cg}.

Example 4.8 Consider the leg shown in Figure 4.47, which is flexed to a right angle. The coordinates of the centers of gravity of the leg between the hip and knee joints (upper leg), the knee and ankle joints, and the foot, as measured from the floor level directly in line with the hip joint, are given in Table 4.2. The weights of the segments of the leg as percentages of the total weight W of the person are also provided in Table 4.2.

Determine the location of the center of gravity of the entire leg.

Solution: The coordinates (x_{cg}, y_{cg}) of the center of gravity of the entire leg can be determined by utilizing Eqs. (4.12) and (4.13). Using Eq. (4.12):

$$x_{cg} = \frac{x_1\, W_1 + x_2\, W_2 + x_3\, W_3}{W_1 + W_2 + W_3}$$
$$x_{cg} = \frac{(17.3)(0.106W) + (42.5)(0.046W) + (45.0)(0.017W)}{0.106W + 0.046W + 0.017W}$$
$$x_{cg} = 26.9 \text{ cm}$$

To determine the y coordinate of the center of gravity of the leg, we must rotate the leg by 90°, as illustrated in Figure 4.48, and apply Eq. (4.13):

$$y_{cg} = \frac{y_1\, W_1 + y_2\, W_2 + y_3\, W_3}{W_1 + W_2 + W_3}$$
$$y_{cg} = \frac{(51.3)(0.106W) + (32.8)(0.046W) + (3.3)(0.017W)}{0.106W + 0.046W + 0.017W}$$
$$y_{cg} = 41.4 \text{ cm}$$

Therefore, the center of gravity of the entire lower extremity when flexed at a right angle is located at a horizontal distance of 26.9 centimeters from the hip joint and at a height of 41.4 centimeters measured from the floor level.

Remarks:

• During standing with upper extremities straight, the center of gravity of a person lies within the pelvis anterior to the second sacral vertebra. If the origin of the rectangular coordinate system is placed at the center of gravity of the person, then the xy-plane corresponds to the *frontal*, *coronal*, or *longitudinal plane*, the yz-plane is called the *sagittal plane*, and the xz-plane is the *horizontal* or *transverse plane*. The frontal plane divides the body into front and back portions, the sagittal plane divides the body into right and left portions, and the horizontal plane divides it into upper and lower portions.

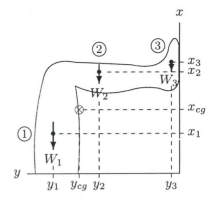

Figure 4.48 ⊗ *is the center of gravity of the leg.*

• The center of gravity of the entire body varies from person to person depending on build. For a given person, the position of the center of gravity can shift depending on the changes in the relative alignment of the extremities during a particular physical activity. Locations of the centers of gravity of the upper and lower extremities can also vary. For example, the center of gravity of a lower limb shifts backward as the knee is flexed. The center of gravity of the entire arm shifts forward as the elbow is flexed. These observations suggest that when the knee is flexed, the leg will tend to move forward bringing the center of gravity of the leg directly under the hip joint. When the elbow is flexed, the arm will tend to move backward bringing its center of gravity under the shoulder joint.

4.13 Exercise Problems

Problem 4.1 As illustrated in Figure 4.49, consider an 80 kg person preparing to dive into a pool. The diving board is represented by a uniform, horizontal beam that is hinged to the ground at A and supported by a frictionless roller at D. B is a point on the board directly under the center of gravity of the person. The distance between A and B is $\ell = 6$ m and the distance between A and D is $d = 2$ m. (Note that one-third of the board is located on the left of the roller support and two-thirds is on the right. Therefore, for the sake of force analyses, one can assume that the board consists of two boards with two different weights connected at D.)

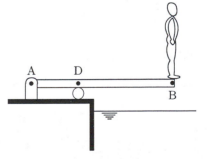

Figure 4.49 *Problem 4.1.*

If the diving board has a total weight of 1500 N, determine the reactions on the beam at A and D.

Answers: $R_A = 2318$ N ($\downarrow$) $R_D = 4602$ N ($\uparrow$)

Problem 4.2 The uniform, horizontal beam shown in Figure 4.50 is hinged to the ground at A and supported by a frictionless roller

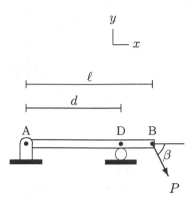

Figure 4.50 *Problem 4.2.*

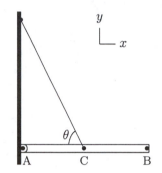

Figure 4.51 *Problem 4.3.*

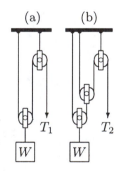

Figure 4.52 *Problem 4.4.*

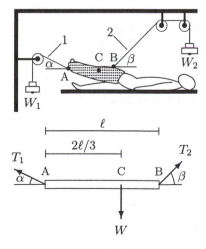

Figure 4.53 *Problem 4.5.*

at D. The distance between A and B is $\ell = 4$ m and the distance between A and D is $d = 3$ m. A force that makes an angle $\beta = 60°$ with the horizontal is applied at B. The magnitude of the applied force is $P = 1000$ N. The total weight of the beam is $W = 400$ N.

By noting that three-quarters of the beam is on the left of the roller support and one-quarter is on the right, calculate the x and y components of reaction forces on the beam at A and D.

Answers: $R_D = 1421$ N ($\uparrow$) $R_{Ax} = 500$ N ($\leftarrow$) $R_{Ay} = 155$ N ($\downarrow$)

Problem 4.3 The uniform, horizontal beam shown in Figure 4.51 is hinged to the wall at A and supported by a cable attached to the beam at C. C also represents the center of gravity of the beam. At the other end, the cable is attached to the wall so that it makes an angle $\theta = 68°$ with the horizontal. If the length of the beam is $\ell = 4$ m and the weight of the beam is $W = 400$ N, calculate the tension T in the cable and components of the reaction force on the beam at A.

Answers: $T = 431$ N $R_{Ax} = 162$ N ($\rightarrow$) $R_{Ay} = 0$

Problem 4.4 Using two different cable-pulley arrangements shown in Figure 4.52, a block of weight W is elevated to a certain height. For each system, determine how much force is applied to the person holding the cable.

Answers: $T_1 = W/2$ $T_2 = W/4$

Problem 4.5 Consider the split Russel traction device and a mechanical model of the leg shown in Figure 4.53. The leg is held in the position shown by two weights that are connected to the leg via two cables. The combined weight of the leg and the cast is $W = 300$ N. ℓ is the horizontal distance between points A and B where the cables are attached to the leg. C is the center of gravity of the leg including the cast, which is located at a distance two-thirds of ℓ as measured from A. The angle cable 2 makes with the horizontal is measured as $\beta = 45°$.

Determine the tensions T_1 and T_2 in the cables, weights W_1 and W_2, and angle α that cable 1 makes with the horizontal, so that the leg remains in equilibrium at the position shown.

Answers: $T_1 = W_1 = 223.6$ N $T_2 = W_2 = 282.8$ N $\alpha = 26.6°$

Problem 4.6 Consider the uniform, horizontal cantilever beam shown in Figure 4.54. The beam is fixed at A and a force that makes an angle $\beta = 63°$ with the horizontal is applied at B. The magnitude of the applied force is $P = 80$ N. C is the center of gravity of the beam and the beam weighs $W = 40$ N and has a length $\ell = 2$ m.

Determine the reactions generated at the fixed end of the beam.

Answers: $R_{Ax} = 36.3\,\text{N}\,(+x)$ $R_{Ay} = 111.3\,\text{N}\,(+y)$ $M_A = 182.3$ N-m (ccw)

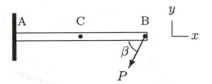

Figure 4.54 *Problem 4.6.*

Problem 4.7 Consider the L-shaped beam illustrated in Figure 4.55. The beam is welded to the wall at A, the arm AB extends in the positive z direction, and the arm BC extends in the negative y direction. A force $\underline{P}$ is applied in the positive x direction at the free-end (B) of the beam. The lengths of arms AB and BC are a and b, respectively, and the magnitude of the applied force is P.

Assuming that the weight of the beam is negligibly small, determine the reactions generated at the fixed end of the beam in terms of a, b, and P.

Answers: The non-zero force and moment components are:

$$R_{Ax} = P\,(-x) \quad M_{Ay} = a\,P\,(-y) \quad M_{Az} = b\,P\,(-z)$$

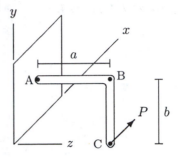

Figure 4.55 *Problems 4.7 and 4.8.*

Problem 4.8 Reconsider the L-shaped beam illustrated in Figure 4.55. This time, assume that the applied force $\underline{P}$ has components in the positive x and positive z directions such that $\underline{P} = P_x \underline{i} + P_y \underline{j}$. Determine the reactions generated at the fixed end of the beam in terms of a, b, P_x, and P_y.

Answers:

$$R_{Ax} = P_x\,(-x) \quad R_{Ay} = 0 \quad R_{Az} = P_y\,(-z)$$

$$M_{Ax} = b\,P_z\,(+y) \quad M_{Ay} = a\,P_x\,(-y) \quad M_{Az} = b\,P_x\,(-z)$$

Problem 4.9 Figure 4.56 illustrates a person who is trying to pull a block on a horizontal surface using a rope. The rope makes an angle θ with the horizontal. If W is the weight of the block and μ is the coefficient of maximum friction between the bottom surface of the block and horizontal surface, show that the magnitude P of minimum force the person must apply in order to overcome the frictional and gravitational effects (to start moving the block) is:

$$P = \frac{\mu\,W}{\cos\theta + \mu\,\sin\theta}$$

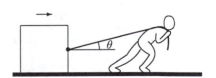

Figure 4.56 *Problem 4.9.*

Problem 4.10 Figure 4.57 illustrates a person trying to push a block up on an inclined surface by applying a horizontal force.

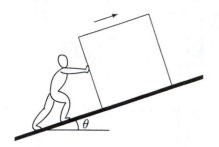

Figure 4.57 *Problem 4.10.*

The weight of the block is W, the coefficient of maximum friction between the block and the incline is μ, and the incline makes an angle θ with the horizontal.

Determine the magnitude P of minimum force the person must apply in order to overcome the frictional and gravitational effects (to start moving the block) in terms of W, μ, and θ.

Answer:

$$P = \frac{\sin\theta + \mu\,\cos\theta}{\cos\theta - \mu\,\sin\theta}\,W$$

Chapter 5

Applications of Statics
to Biomechanics

5.1 Skeletal Joints

The human body is rigid in the sense that it can maintain a posture, and flexible in the sense that it can change its posture and move. The flexibility of the human body is due primarily to the *joints*, or *articulations*, of the skeletal system. The primary function of joints is to provide mobility to the musculoskeletal system. In addition to providing mobility, a joint must also possess a degree of stability. Since different joints have different functions, they possess varying degrees of mobility and stability. Some joints are constructed so as to provide optimum mobility. For example, the construction of the shoulder joint (ball-and -socket) enables the arm to move in all three planes (triaxial motion). However, this high level of mobility is achieved at the expense of reduced stability, increasing the vulnerability of the joint to injuries, such as dislocations. On the other hand, the elbow joint provides movement primarily in one plane (uniaxial motion), but is more stable and less prone to injuries than the shoulder joint. The extreme case of increased stability is achieved at joints that permit no relative motion between the bones constituting the joint. The contacting surfaces of the bones in the skull are typical examples of such joints.

The joints of the human skeletal system may be classified based on their structure and/or function. *Synarthrodial joints*, such as those in the skull, are formed by two tightly fitting bones and do not allow any relative motion of the bones forming them. *Amphiarthrodial joints*, such as those between the vertebrae, allow slight relative motions, and feature an intervening substance (a cartilaginous or ligamentous tissue) whose presence eliminates direct bone-to-bone contact. The third and mechanically most significant type of articulations are called *diarthrodial joints*, which permit varying degrees of relative motion and have articular cavities, ligamentous capsules, synovial membranes, and synovial fluids (Figure 5.1). The *articular cavity* is the space between the articulating bones. The *ligamentous capsule* holds the articulating bones together. The *synovial membrane* is the internal lining of the ligamentous capsule enclosing the *synovial fluid*, which serves as a lubricant. The synovial fluid is a viscous material that functions to reduce friction, reduce wear and tear of the articulating surfaces by limiting direct contact between them, and nourish the articular cartilage lining the surfaces. The *articular cartilage*, on the other hand, is a cartilaginous material whose primary function is to absorb shock. Various diarthrodial joints can be further categorized as gliding (for example, vertebral facets), hinge (elbow and ankle), pivot (proximal radioulnar), condyloid (wrist), saddle (carpometacarpal of thumb), and ball-and -socket (shoulder and hip).

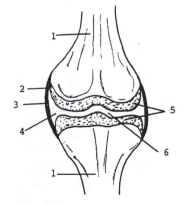

Figure 5.1 *A diarthrodial joint. (1) bone, (2) ligamentous capsule, (3, 4) synovial membrane & fluid, (5, 6) articular cartilage & cavity.*

The nature of motion about a diarthrodial joint and the stability of the joint are dependent upon many factors, including the manner in which the articulating surfaces fit together, the structure of the capsular ligament, the structure and length of the ligaments around the joint, and the degree to which the muscles that cross the joint can be stretched.

5.2 Skeletal Muscles

There are three types of muscles: skeletal, smooth, and cardiac. Movement of human body segments is achieved as a result of forces generated by *skeletal muscles*, which convert chemical energy into mechanical work. The skeletal muscle is composed of muscle fibers and myofibrils. Myofibrils are made up of actin and myosin filaments. Muscles exhibit viscoelastic material behavior, that is, they have both solid and fluid like material properties. Muscles are elastic in the sense that when a muscle is stretched and released, it will resume its original (unstretched) size and shape. Muscles are *viscous* in the sense that there is an internal resistance to motion.

A skeletal muscle is attached, via aponeuroses and/or tendons, to at least two different bones controlling the relative motion of one segment with respect to the other. When its fibers contract under the stimulation of a nerve, the muscle exerts a pulling effect on the bones to which it is attached. *Contraction* is a unique ability of the muscle tissue. In engineering mechanics, contraction implies shortening under compressive forces. In muscle mechanics, contraction can occur as a result of muscle shortening or muscle lengthening, or it can occur without any change in the muscle length. Furthermore, the result of a muscle contraction is always tension: a muscle can only exert a pull. Muscles cannot exert a push.

There are various types of muscle contractions: a *concentric contraction* occurs simultaneously as the length of the muscle decreases (for example, the biceps during flexion of the forearm); a *static contraction* occurs while muscle length remains constant (the biceps when the forearm is flexed and held without any movement); and an *eccentric contraction* occurs as the length of the muscle increases (the biceps during the extension of the forearm). A muscle can cause movement only while its length is shortening. If the length of a muscle increases during a particular activity, then the tension generated by the muscle contraction is aimed at controlling the movement of the body segments associated with that muscle. If a muscle contracts but there is no segmental motion, then the tension in the muscle balances the effects of applied forces such as those due to gravity.

The skeletal muscles can also be named according to the functions they serve during a particular activity. For example, a muscle is called *agonist* if it causes movement through concentric contraction. This type of contraction occurs when the muscles shorten to cause joint movement. An *antagonist* muscle, as a result of eccentric contraction, controls the movement. In this case, the muscle lengthens and the muscle moment occurs opposite to the change in joint angle, thus decelerating the motion of the joint.

5.3 Basic Considerations

In this chapter, we want to apply the principles of statics to investigate the forces involved in various muscle groups and joints for various postural positions of the human body and its segments. Our immediate purpose is to provide answers to questions such as: what tension must the neck extensor muscles exert on the head to support the head in a specified position? When a person bends, what would be the force exerted by the erector spinae on the fifth lumbar vertebra? How does the compression at the elbow, knee, and ankle joints vary with externally applied forces and with different segmental arrangements? How does the force on the femoral head vary with loads carried in the hand? What are the forces involved in various muscle groups and joints during different exercise conditions?

The forces involved in the human body can be grouped as internal and external. Internal forces are those associated with muscles, ligaments, and tendons, and at the joints. Externally applied forces include the effect of gravitational acceleration on the body or on its segments, manually and/or mechanically applied forces on the body during exercise and stretching, and forces applied to the body by prostheses and implements. In general, the unknowns in static problems involving the musculoskeletal system are the joint reaction forces and muscle tensions. Mechanical analysis of a joint requires that we know the vector characteristics of tension in the muscle including the proper locations of muscle attachments, the weights or masses of body segments, the centers of gravity of the body segments, and the anatomical axis of rotation of the joint.

5.4 Basic Assumptions and Limitations

The complete analysis of muscle forces required to sustain various postural positions is difficult because of the complex arrangement of muscles within the human body and because of limited information. In general, the relative motion of body segments about a given joint is controlled by more than one muscle group. To be able to reduce a specific problem of biomechanics to one that is statically determinate and apply the equations of equilibrium, only the muscle group that is the primary source of control over the joint can be taken into consideration. Possible contributions of other muscle groups to the load-bearing mechanism of the joint must be ignored. Note, however, that approximations of the effect of other muscles may be made by considering their cross-sectional areas and their relative positions in relation to the joint. Also, if the phasic activity of muscles is known via some experiments such as the electromyography (EMG) measurements of muscle signals, then the tension in different muscle groups may be estimated.

To apply the principles of statics to analyze the mechanics of human joints, we shall adopt the following assumptions and limitations:

- The anatomical axes of rotation of joints are known.
- The locations of muscle attachments are known.
- The line of action of muscle tension is known.
- Segmental weights and their centers of gravity are known.
- Frictional factors at the joints are negligible.
- Dynamic aspects of the problems will be ignored.
- Only two-dimensional problems will be considered.

These analyses require that the anthropometric data about the segment to be analyzed must be available. For this purpose, there are tables listing anthropometric information including average weights, lengths, and centers of gravity of body segments. See Chaffin and Andersson (1991), Roebuck (1995), and Winter (1990) for a review of the anthropometric data available.

It is clear from this discussion that we shall analyze certain idealized problems of biomechanics. Based on the results obtained and the experience gained, these models may be expanded by taking additional factors into consideration. However, a given problem will become more complex as more factors are considered.

In the following sections, the principles of statics will be applied to analyze forces involved at and around the major joints of the human body. First, a brief functional anatomy of each joint and related muscles will be provided, and specific biomechanical problems will be constructed. For a more complete discussion about the functional anatomy of joints, see texts such as Nordin and Frankel (1989) and Thompson (1989). Next, an analogy will be formed between muscles, bones, and human joints, and certain mechanical elements such as cables, beams, and mechanical joints. This will enable us to construct a mechanical model of the biological system under consideration. Finally, the procedure outlined in Chapter 4.5 will be applied to analyze the mechanical model thus constructed. See LeVeau (1992) for additional examples of the application of the principles of statics to biomechanics.

5.5 Mechanics of the Elbow

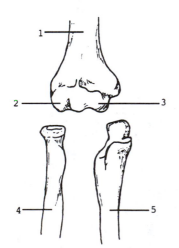

Figure 5.2 *Bones of the elbow: (1) humerus, (2) capitulum, (3) trochlea, (4) radius, (5) ulna.*

The elbow joint is composed of three separate articulations (Figure 5.2). The *humeroulnar joint* is a hinge (ginglymus) joint formed by the articulation between the spool-shaped trochlea of the distal humerus and the concave trochlear fossa of the proximal ulna. The structure of the humeroulnar joint is such that it allows only uniaxial rotations, confining the movements about the elbow joint to flexion (movement of the forearm toward the upper arm) and extension (movement of the forearm away from the upper arm). The *humeroradial joint* is also a hinge joint formed between the capitulum of the distal humerus and the head of the radius. The *proximal radioulnar joint* is a pivot joint formed by the head of the radius and the radial notch of the proximal ulna. This articulation allows the radius and ulna to undergo relative rotation about the longitudinal axis of one or the other bone, giving rise to pronation (the movement experienced while going from the palm-up to the palm-down) or supination (the movement experienced while going from the palm-down to the palm-up).

The muscles coordinating and controlling the movement of the elbow joint are illustrated in Figure 5.3. The biceps brachii muscle is the most powerful flexor of the elbow joint, particularly when the elbow joint is in a supinated position. On the distal side, the biceps is attached to the tuberosity of the radius, and on the proximal side, it has attachments at the top of the coracoid process and upper lip of the glenoid fossa. Another important flexor is the brachialis muscle, which has attachments at the lower half of the anterior portion of the humerus and the coronoid process of the ulna. The most important muscle controlling the extension movement of the elbow is the triceps brachii muscle. It has attachments at the lower head of the glenoid cavity of the scapula, the upper half of the posterior surface of the humerus, the lower two thirds of the posterior surface of the humerus, and the olecranon process of the ulna. Pronation and supination movements of the forearm are performed mainly by the pronator teres and supinator muscles, respectively. The pronator teres is attached to the lower part of the inner condyloid ridge of the humerus, the medial side of the ulna, and the middle third of the outer surface of the radius. The supinator muscle has attachments at the outer condyloid ridge of the humerus, the neighboring part of the ulna, and the outer surface of the upper third of the radius.

Common injuries of the elbow include fractures and dislocations. Fractures usually occur at the epicondyles of the humerus and the olecranon process of the ulna. Another elbow injury is known as "tennis elbow," which occurs as a result of repeated and forceful pronation and supination movements of the elbow.

Example 5.1 Consider the arm shown in Figure 5.4. The elbow is flexed to a right angle and an object is held in the hand. The

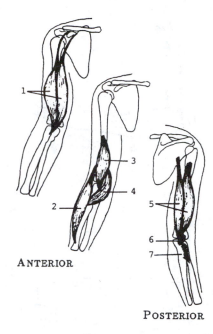

ANTERIOR

POSTERIOR

Figure 5.3 *Muscles of the elbow: (1) biceps, (2) brachioradialis, (3) brachialis, (4) pronator teres, (5) triceps brachii, (6) anconeus, (7) supinator.*

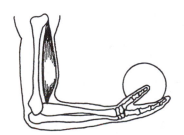

Figure 5.4 *Example 5.1.*

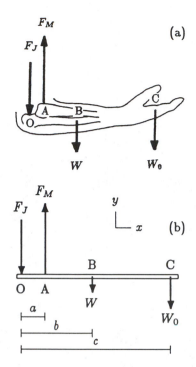

Figure 5.5 *Forces acting on the lower arm.*

forces acting on the forearm are shown in Figure 5.5a, and the free-body diagram of the forearm is shown on a mechanical model in Figure 5.5b. This model assumes that the biceps is the major flexor and that the line of action of the tension (line of pull) in the biceps is vertical.

Point O designates the axis of rotation of the elbow joint, which is assumed to be fixed for practical purposes. Point A is the attachment of the biceps muscle on the radius, B is the center of gravity of the forearm, and C is a point on the forearm that lies along a vertical line passing through the center of gravity of the weight in the hand. The distances between O and A, B, and C are measured as a, b, and c, respectively. W_0 is the weight of the object held in the hand and W is the total weight of the forearm. F_M is the magnitude of the force exerted by the biceps on the radius, and F_J is the magnitude of the reaction force at the elbow joint. Note that the line of action of the muscle force is assumed to be vertical. The gravitational forces are vertical as well. Therefore, for the equilibrium of the lower arm, the line of action of the joint reaction force must also be vertical (a parallel force system).

The task in this example is to determine the magnitudes of the muscle tension and the joint reaction force at the elbow.

Solution: We have a parallel force system, and the unknowns are the magnitudes F_M and F_J of the muscle and joint reaction forces. Considering the rotational equilibrium of the forearm about the elbow joint will yield:

$$\sum M_O = 0: \quad F_M = \frac{1}{a}(b\,W + c\,W_0) \tag{i}$$

For the translational equilibrium of the forearm in the y direction:

$$\sum F_y = 0: \quad F_J = F_M - W - W_0 \tag{ii}$$

For given values of geometric parameters a, b, and c, and weights W and W_0, Eqs. (i) and (ii) can be solved for the magnitudes of the muscle and joint reaction forces. For example, assume that these parameters are given as follows: $a = 4$ cm, $b = 15$ cm, $c = 35$ cm, $W = 20$ N, and $W_0 = 80$ N. Then from Eqs. (i) and (ii):

$$F_M = \frac{1}{0.04}[(0.15)(20) + (0.35)(80)] = 775\text{ N} \quad (+y)$$

$$F_J = 775 - 20 - 80 = 675\text{ N} \quad (-y)$$

Remark:

• The numerical results indicate that the force exerted by the muscle is about ten times larger than the weight of the object held in the position considered. Relative to the axis of the elbow joint, the length a of the lever arm of the muscle force is much smaller than the length c of the lever arm enjoyed by the load held in the hand. The smaller the lever arm, the greater the muscle tension required to balance the clockwise rotational effect of the load about the elbow joint. Therefore, during lifting, it is disadvantageous to have a muscle attachment close to the elbow joint. However, the closer the muscle is to the joint, the larger the range of motion of elbow flexion–extension, and the faster the distal end (hand) of the forearm can reach its goal of moving toward the upper arm or the shoulder.

• The angle (angle of pull) between the line of action of the muscle force and the long axis of the bone upon which the muscle force is exerted is critical in determining the effectiveness of the muscle force. When the lower arm is flexed to a right angle, the muscle tension has only a rotational effect on the forearm about the elbow joint, because the line of action of the muscle force is at right angles with the longitudinal axis of the forearm. For other flexed positions of the forearm, the muscle force can have a translational (stabilizing or sliding) component as well as a rotational component.

Assume that the linkage system shown in Figure 5.6a illustrates the position of the forearm relative to the upper arm. n designates a direction perpendicular (normal) to the long axis of the forearm and t a direction tangent to it. Assuming that the line of action of the muscle force remains parallel to the long axis of the humerus, $\underline{F}_M$ can be decomposed into its rectangular components $\underline{F}_{Mn}$ and $\underline{F}_{Mt}$. In this case, $\underline{F}_{Mn}$ is the *rotational* (rotatory) component of the muscle force because its primary function is to rotate the forearm about the elbow joint. The tangential component $\underline{F}_{Mt}$ of the muscle force acts to compress the elbow joint and is called the *stabilizing* component of the muscle force. As the angle of pull approaches 90°, the magnitude of the rotational component of the muscle force increases while its stabilizing component decreases, and less and less energy is "wasted" to compress the elbow joint. As illustrated in Figure 5.6b, the stabilizing role of $\underline{F}_{Mt}$ changes into a *sliding* or *dislocating* role when the angle between the long axes of the forearm and upper arm becomes less than 90°.

• The elbow is a diarthrodial (synovial) joint. A ligamentous capsule encloses an articular cavity, which is filled with synovial fluid. Synovial fluid is a thick, viscous material. The primary function of this fluid is to lubricate the articulating surfaces, thereby reducing the frictional forces that may develop while one articulating surface slides over the other. The synovial fluid also nourishes the articulating cartilages. Another function of the synovial fluid is to distribute the forces involved at the joint over a relatively large surface area (Figure 5.7). A common

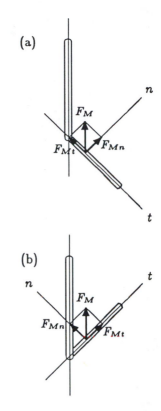

Figure 5.6 *Rotational (F_{Mn}) and stabilizing or sliding (F_{Mt}) components of the muscle force.*

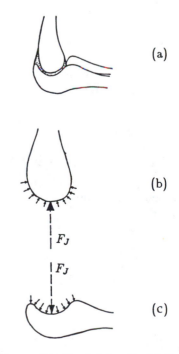

Figure 5.7 *Explaining the joint reaction force at the elbow.*

property of fluids is that they exert pressures (force per unit area) that are distributed over the surfaces they touch. The fluid pressure always acts in a direction toward and perpendicular to the surface it touches having a compressive effect on the surface. Note that in Figure 5.7, the small vectors indicating the fluid pressure have components in the horizontal and vertical directions. We determined that the joint reaction force at the elbow acts vertically downward on the ulna. This implies that the horizontal components of these vectors cancel out (i.e., half pointing to the left and half pointing to the right), but their vertical components (on the ulna, almost all of them are pointing downward) add up to form the resultant force F_J (shown with a dashed arrow in Figure 5.7c). Therefore, the joint reaction force F_J corresponds to the resultant of the distributed force system (pressure) applied through the synovial fluid.

• The most critical simplification made in this example is that the biceps was assumed to be the single muscle group responsible for maintaining the flexed configuration of the forearm. The reason for making such an assumption was to reduce the system under consideration to one that is statically determinate. In reality, in addition to the biceps, the brachialis and the brachioradialis are also primary elbow flexor muscles.

Consider the flexed position of the arm shown in Figure 5.8a. The free-body diagram of the forearm is shown in Figure 5.8b. F_{M1}, F_{M2}, and F_{M3} are the magnitudes of the forces exerted on the forearm by the biceps, the brachialis, and the brachioradialis muscles with attachments at A_1, A_2, and A_3, respectively. Let θ_1, θ_2, and θ_3 be the angles that the biceps, the brachialis, and the brachioradialis muscles make with the long axis of the lower arm. As compared to the single-muscle system that consisted of two unknowns (F_M and F_J), the analysis of this three-muscle system is quite complex. First of all, this is not a simple parallel force system. Even if we assume that the locations of muscle attachments (A_1, A_2, and A_3), their angles of pull (θ_1, θ_2, and θ_3), and the lengths of their moment arms (a_1, a_2, and a_3) as measured from the elbow joint are known, there are still five unknowns in the problem (F_{M1}, F_{M2}, F_{M3}, F_J, and β). The total number of equations available from statics is three:

$$\sum M_O = 0: \quad a_1 F_{M1} + a_2 F_{M2} + a_3 F_{M3} = b\,W + c\,W_0 \quad (iii)$$

$$\sum F_x = 0: \quad F_{Jx} = F_{M1x} + F_{M2x} + F_{M3x} \quad (iv)$$

$$\sum F_y = 0: \quad F_{Jy} = F_{M1y} + F_{M2y} + F_{M3y} - W - W_0 \quad (v)$$

Note that once the muscle forces are determined, Eqs. (iv) and (v) will yield the components of the joint reaction force $\underline{F}_J$. As far as the muscle forces are concerned, we have only Eq. (iii) with three unknowns. In other words, we have a statically

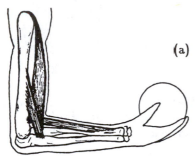

(a)

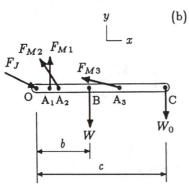

(b)

Figure 5.8 *Three-muscle system.*

indeterminate problem. To obtain a unique solution, we need additional information relating F_{M1}, F_{M2}, and F_{M3}.

There may be several approaches to the solution of this problem. The criterion for estimating the force distribution among different muscle groups may be established by: (1) using cross-sectional areas of muscles, (2) using electromyography (EMG) measurements of muscle signals, and (3) applying certain optimization techniques. It may be assumed that each muscle exerts a force proportional to its cross-sectional area. If A_1, A_2, and A_3 are the cross-sectional areas of the biceps, the brachialis, and the brachioradialis, then this criterion may be applied by expressing muscle forces in the following manner:

$$F_{M2} = k_{21}\, F_{M1} \quad \text{with} \quad k_{21} = \frac{A_2}{A_1} \qquad (vi)$$

$$F_{M3} = k_{31}\, F_{M1} \quad \text{with} \quad k_{31} = \frac{A_3}{A_1} \qquad (vii)$$

If constants k_{21} and k_{31} are known, then Eqs. (vi) and (vii) can be substituted into Eq. (iii), which can then be solved for F_{M1}:

$$F_{M1} = \frac{b\,W + c\,W_0}{a_1 + a_2 k_{21} + a_3 k_{31}}$$

Substituting F_{M1} back into Eqs. (vi) and (vii) will then yield the magnitudes of the forces in the brachialis and the brachioradialis muscles. The values of k_{21} and k_{31} may also be estimated by using the amplitudes of muscle EMG signals.

This statically indeterminate problem may also be solved by considering some optimization techniques. If the purpose is to accomplish a certain task (static or dynamic) in the most efficient manner, then the muscles of the body must act to minimize the forces exerted, the moments about the joints (for dynamic situations), and/or the work done by the muscles. The question is, what force distribution among the various muscles facilitates the maximum efficiency? These concepts and relevant references will be discussed briefly in Section 5.11.

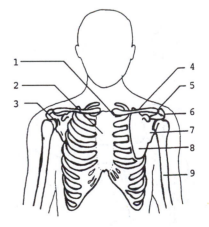

5.6 Mechanics of the Shoulder

The bony structure and the muscles of the shoulder complex are illustrated in Figures 5.9 and 5.10. The shoulder forms the base for all upper extremity movements. The complex structure of the shoulder can be divided into two: the shoulder joint and the shoulder girdle.

Figure 5.9 *The shoulder: (1) sternoclavicular joint, (2) sternum, (3) glenohumeral joint, (4) clavicle, (5) acromioclavicular joint, (6) acromion process, (7) glenoid fossa, (8) scapula, (9) humerus.*

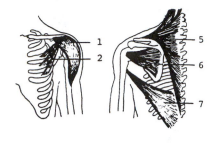

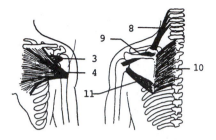

ANTERIOR   POSTERIOR
(FRONT)   (BACK)

Figure 5.10 *Shoulder muscles: (1) deltoideus, (2) pectoralis minor, (3) subscapularis, (4) pectoralis major, (5) trapezius, (6) infraspinatus & teres minor, (7) latissimus dorsi, (8) levator scapulae, (9) supraspinatus, (10) rhomboideus, (11) teres major.*

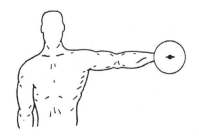

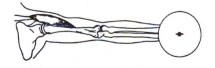

Figure 5.11 *The arm is abducted to horizontal.*

The shoulder joint, also known as the *glenohumeral articulation*, is a ball-and-socket joint between the nearly hemispherical humeral head (ball) and the shallowly concave glenoid fossa (socket) of the scapula. The shallowness of the glenoid fossa allows a significant freedom of movement of the humeral head on the articulating surface of the glenoid. The movements allowed are: flexion (movement of the humerus to the front — a forward upward movement) and extension (return from flexion), abduction (horizontal upward movement of the humerus to the side) and adduction (return from abduction), and outward rotation (movement of the humerus around its long axis to the lateral side) and inward rotation (return from outward rotation). The configuration of the articulating surfaces of the shoulder joint also makes the joint more susceptible to instability and injury, such as dislocation. The stability of the joint is provided by the glenohumeral and coracohumeral ligaments, and by the muscles crossing the joint. The major muscles of the shoulder joint are: deltoideus, supraspinatus, pectoralis major, coracobrachialis, latissimus dorsi, teres major, teres minor, infraspinatus, and subscapularis.

The bony structure of the shoulder girdle consists of the clavicle (collarbone) and the scapula (shoulder blade). The *acromioclavicular joint* is a small synovial articulation between the distal clavicle and the proximal acromion of the scapula. The stability of this joint is reinforced by the coracoclavicular ligaments. The *sternoclavicular joint* is the articulation between the manubrium of the sternum and the proximal clavicle. The stability of this joint is enhanced by the costoclavicular ligament. The acromioclavicular joint and the sternoclavicular joint both have layers of cartilage, called *menisci*, interposed between their bony surfaces.

There are four pairs of scapular movements: elevation (movement of the scapula in the frontal plane) and depression (return from elevation), upward rotation (turning the glenoid fossa upward and the lower medial border of the scapula away from the spinal column) and downward rotation (return from upward rotation), protraction (movement of the distal end of the clavicle forward) and retraction (return from protraction), and forward and backward rotation (rotation of the scapula about the shaft of the clavicle). There are six muscles that control and coordinate these movements: trapezius, levator scapulae, rhomboid, pectoralis minor, serratus anterior, and subclavius.

Example 5.2 Consider a person strengthening the shoulder muscles by means of dumbbell exercises. Figure 5.11 illustrates the position of the left arm when the arm is abducted to horizontal. The free-body diagram of the arm is shown in Figure 5.12

along with a mechanical model of the arm. Also in Figure 5.12, the forces acting on the arm are resolved into their rectangular components along the horizontal and vertical directions. O corresponds to the axis of the shoulder joint, A is where the deltoid muscle is attached to the humerus, B is the center of gravity of the entire arm, and C is the center of gravity of the dumbbell. W is the weight of the arm, W_0 is the weight of the dumbbell, F_M is the magnitude of the tension in the deltoid muscle, and F_J is the joint reaction force at the shoulder. The resultant of the deltoid muscle force makes an angle θ with the horizontal. The distances between O and A, B and C are measured as a, b, and c, respectively.

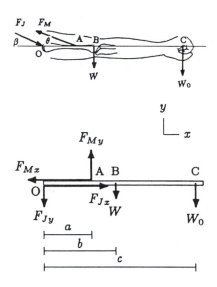

Figure 5.12 *Forces acting on the arm and a mechanical model representing the arm.*

Determine the magnitude F_M of the force exerted by the deltoid muscle to hold the arm at the position shown. Also determine the magnitude and direction of the reaction force at the shoulder joint in terms of specified parameters.

Solution: With respect to the xy coordinate frame, the muscle and joint reaction forces have two components, while the weights of the arm and the dumbbell act in the negative y direction. The components of the muscle force are:

$$F_{Mx} = F_M \cos\theta \quad (-x) \qquad (i)$$
$$F_{My} = F_M \sin\theta \quad (+y) \qquad (ii)$$

Components of the joint reaction force are:

$$F_{Jx} = F_J \cos\beta \quad (+x) \qquad (iii)$$
$$F_{Jy} = F_J \sin\beta \quad (-y) \qquad (iv)$$

β is the angle that the joint reaction force makes with the horizontal. The line of action and direction (in terms of θ) of the force exerted by the muscle on the arm are known. However, the magnitude F_M of the muscle force, the magnitude F_J, and the direction (β) of the joint reaction force are unknowns. We have a total of three unknowns, F_M, F_J, and β (or F_M, F_{Jx}, and F_{Jy}). To be able to solve this two-dimensional problem, we have to utilize all three equilibrium equations.

First, consider the rotational equilibrium of the arm about the shoulder joint at O. The joint reaction force produces no torque about O because its line of action passes through O. For practical purposes, we can neglect the possible contribution of the horizontal component of the muscle force to the moment generated about O by assuming that its line of action also passes through O. Note that this is not a critical or necessary assumption to solve this problem. If we knew the length of its moment arm (i.e., the vertical distance between O and A), we could easily incorporate the torque generated by F_{Mx} about O into the analysis. Under these considerations, there are only three moment-producing forces about O. For the rotational equilibrium of the arm, the net moment about O must be equal to zero. Taking counterclockwise moments to be positive:

$$\sum M_O = 0: \quad a\, F_{My} - b\, W - c\, W_0 = 0$$

$$F_{My} = \frac{1}{a}(b\, W + c\, W_0) \qquad (v)$$

For given a, b, c, W, and W_0, Eq. (v) can be used to determine the vertical component of the force exerted by the deltoid muscle. Eq. (ii) can now be used to determine the total force exerted by the muscle:

$$F_M = \frac{F_{My}}{\sin\theta} \qquad (vi)$$

Knowing F_M, Eq. (i) will yield the horizontal component of the tension in the muscle:

$$F_{Mx} = F_M \cos\theta \qquad (vii)$$

The components of the joint reaction force can be determined by considering the translational equilibrium of the arm in the horizontal and vertical directions:

$$\sum F_x = 0: \quad F_{Jx} = F_{Mx} \qquad (viii)$$

$$\sum F_y = 0: \quad F_{Jy} = F_{My} - W - W_0 \qquad (ix)$$

Knowing the rectangular components of the joint reaction force enables us to compute the magnitude of the force itself and the angle its line of action makes with the horizontal:

$$F_J = \sqrt{(F_{Jx})^2 + (F_{Jy})^2} \qquad (x)$$

$$\beta = \tan^{-1}\left(\frac{F_{Jy}}{F_{Jx}}\right) \qquad (xi)$$

Now consider that $a = 15$ cm, $b = 30$ cm, $c = 60$ cm, $\theta = 15°$, $W = 40$ N, and $W_0 = 60$ N. Then:

$$F_{My} = \frac{1}{0.15}[(0.30)(40) + (0.60)(60)] = 320\ N \quad (+y)$$

$$F_M = \frac{320}{\sin 15°} = 1236\ N$$

$$F_{Mx} = (1236)(\cos 15°) = 1194\ N \quad (-x)$$

$$F_{Jx} = 1194\ N \quad (+x)$$

$$F_{Jy} = 320 - 40 - 60 = 220\ N \quad (-y)$$

$$F_J = \sqrt{(1194)^2 + (220)^2} = 1214\ N$$

$$\beta = \tan^{-1}\left(\frac{220}{1194}\right) = 10°$$

Remark:

• F_{Mx} is the stabilizing component and F_{My} is the rotational component of the deltoid muscle. F_{Mx} is approximately four times larger than F_{My}. A large stabilizing component suggests that the horizontal position of the arm is not stable, and that the muscle needs to exert a high horizontal force to stabilize it.

• It is believed that the joint reaction force at the shoulder is the resultant of two components: an active contraction of the infraspinatous muscle and a force generated at the joint in terms of pressure and friction.

• The human shoulder is very susceptible to injuries. The most common injuries are dislocations of the shoulder joint and the fracture of the humerus. Since the socket of the glenohumeral joint is shallow, the head of the humerus is relatively free to rotate about the articulating surface of the glenoid fossa. This freedom of movement is achieved, however, by reduced joint stability. The humeral head may be displaced in various ways, depending on the strength or weakness of the muscular and ligamentous structure of the shoulder, and depending on the physical activity. Humeral fractures are another common type of injury. The humerus is particularly vulnerable to injuries because of its unprotected configuration.

• Average ranges of motion of the arm about the shoulder joint are 230° during flexion–extension, and 170° in both abduction–adduction and inward–outward rotation.

5.7 Mechanics of the Spinal Column

The human spinal column is the most complex part of the human musculoskeletal system. The principal functions of the spinal column are to protect the spinal cord; to support the head, neck, and upper extremities; to transfer loads from the head and trunk to the pelvis; and to permit a variety of movements. The spinal column consists of the cervical (neck), thoracic (chest), lumbar (lower back), sacral, and coccygeal regions. The thoracic and lumbar sections of the spinal column make up the trunk. The sacral and coccygeal regions are united with the pelvis and can be considered parts of the pelvic girdle.

The vertebral column consists of 24 intricate and complex vertebrae (Figure 5.13). The articulations between the vertebrae are amphiarthrodial joints. A fibrocartilaginous disk is interposed between each pair of vertebrae. The primary functions of these intervertebral disks are to sustain loads transmitted from segments above, act as shock absorbers, eliminate bone-to-bone

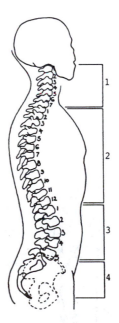

Figure 5.13 *The spinal column: (1) cervical vertabrae, (2) thoracic vertabrae, (3) lumbar vertabrae, (4) sacrum.*

contact, and reduce the effects of impact forces by preventing direct contact between the bony structures of the vertabrae. The articulations of each vertebra with the adjacent vertabrae permit movement in three planes, and the entire spine functions like a single ball-and-socket joint. The structure of the spine allows a wide variety of movements including flexion–extension, lateral flexion, and rotation.

Two particularly important joints of the spinal column are those with the head. The *atlantooccipital joint* is the union between the first cervical vertebra (the atlas) and the occipital bone of the head. This is a double condyloid joint and permits movements of the head in the sagittal and frontal planes. The *atlantoaxial joint* is the union between the atlas and the odontoid process of the head. It is a pivot joint, enabling the head to rotate in the transverse plane. The muscle groups providing, controlling, and coordinating the movement of the head and the neck are the prevertebrals (anterior), hyoids (anterior), sternocleidomastoid (anterior–lateral), scaleni (lateral), levator scapulae (lateral), suboccipitals (posterior), and spleni (posterior).

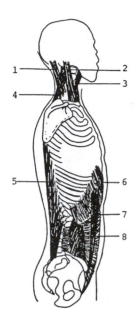

Figure 5.14 *Selected muscles of the neck and spine: (1) splenius, (2) sternocleidomastoid, (3) hyoid, (4) levator scapula, (5) erector spinae, (6) obliques, (7) rectus abdominis, (8) transversus abdominis.*

The spine gains its stability from the intervertebral discs and from the surrounding ligaments and muscles (Figure 5.14). The discs and ligaments provide intrinsic stability, and the muscles supply extrinsic support. The muscles of the spine exist in pairs. The anterior portion of the spine contains the abdominal muscles: the rectus abdominis, external obliques, and internal obliques. These muscles provide the necessary force for trunk flexion and maintain the internal organs in proper position. There are three layers of posterior trunk muscles: erector spinae, semispinalis, and the deep posterior spinal muscle groups. The primary function of the muscles located at the posterior portion of the spine is to provide trunk extension. These muscles also support the spine against the effects of gravity. The quadratus lumborum muscle is important in lateral trunk flexion. It also stabilizes the pelvis and lumbar spine. The lateral flexion of the trunk results from the actions of the abdominal and posterior muscles. The rotational movement of the trunk is controlled by the simultaneous action of anterior and posterior muscles.

The spinal column is vulnerable to various injuries. The most severe injury involves the spinal cord, which is immersed in fluid and protected by the bony structure. Other critical injuries include fractured vertebrae and herniated intervertebral disks. Lower back pain may also result from strains in the lower regions of the spine.

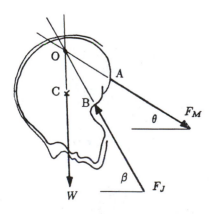

Figure 5.15 *Forces on the skull form a concurrent system.*

Example 5.3 Consider the position of the head and the neck shown in Figure 5.15. Also shown are the forces acting on the head. The head weighs $W = 50$ N and its center of gravity is located at C. F_M is the magnitude of the resultant force exerted by the neck extensor muscles, which is applied on the skull at A. The atlantooccipital joint center is located at B. For this flexed position of the head, it is estimated that the line of action of the neck muscle force makes an angle $\theta = 30°$ and the line of action of the joint reaction force makes an angle $\beta = 60°$ with the horizontal.

What tension must the neck extensor muscles exert to support the head? What is the compressive force applied on the first cervical vertebra at the atlantooccipital joint?

Solution: We have a three-force system with two unknowns: magnitudes F_M and F_J of the muscle and joint reaction forces. Since the problem has a relatively complicated geometry, it is convenient to utilize the condition that for a body to be in equilibrium the force system acting on it must be either concurrent or parallel. In this case, it is clear that the forces involved do not form a parallel force system. Therefore, the system of forces under consideration must be concurrent. Recall that a system of forces is concurrent if the lines of action of all forces have a common point of intersection.

In Figure 5.15, the lines of action of all three forces acting on the head are extended to meet at point O. In Figure 5.16, the forces W, $\underline{F}_M$, and $\underline{F}_J$ acting on the skull are translated to O, which is also chosen to be the origin of the xy coordinate frame. The rectangular components of the muscle and joint reaction forces in the x and y directions are:

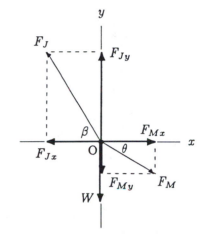

Figure 5.16 *Components of the forces acting on the head.*

$$F_{Mx} = F_M \cos\theta \qquad (i)$$

$$F_{My} = F_M \sin\theta \qquad (ii)$$

$$F_{Jx} = F_J \cos\beta \qquad (iii)$$

$$F_{Jy} = F_J \sin\beta \qquad (iv)$$

The translational equilibrium conditions in the x and y directions will yield:

$$\sum F_x = 0 : \qquad F_{Jx} = F_{Mx} \qquad (v)$$

$$\sum F_y = 0 : \qquad F_{Jy} = W + F_{My} \qquad (vi)$$

Substitute Eqs. (*i*) and (*iii*) into Eq. (*v*):

$$F_J \cos\beta = F_M \cos\theta \qquad (vii)$$

Substitute Eqs. (*ii*) and (*iv*) into Eq. (*vi*):

$$F_J \sin\beta = W + F_M \sin\theta \qquad (viii)$$

Divide Eq. (*viii*) by Eq. (*vii*) so as to eliminate F_J:

$$\tan\beta = \frac{W + F_M \sin\theta}{F_M \cos\theta} \qquad (ix)$$

Eq. (ix) can now be solved for the unknown muscle force F_M:

$$F_M \cos\theta \tan\beta = W + F_M \sin\theta$$
$$F_M (\cos\theta \tan\beta - \sin\theta) = W$$
$$F_M = \frac{W}{\cos\theta \tan\beta - \sin\theta} \qquad (x)$$

Eq. (x) gives the tension in the muscle as a function of the weight W of the head and the angles θ and β that the lines of action of the muscle and joint reaction forces make with the horizontal. Substituting the numerical values of W, θ, and β will yield:

$$F_M = \frac{50}{(\cos 30°)(\tan 60°) - (\sin 30°)} = 50 \text{ N}$$

From Eqs. (i) and (ii):

$$F_{Mx} = (50)(\cos 30°) = 43 \text{ N} \quad (+x)$$
$$F_{My} = (50)(\sin 30°) = 25 \text{ N} \quad (-y)$$

From Eqs. (v) and (vi):

$$F_{Jx} = 43 \text{ N} \quad (-x)$$
$$F_{Jy} = 50 + 25 = 75 \text{ N} \quad (+y)$$

The resultant of the joint reaction force can be computed either from Eq. (iii) or (iv). Using Eq. (iii):

$$F_J = \frac{F_{Jx}}{\cos\beta} = \frac{43}{\cos 60°} = 86 \text{ N}$$

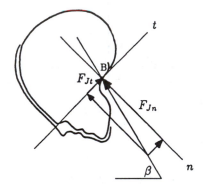

Figure 5.17 *Normal and shear components of the joint reaction force.*

Remark:

• The extensor muscles of the head must apply a force of 50 N to support the head in the position considered. The reaction force developed at the atlantooccipital joint is about 86 N.

• The joint reaction force can be resolved into two rectangular components, as shown in Figure 5.17. F_{Jn} is the magnitude of the normal component of $\underline{F}_J$ compressing the articulating joint surface, and F_{Jt} is the magnitude of its tangential component having a shearing effect on the joint surfaces. Forces in the muscles and ligaments of the neck operate in a manner to counterbalance this shearing effect.

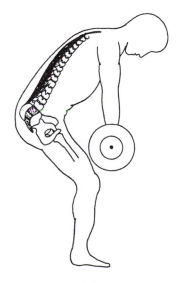

Figure 5.18 *A weight lifter.*

Example 5.4 Consider the weight lifter illustrated in Figure 5.18, who is bent forward and lifting a weight W_0. At the position shown, the athlete's trunk is flexed by an angle θ as measured from the upright (vertical) position.

The forces acting on the lower portion of the athlete's body are shown in Figure 5.19 by considering a section passing through the fifth lumbar vertebra. A mechanical model of the athlete's lower body (the pelvis and legs) is illustrated in Figure 5.20 along with the geometric parameters of the problem under consideration. W is the total weight of the athlete, W_1 is the weight of the legs including the pelvis, $W + W_0$ is the total ground reaction force applied to the athlete through the feet (at C), F_M is the magnitude of the resultant force exerted by the erector spinae muscles supporting the trunk, and F_J is the magnitude of the compressive force generated at the union (O) of the sacrum and the fifth lumbar vertebra. The center of gravity of the legs including the pelvis is located at B. Relative to O, the lengths of the lever arms of the muscle force, lower body weight, and ground reaction force are measured as a, b, and c, respectively.

Assuming that the line of pull of the resultant muscle force exerted by the erector spinae muscles is parallel to the trunk (i.e., making an angle θ with the vertical), determine F_M and F_J in terms of $a, b, c, \theta, W_0, W_1$, and W.

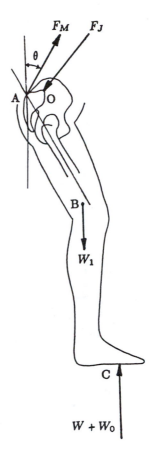

Figure 5.19 *Forces acting on the lower body of the athlete.*

Solution: In this case, there are three unknowns: F_M, F_{Jx}, and F_{Jy}. The lengths of the lever arms of the muscle force, ground reaction force, and the gravitational force of the legs including the pelvis are given as measured from O. Therefore, we can apply the rotational equilibrium condition about O to determine the magnitude F_M of the resultant force exerted by the erector spinae muscles. Considering clockwise moments to be positive:

$$\sum M_O = 0: \qquad a\,F_M + b\,W_1 - c\,(W + W_0) = 0$$

Solving this equation for F_M will yield:

$$F_M = \frac{c\,(W + W_0) - b\,W_1}{a} \qquad (i)$$

For given numerical values of a, b, c, W, W_0, and W_1, Eq. (i) can be used to determine the magnitude of the resultant muscle force. Once F_M is calculated, its components in the x and y directions can be determined using:

$$F_{Mx} = F_M \sin\theta \qquad (ii)$$
$$F_{My} = F_M \cos\theta \qquad (iii)$$

The horizontal and vertical components of the reaction force developed at the sacrum can now be determined by utilizing the translational equilibrium conditions of the lower body of the

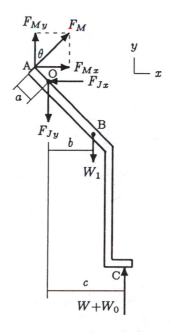

Figure 5.20 *Free-body diagram.*

athlete in the x and y directions:

$$\sum F_x = 0 : \quad F_{Jx} = F_{Mx} \qquad (iv)$$

$$\sum F_y = 0 : \quad F_{Jy} = F_{My} + W + W_0 - W_1 \qquad (v)$$

Assume that at an instant the athlete is bent so that his trunk makes an angle $\theta = 45°$ with the vertical, and that the lengths of the lever arms are measured in terms of the height h of the athlete and the weights are given in terms of the weight W of the athlete as: $a = 0.02h$, $b = 0.08h$, $c = 0.12h$, $W_0 = W$, and $W_1 = 0.4W$. Using Eq. (i):

$$F_M = \frac{(0.12\,h)(W + W) - (0.08\,h)(0.4\,W)}{0.02\,h} = 10.4\ W$$

From Eqs. (ii) and (iii):

$$F_{Mx} = (10.4\ W)(\sin 45°) = 7.4\ W$$
$$F_{My} = (10.4\ W)(\cos 45°) = 7.4\ W$$

From Eqs. (iv) and (v):

$$F_{Jx} = 7.4\ W$$
$$F_{Jy} = 7.4\ W + W + W - 0.4\ W = 9.0\ W$$

Therefore, the magnitude of the resultant force on the sacrum is:

$$F_J = \sqrt{(F_{Jx})^2 + (F_{Jy})^2} = 11.7\ W$$

A remark:

• The results obtained are quite significant. While the athlete is bent forward by 45° and lifting a weight with magnitude equal to his own body weight, the erector spinae muscles exert a force more than 10 times the weight of the athlete and the force applied to the union of the sacrum and the fifth lumbar vertebra is about 12 times that of the body weight.

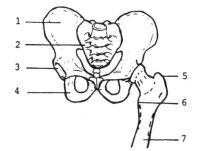

Figure 5.21 *Pelvis and the hip: (1) ilium, (2) sacrum, (3) acetabulum, (4) ischium, (5) greater trochanter, (6) lesser trochanter, (7) femur.*

5.8 Mechanics of the Hip

The articulation between the head of the femur and the acetabulum of the pelvis (Figure 5.21) forms a diarthrodial joint. The stability of the hip joint is provided by its relatively rigid ball-and-socket type of configuration, its ligaments, and by the large and strong muscles crossing it. The femoral head fits well into the deep socket of the acetabulum. The ligaments of the hip joint, such as the transverse and teres femoris ligaments, support and hold the femoral head in the acetabulum as the femoral head moves. The construction of the hip joint is such that it is very stable and has a great deal of mobility, thereby allowing a wide range of motion required for activities such as walking, sitting, and squatting. Movements of the femur about the hip joint include flexion and extension, abduction and adduction, and inward and outward rotation. In some instances, the extent of these movements is constrained by ligaments, muscles, and/or the bony structure of the hip. The articulating surfaces of the femoral head and the acetabulum are lined with hyaline cartilage. Derangements of the hip can produce altered force distributions in the joint cartilage, leading to degenerative arthritis.

The pelvis consists of the ilium, ischium, and pubis bones, and the sacrum. At birth and during growth, the bones of the pelvis are distinct. In adults the bones of the pelvis are fused and form synarthrodial joints, which allow no movement. The pelvis is located between the spine and the two femurs. The position of the pelvis makes it relatively less stable. Movements of the pelvis occur primarily for the purpose of facilitating the movements of the spine or the femurs. There are no muscles whose primary purpose is to move the pelvis. Movements of the pelvis are caused by the muscles of the trunk and the hip.

Based on their primary actions, the muscles of the hip joint can be divided into several groups (Figure 5.22). The psoas, iliacus, rectus femoris, pectineus, and tensor fascia latae are the primary hip flexors. They are also used to carry out activities such as running or kicking. The gluteus maximus and the hamstring muscles (the biceps femoris, semitendinosus, and semimembranosus) are hip extensors. The hamstring muscles also function as knee flexors. The gluteus medius and gluteus minimus are hip abductor muscles providing for the inward rotation of the femur. The gluteus medius is also the primary muscle group stabilizing the pelvis in the frontal plane. The adductor longus, adductor brevis, adductor magnus, and gracilis muscles are the hip adductors. There are also small, deeply placed muscles (outward rotators) that provide for the outward rotation of the femur.

The hip muscles predominantly suffer contusions and strains occurring in the pelvis region.

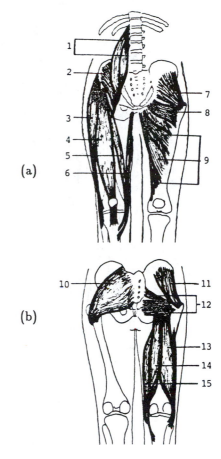

(a)

(b)

Figure 5.22 *Muscles of the hip: (a) anterior & (b) posterior views. (1) psoas, (2) iliacus, (3) tensor fascia latae, (4) rectus femoris, (5) sartorius, (6) gracilis, (7) gluteus minimis, (8) pectineus, (9) adductors, (10), (11) gluteus maximus & medius, (12) lateral rotators, (13) biceps femoris, (14) semitendinosus, (15) semimembranosus.*

Example 5.5 During walking and running, we momentarily put all of our body weight on one leg (the right leg in Figure 5.23). The forces acting on the leg carrying the total body weight are shown in Figure 5.24 during such a single-leg stance. F_M is the magnitude of the resultant force exerted by the hip abductor

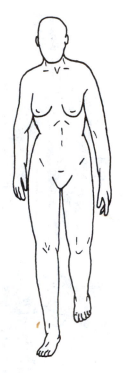

Figure 5.23 *Single-leg stance.*

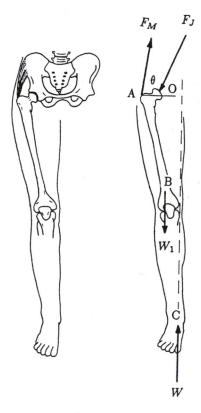

Figure 5.24 *Forces acting on the right leg carrying the entire weight of the body.*

muscles, F_J is the magnitude of the joint reaction force applied by the pelvis on the femur, W_1 is the weight of the leg, W is the total weight of the body applied as a normal force by the ground on the leg. The angle between the line of action of the resultant muscle force and the horizontal is designated by θ.

A mechanical model of the leg, rectangular components of the forces acting on it, and the parameters necessary to define the geometry of the problem are shown in Figure 5.25. O is a point along the instantaneous axis of rotation of the hip joint, A is where the hip abductor muscles are attached to the femur, B is the center of gravity of the leg, and C is where the ground reaction force is applied on the foot. The distances between A and O, B, and C are specified as a, b, and c, respectively. α is the angle of inclination of the femoral neck to the horizontal, and β is the angle that the long axis of the femoral shaft makes with the horizontal. Therefore, $\alpha + \beta$ is approximately equal to the total neck-to-shaft angle of the femur.

Determine the force exerted by the hip abductor muscles and the joint reaction force at the hip to support the leg and the hip in the position shown.

Solution 1:   Utilizing the free-body diagram of the leg.
For the solution of the problem, we can utilize the free-body diagram of the right leg supporting the entire weight of the person. In Figure 5.25a, the muscle and joint reaction forces are shown in terms of their components in the x and y directions. The resultant muscle force has a line of action that makes an angle θ with the horizontal. Therefore:

$$F_{Mx} = F_M \cos\theta \qquad (i)$$
$$F_{My} = F_M \sin\theta \qquad (ii)$$

Since angle θ is specified (given as a measured quantity), the only unknown for the muscle force is its magnitude F_M. For the joint reaction force, neither the magnitude nor the direction is known. With respect to the axis of the hip joint located at O, a_x in Figure 5.25b is the moment arm of the vertical component F_{My} of the muscle force, and a_y is the moment arm of F_{Mx}. Similarly, $(b_x - a_x)$ is the moment arm for W_1 and $(c_x - a_x)$ is the moment arm for the force W applied by the ground on the leg.

From the geometry of the problem:

$$a_x = a \cos\alpha \qquad (iii)$$
$$a_y = a \sin\alpha \qquad (iv)$$
$$b_x = b \cos\beta \qquad (v)$$
$$c_x = c \cos\beta \qquad (vi)$$

Now that the horizontal and vertical components of all forces involved, and their moment arms with respect to O are established, the condition for the rotational equilibrium of the leg about O can be utilized to determine the magnitude of the resultant muscle force applied at A. Assuming that the clockwise moments are positive:

$$\sum M_O = 0: \quad a_x F_{My} - a_y F_{Mx} - (c_x - a_x) W$$
$$+ (b_x - a_x) W_1 = 0$$

Substituting Eqs. (i) through (vi) into the above equation:

$$(a \cos \alpha)(F_M \sin \theta) - (a \sin \alpha)(F_M \cos \theta)$$
$$- (c \cos \beta - a \cos \alpha) W + (b \cos \beta - a \cos \alpha) W_1 = 0$$

Solving this equation for the muscle force:

$$F_M = \frac{(c\,W - b\,W_1) \cos \beta - a\,(W - W_1) \cos \alpha}{a\,(\cos \alpha \sin \theta - \sin \alpha \cos \theta)} \quad (vii)$$

Note that the denominator of Eq. (vii) can be simplified as $a \sin(\theta - \alpha)$. To determine the components of the joint reaction force, we can utilize the horizontal and vertical equilibrium conditions of the leg:

$$\sum F_x = 0: \quad F_{Jx} = F_{Mx} = F_M \cos \theta \quad (viii)$$

$$\sum F_y = 0: \quad F_{Jy} = F_{My} + W - W_1$$

$$F_{Jy} = F_M \sin \theta + W - W_1 \quad (ix)$$

Therefore, the resultant force acting at the hip joint is:

$$F_J = \sqrt{(F_{Jx})^2 + (F_{Jy})^2} \quad (x)$$

Assume that the geometric parameters of the problem and the weight of the leg are measured in terms of the person's height h and total weight W as follows: $a = 0.05h$, $b = 0.20h$, $c = 0.52h$, $\alpha = 45°$, $\beta = 80°$, $\theta = 70°$, and $W_1 = 0.17W$. The solution of the above equations for the muscle and joint reaction forces will yield $F_M = 2.6W$ and $F_J = 3.4W$, the joint reaction force making an angle $\phi = \tan^{-1}(F_{Jy}/F_{Jx}) = 74.8°$.

Solution 2: Utilizing the free-body diagram of the upper body.
Here we have an alternative approach to the solution of the same problem. In this case, instead of the free-body diagram of the right leg, the free-body diagram of the upper body (including the left leg) is utilized. The forces acting on the upper body are shown in Figures 5.26 and 5.27. F_M is the magnitude of the resultant force exerted by the hip abductor muscles applied on the pelvis at D. θ is again the angle between the line of action of the resultant muscle force and the horizontal. F_J is the magnitude of the reaction force applied by the head of the femur on the hip joint at E. $W_2 = W - W_1$ (total body weight

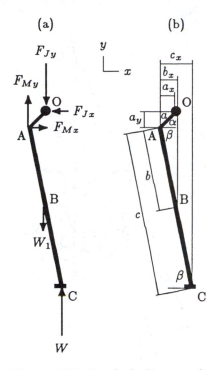

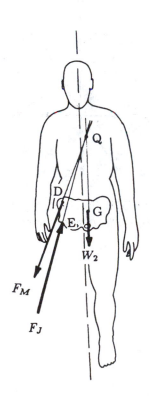

Figure 5.25 *Free-body diagram of the leg (a), and the geometric parameters (b).*

Figure 5.26 *Forces acting on the pelvis during a single-leg (right leg) stance.*

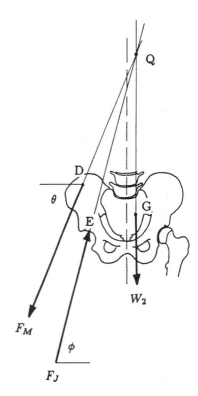

Figure 5.27 *Forces involved form a concurrent system.*

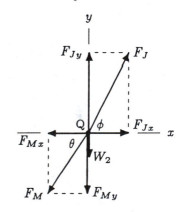

Figure 5.28 *Resolution of the forces into their components.*

minus the weight of the right leg) is the weight of the upper body and the left leg acting as a concentrated force at G. Note that G is not the center of gravity of the entire body. Since the right leg is not included in the free-body, the left-hand side of the body is "heavier" than the right-hand side, and G is located to the left of the original center of gravity (a point along the vertical dashed line in Figure 5.27) of the person. The location of G can be determined utilizing the method provided in Section 4.12.

By combining the individual weights of the segments constituting the body under consideration, the problem is reduced to a three-force system. It is clear from the geometry of the problem that the forces involved do not form a parallel system. Therefore, for the equilibrium of the body, they have to form a concurrent system of forces. This implies that the lines of action of the forces must have a common point of intersection (Q in Figure 5.27), which can be obtained by extending the lines of action of $\underline{W}_2$ and $\underline{F}_M$. A line passing through points Q and E designates the line of action of the joint reaction force $\underline{F}_J$. The angle ϕ that $\underline{F}_J$ makes with the horizontal can now be measured from the geometry of the problem. Since the direction of $\underline{F}_J$ is determined through certain geometric considerations, the number of unknowns is reduced by one. As illustrated in Figure 5.28, the unknown magnitudes F_M and F_J of the muscle and joint reaction forces can now be determined simply by translating $\underline{W}_2$, $\underline{F}_M$, and $\underline{F}_J$ to point Q, and decomposing them into their components along the horizontal (x) and vertical (y) directions:

$$F_{Mx} = F_M \cos\theta$$
$$F_{My} = F_M \sin\theta$$
$$F_{Jx} = F_J \cos\phi$$
$$F_{Jy} = F_J \sin\phi$$

For the translational equilibrium in the x and y directions:

$$\sum F_x = 0: \qquad F_{Jx} = F_{Mx}$$
$$\sum F_y = 0: \qquad F_{Jy} = F_{My} + W_2$$

Simultaneous solutions of these equations will yield:

$$F_M = \frac{\cos\phi \; W_2}{\cos\theta \sin\phi - \sin\theta \cos\phi} \qquad (xi)$$

$$F_J = \frac{\cos\theta \; W_2}{\cos\theta \sin\phi - \sin\theta \cos\phi} \qquad (xii)$$

For example, if $\theta = 70°$, $\phi = 74.8°$, and $W_2 = 0.83W$ (W is the total weight of the person), then Eqs. (xi) and (xii) will yield $F_M = 2.6W$ and $F_J = 3.4W$.

How would the muscle and hip joint reaction forces vary if the person is carrying a load of W_0 in each hand during single-leg stance (Figure 5.29)?

The free-body diagram of the upper body while the person is carrying a load of W_0 in each hand is shown in Figure 5.30. The system to be analyzed consists of the upper body of the person (including the left leg) and the loads carried in each hand. To counterbalance both the rotational and translational (downward) effects of the extra loads, the hip abductor muscles will exert additional forces, and there will be larger compressive forces generated at the hip joint.

In this case, the number of forces is five. The gravitational pull on the upper body (W_2) and on the masses carried in the hands (W_0) form a parallel force system. If these parallel forces can be replaced by a single resultant force, then the number of forces can be reduced to three, and the problem can be solved by applying the same technique explained above (Solution 2). For this purpose, consider the force system shown in Figure 5.31. M and N correspond to the right and left hands of the person where external forces of equal magnitude (W_0) are applied. G is the center of gravity of the upper body including the left leg. The vertical dashed line shows the symmetry axis (midline) of the person in the frontal plane, and G is located to the left of this axis. Note that the distance ℓ_1 between M and G is greater than the distance ℓ_2 between N and G. If ℓ_1, ℓ_2, W_2, and W_0 are given, then a new center of gravity (G') can be determined by applying the technique of finding the center of gravity of a system composed of a number of parts whose centers of gravity are known (see Section 5.14). By intuition, G' is located somewhere between the symmetry axis and G. In other words, G' is closer to the right hip joint, and, therefore, the length of the moment arm of the total weight as measured from the right hip joint is shorter as compared to the case when there is no load carried in the hands. On the other hand, the magnitude of the resultant gravitational force is $W_3 = W_2 + 2W_0$, which overcompensates for the advantage gained by the reduction of the moment arm.

Once the new center of gravity of the upper body is determined, including the left leg and the loads carried in each hand, Eqs. (*xi*) and (*xii*) can be utilized to calculate the resultant force exerted by the hip abductor muscles and the reaction force generated at the hip joint:

$$F_M = \frac{\cos\phi' \, (W_2 + 2W_0)}{\cos\theta \sin\phi' - \sin\theta \cos\phi'}$$

$$F_J = \frac{\cos\theta \, (W_2 + 2W_0)}{\cos\theta \sin\phi' - \sin\theta \cos\phi'}$$

Here, Eqs. (*xi*) and (*xii*) are modified by replacing the weight W_2 of the upper body with the new total weight $W_3 = W_2 + 2W_0$,

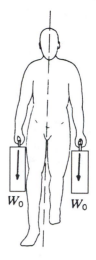

Figure 5.29 *Carrying a load in each hand.*

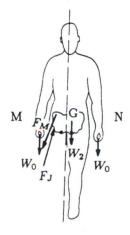

Figure 5.30 *Forces acting on the upper body.*

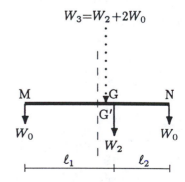

Figure 5.31 $\underline{W_3}$ *is the resultant of the three-force system.*

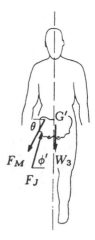

Figure 5.32 *The problem is reduced to a three-force concurrent system.*

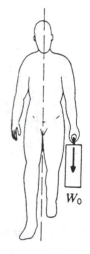

Figure 5.33 *Carrying a load in one hand.*

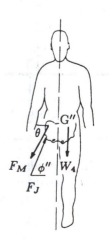

Figure 5.34 *Forces acting on the upper body.*

and by replacing the angle ϕ that the line of action of the joint reaction force makes with the horizontal with the new angle ϕ' (Figure 5.32). ϕ' is slightly larger than ϕ because of the shift of the center of gravity from G to G' toward the right of the person. Also, it is assumed that the angle θ between the line of action of the muscle force and the horizontal remains unchanged.

What happens if the person is carrying a load of W_0 in the left hand during a right-leg stance (Figure 5.33)?

Assuming that the system we are analyzing consists of the upper body, left leg, and the load in hand, the extra load W_0 carried in the left hand will shift the center of gravity of the system from G to G'' toward the left of the person. Consequently, the length of the lever arm of the total gravitational force $W_4 = W_2 + W_0$ as measured from the right hip joint (Figure 5.34) will increase. This will require larger hip abductor muscle forces to counterbalance the clockwise rotational effect of W_4 and also increase the compressive forces at the right hip joint.

It can be observed from the geometry of the system analyzed that a shift in the center of gravity from G to G'' toward the left of the person will decrease the angle between the line of action of the joint reaction force and the horizontal from ϕ to ϕ''. For the new configuration of the free-body shown in Figure 5.34, Eqs. (*xi*) and (*xii*) can again be utilized to calculate the required hip abductor muscle force and joint reaction force produced at the right hip (opposite to the side where the load is carried):

$$F_M = \frac{\cos \phi'' \, (W_2 + W_0)}{\cos \theta \sin \phi'' - \sin \theta \cos \phi''}$$

$$F_J = \frac{\cos \theta \, (W_2 + W_0)}{\cos \theta \sin \phi'' - \sin \theta \cos \phi''}$$

Remarks:

• When the body weight is supported equally on both feet, half of the supra-femoral weight falls on each hip joint. During walking and running, the entire mass of the body is momentarily supported by one joint, and we have analyzed some of these cases.

• The above analyses indicate that the supporting forces required at the hip joint are greater when a load is carried on the opposite side of the body as compared to the forces required to carry the load when it is distributed on either side. Carrying loads by using both hands and by bringing the loads closer to the midline of the body is effective in reducing required musculoskeletal forces.

• While carrying a load on one side, people tend to lean toward the other side. This brings the center of gravity of the upper body and the load being carried in the hand closer to the midline of the body, thereby reducing the length of the moment arm of the resultant gravitational force as measured from the hip joint distal to the load.

- People with weak hip abductor muscles and/or painful hip joints usually lean toward the weaker side and walk with a so-called abductor gait. Leaning the trunk sideways toward the affected hip shifts the center of gravity of the body closer to that hip joint, and consequently reduces the rotational action of the moment of the body weight about the hip joint by reducing its moment arm. This in return reduces the magnitude of the forces exerted by the hip abductor muscles required to stabilize the pelvis.

- Abductor gait can be corrected more effectively with a cane held in the hand opposite to the weak hip, as compared to the cane held in the hand on the same side as the weak hip.

5.9 Mechanics of the Knee

The knee is the largest joint in the body. It is a modified hinge joint. In addition to flexion and extension action of the leg in the sagittal plane, the knee joint permits some inward and outward rotation. The knee joint is designed to sustain large loads. It is an essential component of the linkage system responsible for human locomotion. The knee is extremely vulnerable to injuries.

The knee is a two-joint structure composed of the tibiofemoral joint and the patellofemoral joint (Figure 5.35). The *tibiofemoral joint* has two distinct articulations between the medial and lateral condyles of the femur and the tibia. These articulations are separated by layers of cartilage, called *menisci*. The lateral and medial menisci eliminate bone-to-bone contact between the femur and the tibia, and function as shock absorbers. The *patellofemoral joint* is the articulation between the patella and the anterior end of the femoral condyles. The patella is a "floating" bone kept in position by the quadriceps tendon and the patellar ligament. It protects the knee from impact-related injuries and improves the pulling effect of the quadriceps muscles on the tibia via the patellar tendon. The stability of the knee is provided by an intricate ligamentous structure, the menisci and the muscles crossing the joint. Most knee injuries are characterized by ligament and cartilage damage occurring on the medial side.

The muscles crossing the knee protect it, provide internal forces for movement, and/or control its movement. The muscular control of the knee is produced primarily by the quadriceps muscles and the hamstring muscle group (Figure 5.36). The quadriceps muscle group is composed of the rectus femoris, vastus lateralis,

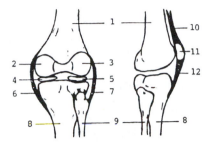

POSTERIOR   LATERAL
(BACK)   (SIDE)

Figure 5.35 *The knee: (1) femur, (2), medial condyle, (3) lateral condyle, (4) medial meniscus, (5) lateral meniscus, (6) tibial collateral ligament, (7) fibular collateral ligament, (8) tibia, (9) fibula, (10) quadriceps tendon, (11) patella, (12) patellar ligament.*

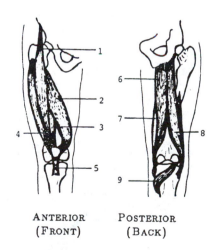

ANTERIOR   POSTERIOR
(FRONT)   (BACK)

Figure 5.36 *Muscles of the knee: (1) rectus femoris, (2) vastus medialis, (3) vastus intermedius, (4) vastus lateralis, (5) patellar ligament, (6) semitendinosus, (7) semimembranosus, (8) biceps femoris, (9) gastrocnemius.*

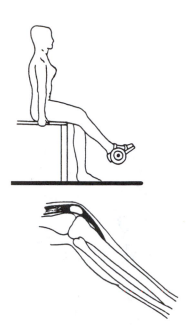

Figure 5.37 *Exercising the muscles around the knee joint.*

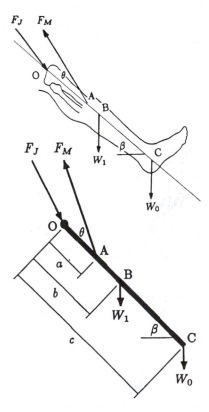

Figure 5.38 *Forces acting on the lower leg.*

vastus medialis, and vastus intermedius muscles. The rectus femoris muscle has attachments at the anterior–inferior iliac spine and the patella, and its primary actions are the flexion of the hip and the extension of the knee. The vastus lateralis, medialis, and intermedius muscles connect the femur and tibia through the patella, and they are all knee extensors. The biceps femoris, semitendinosus, and semimembranosus muscles make up the hamstring muscle group, which help control the extension of the hip, flexion of the knee, and some inward–outward rotation of the tibia. Semitendinosus and semimembranosus muscles have proximal attachments on the pelvic bone and distal attachments on the tibia. The biceps femoris has proximal attachments on the pelvic bone and the femur, and distal attachments on the tibia and fibula. There is also the popliteus muscle that has attachments on the femur and tibia. The primary function of this muscle is knee flexion. The other muscles of the knee are sartorius, gracilis, gastrocnemius, and plantaris.

Example 5.6 Consider a person wearing a weight boot, and from a sitting position, doing lower leg flexion/extension exercises to strengthen the quadriceps muscles (Figure 5.37).

Forces acting on the lower leg and a simple mechanical model of the leg are illustrated in Figure 5.38. W_1 is the weight of the lower leg, W_0 is the weight of the boot, F_M is the magnitude of the tensile force exerted by the quadriceps muscle on the tibia through the patellar tendon, and F_J is the magnitude of the tibiofemoral joint reaction force applied by the femur on the tibial plateau. The tibiofemoral joint center is located at O, the patellar tendon is attached to the tibia at A, the center of gravity of the lower leg is located at B, and the center of gravity of the weight boot is located at C. The distances between O and A, B, and C are measured as a, b, and c, respectively. For the position of the lower leg shown, the long axis of the tibia makes an angle β with the horizontal, and the line of action of the quadriceps muscle force makes an angle θ with the long axis of the tibia.

Assuming that points O, A, B, and C all lie along a straight line, determine F_M and F_J in terms of a, b, c, θ, β, W_1, and W_0.

Solution: Horizontal (x) and vertical (y) components of the forces acting on the trunk and their lever arms as measured from the knee joint located at O are shown in Figure 5.39. The components of the muscle force are:

$$F_{Mx} = F_M \cos(\theta + \beta) \qquad (i)$$
$$F_{My} = F_M \sin(\theta + \beta) \qquad (ii)$$

There are three unknowns, namely, F_M, F_{Jx}, and F_{Jy}. For the solution of this two-dimensional (plane) problem, all three equilibrium conditions must be utilized. Assuming that the counter-clockwise moments are positive, consider the rotational equilibrium of the lower leg about O:

$$\sum M_O = 0: \quad (a \cos \beta) F_{My} - (a \sin \beta) F_{Mx}$$
$$- (b \cos \beta) W_1 - (c \cos \beta) W_0 = 0$$

Substituting Eqs. (i) and (ii) into the above equation, and solving it for F_M will yield:

$$F_M = \frac{(b W_1 + c W_0) \cos \beta}{a [\cos \beta \sin(\theta + \beta) - \sin \beta \cos(\theta + \beta)]} \qquad (iii)$$

Note that this equation can be simplified by considering that $[\cos \beta \sin(\theta + \beta) - \sin \beta \cos(\theta + \beta)] = \sin \theta$.

Equation (iii) yields the magnitude of the force that must be exerted by the quadriceps muscles to support the leg when it is extended forward making an angle β with the horizontal. Once F_M is determined, the components of the reaction force developed at the knee joint along the horizontal and vertical directions can also be evaluated by considering the translational equilibrium of the lower leg in the x and y directions:

$$\sum F_x = 0: \quad F_{Jx} = F_{Mx} = F_M \cos(\theta + \beta)$$
$$\sum F_y = 0: \quad F_{Jy} = F_{My} - W_0 - W_1$$
$$F_{Jy} = F_M \sin(\theta + \beta) - W_0 - W_1$$

The magnitude of the resultant compressive force applied on the tibial plateau at the knee joint is:

$$F_J = \sqrt{(F_{Jx})^2 + (F_{Jy})^2} \qquad (iv)$$

Assume that the geometric parameters and the weights involved are given as: $a = 12$ cm, $b = 22$ cm, $c = 50$ cm, $W_1 = 150$ N, $W_0 = 100$ N, $\theta = 15°$, and $\beta = 45°$, then by using Eqs. (iii) and (iv):

$$F_M = 1381 N, \quad F_J = 1171 N$$

Remarks:

• The force $\underline{F}_M$ exerted by the quadriceps muscle on the tibia through the patellar tendon can be expressed in terms of two components normal and tangential to the long axis of the tibia (Figure 5.40). The primary function of the normal component $\underline{F}_{Mn}$ of the muscle force is to rotate the tibia about the knee joint, while its tangential component $\underline{F}_{Mt}$ tends to translate the lower leg in a direction collinear with the long axis of the tibia and applies a compressive force on the articulating surfaces of the tibiofemoral joint. Since the normal component of $\underline{F}_M$ is a sine function of

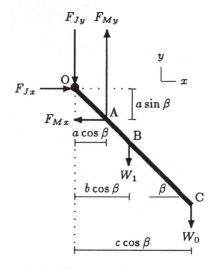

Figure 5.39 *Force components, and their lever arms.*

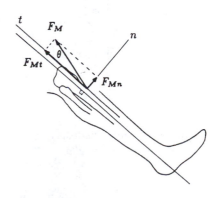

Figure 5.40 *Rotational and translatory components of $\underline{F}_M$.*

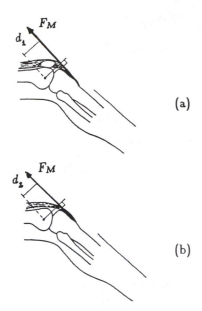

Figure 5.41 *Patella increases the length of the lever arm.*

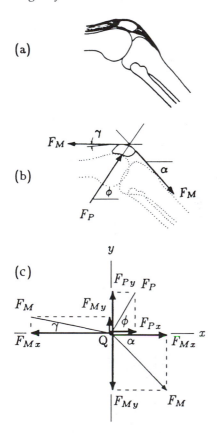

Figure 5.42 *Static analysis of the forces acting on the patella.*

angle θ, a larger angle between the patellar tendon and the long axis of the tibia indicates a larger rotational effect of the muscle exertion. This implies that for large θ, less muscle force is wasted to compress the knee joint, and a larger portion of the muscle tension is utilized to rotate the lower leg about the knee joint.

• One of the most important biomechanical functions of the patella is to provide anterior displacement of the quadriceps and patellar tendons, thus lengthening the lever arm of the knee extensor muscle forces with respect to the center of rotation of the knee by increasing angle θ (Figure 5.41a). Surgical removal of the patella brings the patellar tendon closer to the center of rotation of the knee joint (Figure 5.41b), which causes the length of the lever arm of the muscle force to decrease ($d_2 < d_1$). Losing the advantage of having a relatively long lever arm, the quadriceps muscle has to exert more force than normal to rotate the lower leg about the knee joint.

• The human knee has a two-joint structure composed of the tibiofemoral and patellofemoral joints. Note that the quadriceps muscle goes over the patella, and the patella and the muscle form a pulley–rope arrangement. The higher the tension in the muscle, the larger the compressive force (pressure) the patella exerts on the patellofemoral joint.

We have analyzed the forces involved around the tibiofemoral joint by considering the free-body diagram of the lower leg. Having determined the tension in the patellar tendon, and assuming that the tension is uniform throughout the quadriceps, we can calculate the compressive force applied on the patellofemoral joint by considering the free-body diagram of the patella (Figure 5.42a). Let F_M be the uniform magnitude of the tensile force in the patellar and quadriceps tendons, F_P be the magnitude of the force exerted on the patellofemoral joint, α be the angle between the patellar tendon and the horizontal, γ be the angle between the quadriceps tendon and the horizontal, and ϕ be the unknown angle between the line of action of the compressive reaction force at the joint (Figure 5.42b). We have a three-force system and for the equilibrium of the patella it has to be concurrent. We can first determine the common point of intersection Q by extending the lines of action of patellar and quadriceps tendon forces. A line connecting point Q and the point of application of $\underline{F}_P$ will correspond to the line of action of $\underline{F}_P$. The forces can then be translated to Q (Figure 5.42c), and the equilibrium equations can be applied. For the equilibrium of the patella in the x and y directions:

$$\sum F_x = 0: \quad F_P \cos\phi = F_M (\cos\gamma - \cos\alpha)$$

$$\sum F_y = 0: \quad F_P \sin\phi = F_M (\sin\alpha - \sin\gamma)$$

These equations can be solved simultaneously for angle ϕ and the magnitude F_P of the compressive force applied by the femur on the patella at the patellofemoral joint:

$$F_P = \left(\frac{\cos\gamma - \cos\alpha}{\cos\phi}\right) F_M$$

$$\phi = \tan^{-1}\left(\frac{\sin\alpha - \sin\gamma}{\cos\gamma - \cos\alpha}\right)$$

5.10 Mechanics of the Ankle

The ankle is the union of three bones: the tibia, the fibula, and the talus of the foot (Figure 5.43). Like other major joints in the lower extremity, the ankle is responsible for load-bearing and kinematic functions. The anatomical configuration of the ankle joints is similar to that of the hip. The ankle joint is inherently more stable than the knee joint, which requires ligamentous and muscular restraints for its stability.

The ankle joint complex consists of the tibiotalar, fibulotalar, and distal tibiofibular articulations. The *ankle (tibiotalar) joint* is a hinge or ginglymus-type articulation between the spool-like convex surface of the trochlea of the talus and the concave distal end of the tibia. Being a hinge joint, the ankle permits only flexion–extension (dorsiflexion–plantar flexion) movement of the foot in the sagittal plane. Other foot movements include inversion and eversion, inward and outward rotation, and pronation and supination. These movements occur about the foot joints such as the subtalar and transverse tarsal joints between the talus and calcaneus.

The ankle mortise is maintained by the shape of the three articulations, and the ligaments and muscles crossing the joint. The integrity of the ankle joint is improved by the medial (deltoid) and lateral collateral ligament systems, and the interosseous ligaments. There are numerous muscle groups crossing the ankle. The most important ankle plantar flexors are the gastrocnemius and soleus muscles (Figure 5.44). Both the gastrocnemius and soleus muscles have attachments to the posterior surface of the calcaneus via the Achilles tendon. The gastrocnemius muscle is more effective as a knee flexor if the foot is elevated, and more effective as a plantar flexor of the foot if the knee is held in extension. The plantar extensors are posterior muscles. There are also anterior (tibialis anterior, extensor digitorum longus, extensor hallucis longus, peroneus tertius) and lateral (peroneus longus, peroneus brevis) muscles whose primary function is to provide pronation and supination, and inward and outward rotation of the foot.

The ankle joint responds poorly to small changes in its anatomical configuration. Loss of kinematic and structural restraints

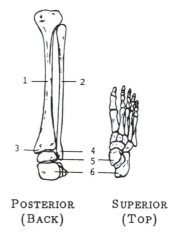

POSTERIOR   SUPERIOR
(BACK)   (TOP)

Figure 5.43 *The ankle and foot: (1) tibia, (2) fibula, (3) medial malleolus, (4) lateral malleolus, (5) talus, (6) calcaneus.*

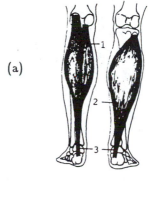

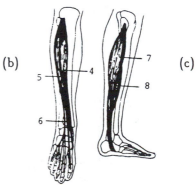

Figure 5.44 *Ankle muscles. (a) posterior, (b) anterior and (c) lateral views. (1) gastrocnemius, (2) soleus, (3) Achilles tendon, (4) tibialis antherior, (5) extensor digitorum longus, (6) extensor hallucis longus, (7) peroneus longus, (8) peroneus brevis.*

due to severe sprains can seriously affect ankle stability and can produce malalignment of the ankle joint surfaces. The most common ankle injury, inversion sprain, occurs when the body weight is forcefully transmitted to the ankle while the foot is inverted (the sole of the foot facing inward).

Example 5.7 Consider a person standing on tiptoe on one foot (a strenuous position illustrated). The forces acting on the foot during this instant are shown in Figure 5.45. W is the person's weight applied on the foot as the ground reaction force, F_M is the magnitude of the tensile force exerted by the gastrocnemius and soleus muscles on the calcaneus through the Achilles tendon, and F_J is the magnitude of the ankle joint reaction force applied by the tibia on the dome of the talus. The weight of the foot is small compared to the weight of the body and is therefore ignored. The Achilles tendon is attached to the calcaneus at A, the ankle joint center is located at B, and the ground reaction force is applied on the foot at C. For this position of the foot, it is estimated that the line of action of the tensile force in the Achilles tendon makes an angle θ with the horizontal, and the line of action of the ankle joint reaction force makes an angle β with the horizontal.

Assuming that the relative positions of A, B, and C are known, determine expressions for the tension in the Achilles tendon and the magnitude of the reaction force at the ankle joint.

Solution: We have a three-force system composed of muscle force $\underline{F}_M$, joint reaction force $\underline{F}_J$, and the ground reaction force $\underline{W}$. From the geometry of the problem, it is obvious that for the position of the foot shown, the forces acting on the foot do not form a parallel force system. Therefore, the force system must be a concurrent one. The common point of intersection (O in Figure 5.45) of these forces can be determined by extending the lines of action of $\underline{W}$ and $\underline{F}_M$. A straight line passing through both points O and B represents the line of action of the joint reaction force. Assuming that the relative positions of points A, B, and C are known (as stated in the problem), the angle (say β) of the line of action of the joint reaction force can be measured.

Once the line of action of the joint reaction force is determined by graphical means, the magnitudes of the joint reaction and muscle forces can be calculated by translating all three forces involved to the common point of intersection at O (Figure 5.46). The two unknowns F_M and F_J can now be determined by applying the translational equilibrium conditions in the horizontal (x) and vertical (y) directions. For this purpose, the joint reaction and muscle forces must be decomposed into their rectangular

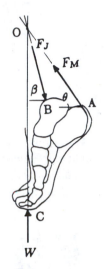

Figure 5.45 *Forces acting on the foot form a concurrent system of forces.*

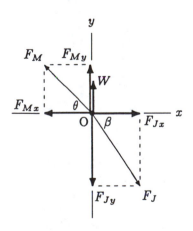

Figure 5.46 *Components of the forces acting on the foot.*

components first:

$$F_{Mx} = F_M \cos\theta$$
$$F_{My} = F_M \sin\theta$$
$$F_{Jx} = F_J \cos\beta$$
$$F_{Jy} = F_J \sin\beta$$

For the translational equilibrium of the foot in the horizontal and vertical directions:

$$\sum F_x = 0: \qquad F_{Jx} = F_{Mx}$$
$$\sum F_y = 0: \qquad F_{Jy} = F_{My} + W$$

Simultaneous solutions of these equations will yield:

$$F_M = \frac{W > \cos\beta}{\cos\theta \sin\beta - \sin\theta \cos\beta}$$
$$F_J = \frac{W > \cos\theta}{\cos\theta \sin\beta - \sin\theta \cos\beta}$$

For example, assume that $\theta = 45°$ and $\beta = 60°$. Then:

$$F_M = 1.93\ W \qquad F_J = 2.73\ W$$

5.11 Discussion

Analyses of posture and movement are based on an understanding of the leverage afforded gravitational force by the body position. Lifting while the arms are flexed requires less effort than lifting with arms extended. This knowledge enables us not only to grade exercises effectively but also to minimize gravitational effects in our daily activities. Low back injuries usually result from ignorance of safe lifting techniques that minimize gravitational effects. Arm and shoulder strain may result from prolonged use of hands away from the midline of the body where the center of gravity is located. It should be noted here that there are other factors that influence the level of difficulty of performing certain tasks. For example, we must consider the length of the antagonist muscles and relative size of the various body segments.

The point of insertion of a muscle or tendon varies from person to person. A slight variation in the point of insertion of a tendon can change the force distribution between the muscle and related joint considerably, as in the case of the biceps muscle and elbow joint. The farther a tendon lies from the axis of the joint, the better its ability to handle the turning effect of the segment about the joint, which can be a considerable advantage for lifting and throwing.

While analyzing the gravitational forces in body movement, consider the fact that the downward gravitational pull of the body is opposed by an equal upward push of the supporting surface regardless of whether the person is reclining, sitting, or standing. Sufficient floor reactions in terms of frictional forces are necessary for the stabilization of the body. Frictional forces are also important for walking and running. The joints of the body sustain compressive forces. Both the superincumbent and the supporting segment of the joint resist equal and opposite forces applied to their respective articular surfaces. The shape of the joints and the arrangement of reinforcing ligaments and tendons are such that the upright posture is maintained with an economy of muscular effort.

Researchers in the field of biomechanics have suggested various mathematical models to evaluate the forces involved in different muscles, bones, and joints of the human body during its various activities and static postures. Some of these models attempt to take into consideration the contributions of more than one muscle group. Different researchers utilize different formulation techniques and different criteria for the prediction of muscle forces. Some classic papers have been written describing these new approaches to the study of skeletal muscle, and many researchers have presented the application and refinement of these methods in more recent publications.

Seireg and Arvikar (1973) developed a mathematical model for evaluating the muscle forces necessary to maintain the human body equilibrium in standing, leaning, and stooping. These researchers utilized linear optimization techniques in the prediction of muscle forces. Penrod et al. (1974) investigated the problem of distribution of forces among muscles at a joint. In order to arrive at a unique solution for the redundant (statically indeterminate) system, they suggested that the solution must be based on optimizing the total muscle effort. They applied the resulting mathematical formulation to a two tendon model. Crowninshield and Brand (1981) developed a nonlinear model for predicting the muscle activity during locomotion. These investigators utilized the criterion of maximum endurance of musculoskeletal function to develop a model based on the inversely nonlinear relationship of muscle force and contraction endurance. More recently, Yamaguchi et al. (1995) refined the mathematical modeling for this problem.

In a review paper, Dul et al. (1984a) compared the characteristics and performance of several linear and nonlinear criteria for load sharing between synergistic muscles reported in the literature. Based on the assumption that during a physical activity the muscular fatigue is minimized, the same group of researchers

(Dul et al., 1984b) developed a model for predicting the load-sharing mechanisms between muscles. In 1995, Cholewicki et al. proposed a new approach to this model.

5.12 Suggested Reading

Chaffin, D.B., and Andersson, G.B.J. 1991. *Occupational Biomechanics*. 2nd ed. New York: John Wiley & Sons.

Cholewicki, J., McGill, S.M., and Norman, R.W. 1995. Comparison of muscle force and joint load form an optimization and EMG-assisted lumbar spine model: Towards development of a hybrid approach. *J. Biomechanics* 28:321–31.

Crowninshield, R.D., and Brand, R.A. 1981. A physiologically based criterion on muscle force prediction in locomotion. *J. Biomechanics* 14:793–801.

Dul, J., Townsend, M.A., Shiavi, R., and Johnson, G.E. 1984a. Muscular synergism—I. On the criteria for load sharing between synergistic muscles. *J. Biomechanics* 17:663–673.

Dul, J., Johnson, G.E., Shiavi, R., and Townsend, M.A. 1984b. Muscular synergism—II. A minimum-fatigue criterion for load sharing between synergistic muscles. *J. Biomechanics* 17:663–673.

Goel, V.K., Weinstein, J.N. 1990. *Biomechanics of the Spine: Clinical and Surgical Perspective*. Boston: CRC Press, Inc.

Kroemer, K.H.E., Marras, W.S., McGlothlin, J.D., McIntyre, D.R., and Nordin, M. 1990. On the measurement of human strength. *Int. J. Industrial Ergonomics* 6:199–210.

LeVeau, B.F. 1992. *Williams & Lissner's Biomechanics of Human Motion*. 2nd ed. Philadelphia: W.B. Saunders Company.

McMinn, R.M.H., Hutchings, R.T. 1988. *Color Atlas of Human Anatomy*. 2nd ed. Chicago: Year Book Medical Publishers Inc.

Nordin, M., Andersson, G.B.J., Pope, M.H. (Eds.) 1997. *Musculoskeletal Disorders in the Workplace: Principles & Practice*. Philadelphia: Mosby-Year Book, Inc.

Nordin, M., and Frankel, V.H. 1989. *Basic Biomechanics of the Musculoskeletal System*. 2nd ed. Philadelphia: Lea & Febiger.

Penrod, D.D, Davy, D.T., and Singh, D.P. 1974. An optimization approach to tendon force analysis. *J. Biomechanics* 7:123–129.

Roebuck, J.A. 1995. *Anthropometric Methods: Designing to Fit the Human Body*. Monographs in Human Factors and Ergonomics: Alphonse Chapanis, Series Editor. Santa Monica: Human Factors & Ergonomics Society.

Seireg, A., and Arvikar, R.J. 1973. A mathematical model for evaluation of force in lower extremities of the musculoskeletal system. *J. Biomechanics* 6:313–326.

Simon, S.R. (Ed.) 1994. *Orthopaedic Basic Science.* Rosemont, IL: American Academy of Orthopaedic Surgeons.

Thompson, C.W. 1989. *Manual of Structural Kinesiology.* 11th ed. St. Louis, MO: Times Mirror/Mosby.

Wilson, J.R., Corlett, E. N. (Eds.) 1995. *Evaluation of Human Work: A Practical Ergonomics Methodology.* 2nd ed. Bristol, UK: Taylor & Francis.

Winter, D.A. 1990. *Biomechanics and Motor Control of Human Behavior.* 2nd ed. New York: John Wiley & Sons.

Yamaguchi, G.T., Moran, D.W., and Si, J. 1995. A computationally efficient method for solving the redundant problem in biomechanics. *J. Biomechanics* 28:999–1005.

Chapter 6

Introduction to Deformable Body Mechanics

6.1 Overview

Basic concepts of statics were introduced in Chapters 4 and 5 along with some of their applications. The field of statics is based on Newton's laws (Newtonian mechanics). It constitutes one of the two main branches of the more general field of rigid body mechanics, dynamics being the other branch. The basic assumption in rigid body mechanics is that the bodies involved do not deform under applied loads. This idealization is necessary to simplify the problem under investigation for the sake of analyzing external forces and moments. The field of deformable body mechanics, on the other hand, does not treat the body as rigid, but incorporates the deformability (ability to undergo shape change) and the material properties of the body into the analyses. This field of applied mechanics utilizes the experimentally determined and/or verified relationships between applied forces and corresponding deformations.

Rigid body mechanics has its limitations. One of these limitations was discussed in Section 4.6 where the concept of statically indeterminant systems was introduced. A system for which the equations of equilibrium are not sufficient to determine the unknown forces is called *statically indeterminate*. For the analyses of such systems, there is a need for equations in addition to those provided by the conditions of static equilibrium. These additional equations can be derived by considering the material properties of the parts constituting a system and by relating forces to deformations, which is the focus of deformable body mechanics.

The desire to analyze statically indeterminate systems is only one of the reasons why deformable body mechanics is important. The applications of this field extend to almost all branches of engineering by providing essential design and analysis tools. The task of an engineer–mechanical, civil, electrical, or biomedical–is to determine the safest and most efficient operating condition for a machine, a structure, a piece of equipment, or a prosthetic device. A design engineer can accomplish this task by first assessing the proper operational environment through force analyses, making the correct structural design, and choosing the material that can sustain the forces involved in that environment. The primary concern of a design engineer is to make sure that when loaded, a machine part, a structure, a piece of equipment, or a device will not break or deform excessively.

6.2 Applied Forces and Deformations

Mechanics is concerned with forces and motions. It is possible to distinguish two types of motion. If the resultant of external forces or moments applied on a body is not zero, then the body will undergo gross overall motion (translation and/or rotation).

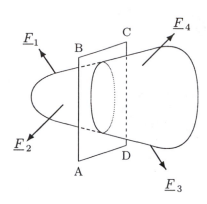

Figure 6.1 *An object subjected to externally applied forces.*

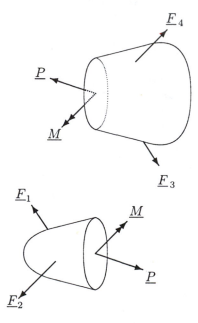

Figure 6.2 *Method of sections.*

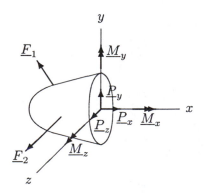

Figure 6.3 *Internal forces and moments.*

In other words, the position of the body as a whole will change over time. Such movements are studied within the field of dynamics. The second type of motion involves local changes of shape within a body, called *deformations*, which are the primary concern of the field of deformable body mechanics. If a body is subjected to externally applied forces and moments but remains in static equilibrium, then it is most likely that there is some local shape change within the body. The extent of the shape change may depend upon the magnitude, direction, and duration of the applied forces, material properties of the body, and environmental conditions such as heat and humidity.

6.3   Internal Forces and Moments

Consider the arbitrarily shaped object illustrated in Figure 6.1, which is subjected to a number of externally applied forces. Assume that the resultant of these forces and the net moment acting on the object are equal to zero. That is, the object is in static equilibrium. Also assume that the object is fictitiously separated into two parts by passing an arbitrary plane ABCD through the object. If the object as a whole is in equilibrium, then its individual parts must be in equilibrium as well. If one of these two parts is considered, then the equilibrium condition requires that there is a force vector and/or a moment vector acting on the cut section to counterbalance the effects of the external forces and moments applied on that part. These are called the *internal force* and *internal moment* vectors. Of course, the same argument is true for the other part of the object. Furthermore, for the overall equilibrium of the object, the force vectors and moment vectors on either surface of the cut section must have equal magnitudes and opposite directions (Figure 6.2).

For a three-dimensional object, the internal forces and moments can be resolved into their components along three mutually perpendicular directions, as illustrated in Figure 6.3. The force and moment vector components measured at the cut sections take special names reflecting their orientation and effects on the cut sections. Assuming that x is the direction normal (perpendicular) to the cut section, the force component P_x in Figure 6.3 is called the *axial* or *normal force*, and it is a measure of the pulling or pushing action of the externally applied forces in a direction perpendicular to the cut section. It is called a *tensile force* if it has a pulling action trying to elongate the part, or a *compressive force* if it has a pushing action tending to shorten the part. The force components P_y and P_z are called *shear forces*, and they are measures of resistance to the sliding action of one cut section over the other. Their subscripts indicate their lines of action. The moment component M_x is also called *twisting torque*, and it is a measure of the twisting action of the externally applied forces along an axis normal to the plane of the cut section (in this case, in the x direction). The components M_y and M_z of the moment vector are called the *bending moments*, and they respectively indicate the extent of bending action to which the cut part is subjected in the y and z directions.

Note here that it may be more informative to refer to forces and moments with double subscripted symbols. For example, using P_{xy} instead of P_y would indicate that the force component is acting in the y direction (second subscript) on a section whose normal is in the x direction (first subscript). Similarly, M_{xz} would refer to the component of the moment vector in the z direction measured on the same section.

6.4 Stress and Strain

The purpose of studying the mechanics of deformable bodies or strength of materials is to make sure that the design of a structure is safe against the combined effects of applied forces and moments. The idea is to select the proper material for the structure, or if there is an existing structure, to determine the loading conditions under which the structure can operate safely and efficiently. To make a selection, however, one needs to know the mechanical properties of materials under different loading conditions.

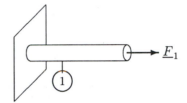

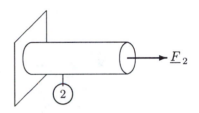

Consider the two bars shown in Figure 6.4, which are made of the same material, and have the same length but different sizes. The cross-sectional area A_1 of bar 1 is less than the cross-sectional area A_2 of bar 2. Assume that these bars are subjected to successively increasing forces until they break. If the forces F_1 and F_2 at which the bars 1 and 2 break were recorded, it would be observed that the force F_2 required to break bar 2 is greater than the force F_1 required to break bar 1 because bar 2 has a larger cross-sectional area and volume than bar 1. These forces might be an indication of the strength of the bars. However, the fact that the force-to-failure depends on the cross-sectional area of the specimen (in addition to some other factors) makes force an impractical measure of the strength of a material. To eliminate this inconvenience, a concept called *stress* is defined by dividing force with the cross-sectional area.

Figure 6.4 *Two bars made of the same material, have the same length, but different cross-sectional areas.*

Although the bars in Figure 6.4 have different cross-sectional areas and require different forces-to-failure, since they are made of the same material, their stress measurement at failure would be equal.

As stated earlier, the mechanics of deformable bodies is concerned with applied forces and their internal effects on bodies. One of these effects is shape change or deformation. The amount of deformation an object will undergo depends on its size, material properties, and the magnitude and duration of applied forces. Consider the two bars shown in Figure 6.5, which are made of the same material and have the same cross-sectional

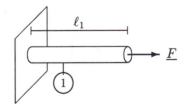

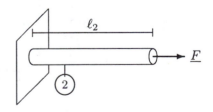

Figure 6.5 *Two bars made of the same material, have the same cross-sectional area, but different lengths.*

area, but have different lengths. The length ℓ_1 of bar 1 is less than the length ℓ_2 of bar 2. Assume that the same force F is applied to both bars, and the elongation of each bar is measured. It would be observed that the increase of length in bar 2 is greater than the increase of length in bar 1, indicating that the amount of elongation depends on the original length of the specimen. To eliminate the size dependence of deformation measurements, another concept called *strain* is defined by dividing the amount of elongation with the original length of the specimen in the direction of elongation.

Broad definitions of stress and strain are introduced here. More detailed descriptions of these concepts will be provided in the following sections.

6.5 General Procedure

A general procedure for analyzing problems in deformable body mechanics, including the purpose of these analyses, is provided below.

• Static analyses. At this first stage, the analytical methods of statics are employed to determine the external reaction forces and moments. This stage involves drawing free-body diagrams and applying the conditions of equilibrium to determine the unknown reaction forces and moments by utilizing concepts such as equivalent force systems.

• Analyses of internal forces and moments. The internal forces and moments can be determined by the *method of sections*. As discussed briefly in Section 6.3, this can be done by separating the body into two sections at the location where the forces and moments need to be calculated. Here, the concern is to determine critical load conditions that correspond to maximum stress levels. These critical loads can be determined by drawing the shear and bending diagrams of the body, which essentially requires the application of the method of sections throughout the body.

• Stress analyses. This stage involves the conversion of internal forces and moments, in particular the critical forces and moments, into corresponding stresses by using formulas that also incorporate the material and geometric properties of the problem into the analyses.

• Material selection. Materials can be distinguished by their physical and mechanical properties. At this final stage of analysis, a material must be selected for the safe operation of the structure based on the maximum stresses calculated. If the material is already selected and the design is already made, then the maximum stresses calculated are used to set the allowable load conditions.

Note that the prerequisite for analyses in deformable body mechanics is statics. However, the procedure outlined above is not limited to analyzing systems in equilibrium. Under the effect of externally applied forces, a body may deform and undergo overall motion simultaneously. Such a problem can also be analyzed with the procedure outlined above by utilizing the *d'Alembert principle*. This principle is applied by treating the inertial effects due to the acceleration of the body as another external force acting at the center of gravity of the body in a direction opposite to the direction of acceleration.

6.6 Mathematics Involved

The analyses in this part of the text (Chapters 6 through 10) will utilize vector algebra and differential and integral calculus as computational tools. Therefore, the reader is advised to review Appendices A through C. Some of the analyses may also require familiarity with ordinary differential equations.

6.7 Topics to Be Covered

At the beginning of Chapter 7, detailed definitions of stress and strain will be provided. Based on the stress–strain diagrams, material properties such as ductility, stiffness, and brittleness will be discussed. Elastic and plastic deformations, Hooke's law, the necking phenomenon, and the concepts of work and strain energy will be explained. The analyses in Chapter 7 will be limited to uniaxial deformations. The concepts introduced will be applied to analyze relatively simple systems.

In Chapter 8, more advanced topics in stress–strain analyses will be introduced. Two- and three-dimensional stress analyses, techniques of transforming stresses from one plane to another, methods for finding critical stresses, reasons why stress analyses are important in the design of structures, failure theories, concepts such as fatigue, endurance, and stress concentration will also be discussed in Chapter 8. Also in Chapter 8, analyses of bodies subjected to torsion, bending, and combined loading will be explained.

In Chapter 9, the viscoelastic behavior of materials and emprical models of viscoelasticity will be reviewed, and elasticity and viscoelasticity will be compared. Also in Chapter 9, the mechanical properties of biological tissues including bone, tendons, ligaments, muscles, and articular cartilage will be discussed and the relevance of mechanical concepts introduced earlier to orthopedics will be demonstrated.

6.8 Suggested Reading

The field of deformable body mechanics has been studied under various titles such as solid mechanics, mechanics of materials, and strength of materials. The subjects covered within deformable body mechanics form the basis for the study of more advanced topics in elasticity, inelasticity, and continuum mechanics. The following books can be reviewed to gain more detailed information on the principles of deformable body mechanics:

Crandall, S.H., Dahl, N.C. and Lardner, T.J. 1978. *An Introduction to the Mechanics of Solids.* 2nd ed. New York: McGraw-Hill.

Popov, E.P. 1978. *Mechanics of Materials.* 2nd ed. Englewood Cliffs, NJ: Prentice-Hall.

Pytel, A. and Singer, F.L. 1987. *Strength of Materials.* 4th ed. New York: Harper & Row.

Chapter 7

Stress and Strain

7.1 Basic Loading Configurations

An object subjected to an external force will move in the direction of the applied force. The object will deform if its motion is constrained in the direction of the applied force. Deformation implies relative displacement of any two points within the object. The extent of deformation will be dependent upon many factors including the magnitude, direction, and duration of the applied force, the material properties of the object, the geometry of the object, and environmental factors such as heat and humidity.

In general, materials respond differently to different loading configurations. For a given material, there may be different mechanical properties that must be considered while analyzing its response to, for example, tensile loading as compared to loading that may cause bending or torsion. Figure 7.1 is drawn to illustrate different loading conditions, in which an L-shaped beam is subjected to forces F_1, F_2, and F_3. The force F_1 subjects the arm AB of the beam to tensile loading. The force F_2 tends to bend the arm AB. The force F_3 has a bending effect on arm BC and a twisting (torsional) effect on arm AB. Furthermore, all of these forces are subjecting different sections of the beam to shear loading.

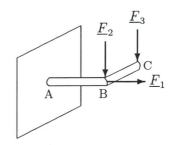

Figure 7.1 *Loading modes.*

7.2 Uniaxial Tension Test

The mechanical properties of materials are established by subjecting them to various experiments. The mechanical response of materials under tensile loading is analyzed by the *uniaxial* or *simple tension test* that will be discussed next. The response of materials to forces that cause bending and torsion will be reviewed in the following chapter.

The experimental setup for the uniaxial tension test is illustrated in Figure 7.2. It consists of one fixed and one moving head with attachments to grip the test specimen. A specimen is placed and firmly fixed in the equipment, a tensile force of known magnitude is applied through the moving head, and the corresponding elongation of the specimen is measured. A general understanding of the response of the material to tensile loading is obtained by repeating this test for a number of specimens made of the same material, but with different lengths, cross-sectional areas, and under tensile forces with different magnitudes.

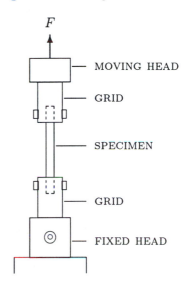

Figure 7.2 *Uniaxial tension test.*

7.3 Load-Elongation Diagrams

Consider the three bars shown in Figure 7.3. Assume that these bars are made of the same material. The first and the second bars have the same length but different cross-sectional areas, and the second and third bars have the same cross-sectional area but

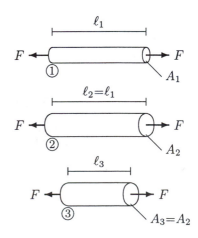

Figure 7.3 *Specimens.*

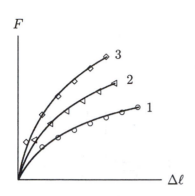

Figure 7.4 *Load-elongation diagrams.*

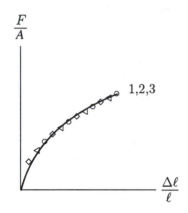

Figure 7.5 *Load over area versus elongation over length diagram.*

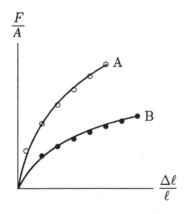

Figure 7.6 *Material A is stiffer than material B.*

different lengths. Each of these bars can be subjected to a series of uniaxial tension tests by gradually increasing the applied forces and measuring corresponding increases in their lengths. If F is the magnitude of the applied force and $\Delta\ell$ is the increase in length, then the data collected can be plotted to obtain a load versus elongation diagram for each specimen. Effects of geometric parameters (cross-sectional area and length) on the load-bearing ability of the material can be judged by drawing the curves obtained for each specimen on a single graph (Figure 7.4) and comparing them. At a given force magnitude, the comparison of curves 1 and 2 indicate that the larger the cross-sectional area, the more difficult it is to deform the specimen in a simple tension test, and the comparison of curves 2 and 3 indicate that the longer the specimen, the larger the deformation in tension.

Note that instead of applying a series of tensile forces to a single specimen, it is preferable to have a number of specimens with almost identical geometries and apply one force to one specimen only once. As will be discussed later, a force applied on an object may alter its mechanical properties.

Another method of representing the results obtained in a uniaxial tension test is by first dividing the magnitude of the applied force F with the cross-sectional area A of the specimen, normalizing the amount of deformation by dividing the measured elongation with the original length of the specimen, and then plotting the data on a F/A versus $\Delta\ell/\ell$ graph as shown in Figure 7.5. The three curves in Figure 7.4, representing three specimens made of the same material, are represented by a single curve in Figure 7.5. It is obvious that some of the information provided in Figure 7.4 is lost in Figure 7.5. That is why the representation in Figure 7.5 is more advantageous than that in Figure 7.4. The single curve in Figure 7.5 is unique for a particular material, independent of the geometries of the specimens used during the experiments. This type of representation eliminates geometry as one of the variables, and makes it possible to focus attention on the mechanical properties of different materials. For example, consider the curves in Figure 7.6, representing the mechanical behavior of materials A and B in simple tension. It is clear that material B can be deformed more easily than material A in a uniaxial tension test, or material A is "stiffer" than material B.

7.4 Simple Stress

Consider the cantilever beam shown in Figure 7.7a. The beam has a circular cross-section, cross-sectional area A, welded to the wall at one end, and is subjected to a tensile force with magnitude F at the other end. The bar does not move, so it is in equilibrium. To analyze the forces induced within the beam, the method of sections can be applied by hypothetically cutting the beam into two pieces through a plane ABCD perpendicular to the centerline of the beam. Since the beam as a whole is in equilibrium, the two pieces must individually be in equilibrium as well. This requires the presence of an internal force collinear with the externally applied force at the cut section of each piece. To satisfy the condition of equilibrium, the internal forces must have the same magnitude as the external force (Figure 7.7b). The internal force at the cut section represents the resultant of a force system distributed over the cross-sectional area of the beam (Figure 7.7c). The intensity of the internal force over the cut section (force per unit area) is known as the *stress*. For the case shown in Figure 7.7, since the force resultant at the cut section is perpendicular (normal) to the plane of the cut, the corresponding stress is called a *normal stress*. It is customary to use the symbol σ (sigma) to refer to normal stresses. The intensity of this distributed force may or may not be uniform (constant) throughout the cut section. Assuming that the intensity of the distributed force at the cut section is uniform over the cross-sectional area A, the normal stress can be calculated using:

$$\sigma = \frac{F}{A} \qquad (7.1)$$

If the intensity of the stress distribution over the area is not uniform, then Eq. (7.1) will yield an *average normal stress*. It is customary to refer to normal stresses that are associated with tensile loading as *tensile stresses*. On the other hand, *compressive stresses* are those associated with compressive loading. It is also customary to treat tensile stresses as positive and compressive stresses as negative.

The other form of stress is called *shear stress*, which is a measure of the intensity of internal forces acting parallel or tangent to a plane of cut. To get a sense of shear stresses, hold a stack of paper with both hands such that one hand is under the stack while the other hand is above it. First, press the stack of papers together. Then, slowly slide one hand in the direction parallel to the surface of the papers while sliding the other hand in the opposite direction. This will slide individual papers relative to one another and generate frictional forces on the surfaces of individual papers. The shear stress is comparable to the intensity of the frictional force over the surface area upon which it is applied. Now, consider the cantilever beam illustrated in Figure 7.8a. A downward force with magnitude F is applied to its free end. To analyze internal forces and moments, the method of sections can be applied by fictitiously cutting the beam through a plane ABCD that is perpendicular to the centerline of the beam. Since the beam as a whole is in equilibrium, the two pieces thus obtained must individually be in equilibrium as well. The free-body diagram of the right-hand piece of the beam is illustrated in Figure 7.8b along with the internal force and moment on the left-hand piece. For the equilibrium of the right-hand piece, there

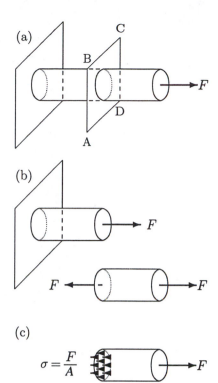

Figure 7.7 *Normal stress.*

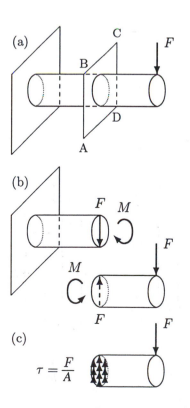

Figure 7.8 *Shear stress.*

has to be an upward force resultant and an internal moment at the cut surface. Again for the equilibrium of this piece, the internal force must have a magnitude F. This is known as the *internal shearing force* and is the resultant of a distributed load over the cut surface (Figure 7.8c). The intensity of the shearing force over the cut surface is known as the *shear stress*, and is commonly denoted with the symbol τ (tau). If the area of this surface (in this case, the cross-sectional area of the beam) is A, then:

$$\tau = \frac{F}{A} \qquad (7.2)$$

The underlying assumption in Eq. (7.2) is that the shear stress is distributed uniformly over the area. For some cases this assumption may not be true. In such cases, the shear stress calculated by Eq. (7.2) will represent an average value.

The dimension of stress can be determined by dividing the dimension of force $[F] = [M][L]/[T^2]$ with the dimension of area $[L^2]$. Therefore, stress has the dimension of $[M]/[L][T^2]$. The units of stress in different unit systems are listed in Table 7.1. Note that stress has the same dimension and units as pressure.

Table 7.1 *Units of stress.*

SYSTEM	UNITS OF STRESS	SPECIAL NAME
SI	N/m^2	Pascal (Pa)
c-g-s	dyn/cm^2	
British	lb/ft^2 or lb/in^2	psf or psi

7.5 Simple Strain

Strain, which is also known as unit deformation, is a measure of the degree or intensity of deformation. Consider the bar in Figure 7.9. Let A and B be two points on the bar located at a distance ℓ_1, and C and D be two other points located at a distance ℓ_2 from one another, such that $\ell_1 > \ell_2$. ℓ_1 and ℓ_2 are called *gage lengths*. The bar will elongate when it is subjected to tensile loading. Let $\Delta\ell_1$ be the amount of elongation measured between A and B, and $\Delta\ell_2$ be the increase in length between C and D. $\Delta\ell_1$ and $\Delta\ell_2$ are certainly some measures of deformation. However, they depend on the respective gage lengths, such that $\Delta\ell_1 > \Delta\ell_2$. On the other hand, if the ratio of the amount of elongation to the gage length is calculated for each case and compared, it would be observed that $(\Delta\ell_1/\ell_1) \simeq (\Delta\ell_2/\ell_2)$. Elongation per unit gage length is known as *strain* and is a more fundamental means of measuring deformation.

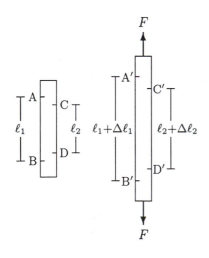

Figure 7.9 *Normal strain.*

As in the case of stress, two types of strains can be distinguished. The *normal* or *axial strain* is associated with axial forces and defined as the ratio of the change (increase or decrease) in length, $\Delta \ell$, to the original gage length, ℓ, and is denoted with the symbol ϵ (epsilon):

$$\epsilon = \frac{\Delta \ell}{\ell} \tag{7.3}$$

When a body is subjected to tension, its length increases, and both $\Delta \ell$ and ϵ are positive. The length of a specimen under compression decreases, and both $\Delta \ell$ and ϵ become negative.

The second form of strain is related to distortions caused by shearing forces. Consider the rectangle (ABCD) shown in Figure 7.10, which is acted upon by a pair of shearing forces. Shear forces deform the rectangle into a parallelogram (AB'C'D). If the relative horizontal displacement of the top and the bottom of the rectangle is d and the height of the rectangle is ℓ, then the *average shear strain* is defined as the ratio of d and ℓ, which is equal to the tangent of angle γ (gamma). This angle is usually very small. For small angles, the tangent of the angle is approximately equal to the angle itself. Hence, the average shear strain is equal to angle γ (measured in radians), which can be calculated using:

$$\gamma = \frac{d}{\ell} \tag{7.4}$$

Strains are calculated by dividing two quantities having the dimension of length. Therefore, they are dimensionless quantities and there is no unit associated with them. For most applications, the deformations and, consequently, the strains involved are very small, and the precision of the measurements taken is very important. To indicate the type of measurements taken, it is not unusual to attach units such as cm/cm or mm/mm next to a strain value. Strains can also be given in percent. In engineering applications, the strains involved are of the order of magnitude 0.1 percent or 0.001.

Figure 7.11 is drawn to compare the effects of tensile, compressive, and shear loading. Figure 7.11a shows a square object (in a two-dimensional sense) under no load. The square object is ruled into 16 smaller squares to illustrate different modes of deformation. In Figure 7.11b, the object is subjected to a pair of tensile forces. Tensile forces distort squares into rectangles such that the dimension of each square in the direction of applied force (axial dimension) increases while its dimension perpendicular to the direction of the applied force (transverse dimension) decreases. In Figure 7.11c, the object is subjected to a pair of compressive forces that distort squares into rectangles such that the axial dimension of each square decreases while its transverse dimension

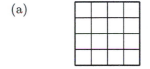

Figure 7.10 *Shear strain.*

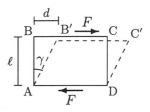

(a)

(b)

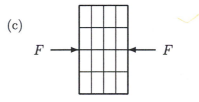

(c)

(d)

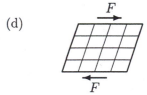

Figure 7.11 *Distorsions under tensile (b), compressive (c), and shear (d) loading.*

increases. In Figure 7.11d, the object is subjected to a pair of shear forces that distort the squares into diamonds.

7.6   Stress–Strain Diagrams

It was demonstrated in Section 7.3 that the results of uniaxial tension tests can be used to obtain a unique curve representing the relationship between the applied load and corresponding deformation for a material. This can be achieved by dividing the applied load with the cross-sectional area (F/A) of the specimen, dividing the amount of elongation measured with the gage length ($\Delta\ell/\ell$), and plotting a F/A versus $\Delta\ell/\ell$ graph. Note, however, that for a specimen under tension, F/A is the average tensile stress σ and $\Delta\ell/\ell$ is the average tensile strain ϵ. Therefore, the F/A versus $\Delta\ell/\ell$ graph of a material is essentially the *stress–strain diagram* of that material.

Different materials demonstrate different stress–strain relationships, and the stress–strain diagrams of two or more materials can be compared to determine which material is relatively stiffer, harder, tougher, more ductile, and/or more brittle. Before explaining these concepts related to the strength of materials, it is appropriate to first analyze a typical stress–strain diagram in detail.

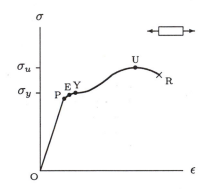

Figure 7.12 *Stress–strain diagram for axial loading.*

Consider the stress–strain diagram shown in Figure 7.12. There are six distinct points on the curve that are labeled as O, P, E, Y, U, and R. Point O is the origin of the σ-ϵ diagram, which corresponds to the initial no load, no deformation stage. Point P represents the *proportionality limit*. Between O and P, stress and strain are linearly proportional, and the σ-ϵ curve is a straight line. Point E represents the *elastic limit*. The stress corresponding to the elastic limit is the greatest stress that can be applied to the material without causing any permanent deformation within the material. The material will not resume its original size and shape upon unloading if it is subjected to stress levels beyond the elastic limit. Point Y is the *yield point*, and the stress σ_y corresponding to the yield point is called the *yield strength* of the material. At this stress level, considerable elongation (yielding) can occur without a corresponding increase of load. U is the highest stress point on the σ-ϵ curve. The stress σ_u is the *ultimate strength* of the material. For some materials, once the ultimate strength is reached, the applied load can be decreased and continued yielding may be observed. This is due to the phenomena called *necking* that will be discussed later. The last point on the σ-ϵ curve is R, which represents the *rupture* or *failure point*. The stress at which the rupture occurs is called the *rupture strength* of the material.

For some materials, it may not be easy to determine or distinguish the elastic limit and the yield point. The yield strength of such materials is determined by the *offset method*, illustrated in Figure 7.13. The offset method is applied by drawing a line

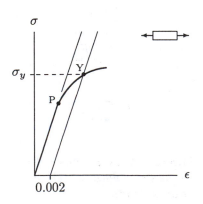

Figure 7.13 *Offset method.*

parallel to the linear section of the stress–strain diagram and passing through a strain level of about 0.2 percent (0.002). The intersection of this line with the σ-ϵ curve is taken to be the yield point, and the stress corresponding to this point is called the *apparent yield strength* of the material.

Note that a given material may behave differently under different load and environmental conditions. If the curve shown in Figure 7.12 represents the stress–strain relationship for a material under tensile loading, there may be a similar but different curve representing the stress–strain relationship for the same material under compressive or shear loading. Also, temperature is known to alter the relationship between stress and strain. For a given material and fixed mode of loading, different stress–strain diagrams may be obtained under different temperatures. Furthermore, the data collected in a particular tension test may depend on the rate at which the tension is applied on the specimen. Some of these factors affecting the relationship between stress and strain will be discussed later.

7.7 Elastic Deformations

Consider the partial stress–strain diagram shown in Figure 7.14. Y is the yield point, and in this case, it also represents the proportionality and elastic limits. σ_y is the yield strength and ϵ_y is the corresponding strain. (The σ-ϵ curve beyond the elastic limit is not shown.) The straight line in Figure 7.14 represents the stress–strain relationship in the elastic region. *Elasticity* is defined as the ability of a material to resume its original (stress-free) size and shape upon removal of applied loads. In other words, if a load is applied on a material such that the stress generated in the material is equal to or less than σ_y, then the deformations that took place in the material will be completely recovered once the applied loads are removed (the material is unloaded).

An elastic material whose σ-ϵ diagram is a straight line is called a *linearly elastic* material. For such a material, the stress is linearly proportional to strain, and the constant of proportionality is called the *elastic* or *Young's modulus* of the material. Denoting the elastic modulus with E:

$$\sigma = E \epsilon \qquad (7.5)$$

The elastic modulus, E, is equal to the slope of the σ-ϵ diagram in the elastic region, which is constant for a linearly elastic material. E represents the *stiffness* of a material, such that the higher the elastic modulus, the stiffer the material.

The distinguishing factor in linearly elastic materials is their elastic moduli. That is, different linearly elastic materials have different elastic moduli. If the elastic modulus of a material is known, then the mathematical definitions of stress and strain ($\sigma = F/A$ and $\epsilon = \Delta\ell/\ell$) can be substituted into Eq. (7.5) to derive a relationship between the applied load and corresponding

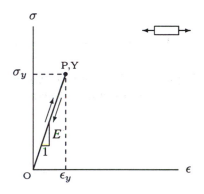

Figure 7.14 *Stress–strain diagram for a linearly elastic material.* ($\nearrow$: loading, $\swarrow$: unloading)

deformation:

$$\Delta \ell = \frac{F \ell}{E A} \qquad (7.6)$$

In Eq. (7.6), F is the magnitude of the tensile or compressive force applied to the material, E is the elastic modulus of the material, A is the area of the surface that cuts the line of action of the applied force at right angles, ℓ is the length of the material measured along the line of action of the applied force, and $\Delta \ell$ is the amount of elongation or shortening in ℓ due to the applied force. For a given linearly elastic material (or any material for which the deformations are within the linearly elastic region of the σ-ϵ diagram) and applied load, Eq. (7.6) can be used to calculate the corresponding deformation. This equation can be used when the object is under a tensile or compressive force.

Not all elastic materials demonstrate linear behavior. As illustrated in Figure 7.15, the stress–strain diagram of a material in the elastic region may be a straight line up to the proportionality limit followed by a curve. A curve implies varying slope and nonlinear behavior. Materials for which the σ-ϵ curve in the elastic region is not a straight line are known as *nonlinear elastic materials*. For a nonlinear elastic material, there is not a single elastic modulus because the slope of the σ-ϵ curve is not constant throughout the elastic region. Therefore, the stress–strain relationships for nonlinear materials take more complex forms. Note, however, that even nonlinear materials may have a linear elastic region in their σ-ϵ diagrams at low stress levels (region between O and P in Figure 7.15).

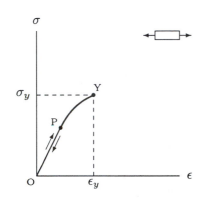

Figure 7.15 *Stress–strain diagram for a nonlinearly elastic material.*

Some materials may exhibit linearly elastic behavior when they are subjected to shear loading (Figure 7.16). For such materials, the shear stress, τ, is linearly proportional to the shear strain, γ, and the constant of proportionality is called the *shear modulus* or the *modulus of rigidity*, which is commonly denoted with the symbol G:

$$\tau = G \gamma \qquad (7.7)$$

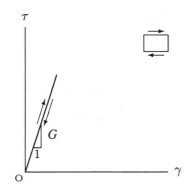

Figure 7.16 *Shear stress versus shear strain diagram for a linearly elastic material.*

The shear modulus of a given linear material is equal to the slope of the τ-γ curve in the elastic region. The higher the shear modulus, the more rigid the material.

Note that Eqs. (7.5) and (7.6) relate stresses to strains for linearly elastic materials, and are called *material functions*. Obviously, for a given material, there may exist different material functions for different modes of deformation. There are also *constitutive equations* that incorporate all material functions.

7.8 Hooke's Law

The load-bearing characteristics of elastic materials are similar to those of springs, which was first noted by Robert Hooke.

Like springs, elastic materials have the ability to store potential energy when they are subjected to externally applied loads. During unloading, it is the release of this energy that causes the material to resume its undeformed configuration. A linear spring subjected to a tensile load will elongate, the amount of elongation being linearly proportional to the applied load (Figure 7.17). The constant of proportionality between the load and the deformation is usually denoted with the symbol k, which is called the *spring constant* or *stiffness* of the spring. For a linear spring with a spring constant k, the relationship between the applied load F and the amount of elongation d is:

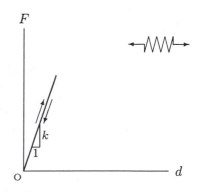

Figure 7.17 *Load-elongation diagram for a linear spring.*

$$F = k\,d \qquad (7.8)$$

By comparing Eqs. (7.5) and (7.8), it can be observed that stress in an elastic material is analogous to the force applied to a spring, strain in an elastic material is analogous to the amount of deformation of a spring, and the elastic modulus of an elastic material is analogous to the spring constant of a spring. This analogy between elastic materials and springs is known as *Hooke's Law*.

7.9 Plastic Deformations

We have defined elasticity as the ability of a material to regain completely its original dimensions upon removal of the applied forces. Elastic behavior implies the absence of permanent deformation. On the other hand, *plasticity* implies permanent deformation. In general, materials undergo plastic deformations following elastic deformations when they are loaded beyond their elastic limits or yield points.

Consider the stress–strain diagram of a material shown in Figure 7.18. Assume that a specimen made of the same material is subjected to a tensile load and the stress, σ, in the specimen is brought to such a level that $\sigma > \sigma_y$. The corresponding strain in the specimen is measured as ϵ. Upon removal of the applied load, the material will recover the elastic deformation that had taken place by following an unloading path parallel to the initial linearly elastic region (straight line between O and P). The point where this path cuts the strain axis is called the plastic strain, ϵ_p, that signifies the extent of permanent (unrecoverable) shape change that has taken place in the specimen.

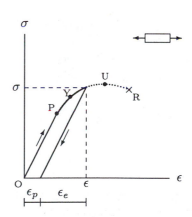

Figure 7.18 *Plastic deformation.*

The difference in strains between when the specimen is loaded and unloaded ($\epsilon - \epsilon_p$) is equal to the amount of elastic strain, ϵ_e, that had taken place in the specimen and that was recovered upon unloading. Therefore, for a material loaded to a stress level beyond its elastic limit, the total strain is equal to the sum of the elastic and plastic strains:

$$\epsilon = \epsilon_e + \epsilon_p \qquad (7.9)$$

The elastic strain, ϵ_e, is completely recoverable upon unloading, whereas the plastic strain, ϵ_p, is a permanent residue of the deformations.

7.10 Necking

As defined in Section 7.6, the largest stress a material can endure is called the ultimate strength of that material. Once a material is subjected to a stress level equal to its ultimate strength, an increased rate of deformation can be observed, and in most cases, continued yielding can occur even by reducing the applied load. The material will eventually fail to hold any load, and rupture. The stress at failure is called the rupture strength of the material, which may be lower than its ultimate strength. Although this may seem to be unrealistic, the reason is due to a phenomenon called *necking* and because of the manner in which stresses are calculated.

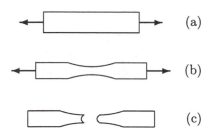

Figure 7.19 *Necking.*

Stresses are usually calculated on the basis of the original cross-sectional area of the material. Such stresses are called *conventional stresses*. Calculating a stress by dividing the applied force with the original cross-sectional area is convenient but not necessarily accurate. The *true* or *actual stress* calculations must be made by taking the cross-sectional area of the deformed material into consideration. As illustrated in Figure 7.19, under a tensile load a material may elongate in the direction of the applied load but contract in the transverse directions. At stress levels close to the breaking point, the elongation may occur very rapidly and the material may narrow simultaneously. The cross-sectional area at the narrowed section decreases, and although the force required to further deform the material may decrease, the force per unit area (stress) may increase. As illustrated with the dotted curve in Figure 7.20, the actual stress–strain curve may continue having a positive slope, which indicates increasing strain with increasing stress rather than a negative slope, which implies increasing strain with decreasing stress. Also the rupture point and the point corresponding to the ultimate strength of the material may be the same.

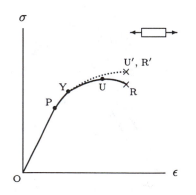

Figure 7.20 *Conventional (solid curve) and actual (dotted curve) stress–strain diagrams.*

7.11 Work and Strain Energy

In dynamics, *work done* is defined as the product of force and the distance traveled in the direction of applied force, and *energy* is the capacity of a system to do work. Stress and strain in deformable body mechanics are respectively related to force and displacement. Stress multiplied by area is equal to force, and strain multiplied by length is displacement. Therefore, the product of stress and strain is equal to the work done on a body per unit volume of that body, or the *internal work* done on the body by the externally applied forces. For an elastic body, this work is stored as an internal *elastic strain energy*, and it is the release of this energy that brings the body back to its original shape upon unloading. The maximum elastic strain energy (per unit volume) that can be stored in a body is equal to the total area under the σ-ϵ diagram in the elastic region (Figure 7.21). There is also a *plastic strain energy* that is dissipated as heat while deforming the body.

7.12   Strain Hardening

Figure 7.22 represents the σ-ϵ diagram of a material. Assume that the material is subjected to a tensile force such that the stress generated is beyond the elastic limit (yield point) of the material. The stress level in the material is indicated with point A on the σ-ϵ diagram. Upon removal of the applied force, the material will follow the path AB, which is almost parallel to the initial, linear section OP of the σ-ϵ diagram. The strain at B corresponds to the amount of plastic deformation in the material. If the material is reloaded, it will exhibit elastic behavior between B and A, the stress at A being the new yield strength of the material. This technique of changing the yield point of a material is called *strain hardening*. Since the stress at A is greater than the original yield strength of the material, strain hardening increases the yield strength of the material. Upon reloading, if the material is stressed beyond A, then the material will deform according to the original σ-ϵ curve for the material.

7.13   Hysteresis Loop

Consider the σ-ϵ diagram shown in Figure 7.23. Between O and A a tensile force is applied on the material and the material is deformed beyond its elastic limit. At A, the tensile force is removed, and the line AB represents the unloading path. At B the material is reloaded, this time with a compressive force. At C, the compressive force applied on the material is removed. Between C and O, a second unloading occurs, and finally the material resumes its original shape. The loop OABCO is called the *hysteresis loop*, and the area enclosed by this loop is equal to the total strain energy dissipated as heat to deform the body in tension and compression.

7.14   Properties Based on Stress–Strain Diagrams

As defined earlier, the elastic modulus of a material is equal to the slope of its stress–strain diagram in the elastic region. The elastic modulus is a relative measure of the *stiffness* of one material with respect to another. The higher the elastic modulus, the stiffer the material and the higher the resistance to deformation. For example, material 1 in Figure 7.24 is stiffer than material 2.

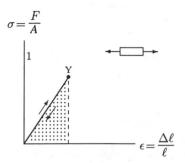

Figure 7.21 *Internal work done and elastic strain energy per unit volume.*

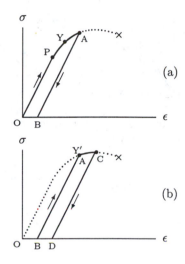

Figure 7.22 *Strain hardening.*

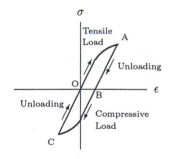

Figure 7.23 *Hysteresis loop.*

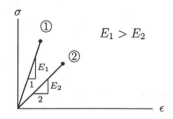

Figure 7.24 *Material 1 is stiffer than material 2.*

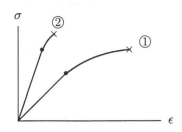

Figure 7.25 *Material 1 is more ductile and less brittle than 2.*

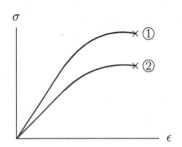

Figure 7.26 *Material 1 is tougher than material 2.*

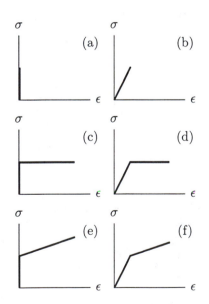

Figure 7.27 *Idealized models of material behavior.*

A *ductile* material is one that exhibits a large plastic deformation prior to failure. For example, material 1 in Figure 7.25 is more ductile than material 2. A *brittle* material, on the other hand, shows a sudden failure (rupture) without undergoing a considerable plastic deformation. Glass is a typical example of a brittle material.

Toughness is a measure of the capacity of a material to sustain permanent deformation. The toughness of a material is measured by considering the total area under its stress–strain diagram. The larger this area, the tougher the material. For example, material 1 in Figure 7.26 is tougher than material 2.

The ability of a material to store or absorb energy without permanent deformation is called the *resilience* of the material. The resilience of a material is measured by its *modulus of resilience*, which is equal to the area under the stress–strain curve in the elastic region. The modulus of resilience is equal to $\sigma_y \epsilon_y / 2$ or $\sigma_y^2 / 2E$ for linearly elastic materials.

Although they are not directly related to the stress–strain diagrams, there are other important concepts used to describe material properties. A material is called *homogeneous* if its properties do not vary from location to location within the material. A material is called *isotropic* if its properties are independent of direction or orientation. A material is called *incompressible* if it has a constant density.

7.15   Idealized Models of Material Behavior

Stress–strain diagrams are most useful when they are represented by mathematical functions. The stress–strain diagrams of materials may come in various forms, and it may not be possible to find a single mathematical function to represent them. For the sake of mathematical modeling and the analytical treatment of material behavior, these diagrams can be simplified. Some of these diagrams representing certain idealized material behavior are illustrated in Figure 7.27.

A *rigid* material is one that cannot be deformed even under very large loads (Figure 7.27a). A *linearly elastic* material is one for which the stress and strain are linearly proportional, with the modulus of elasticity being the constant of proportionality (Figure 7.27b). A *rigid-perfectly plastic* material does not exhibit any elastic behavior, and once a critical stress level is reached, it will deform continuously and permanently until failure (Figure 7.27c). After a linearly elastic response, a *linearly elastic-perfectly plastic* material is one that deforms continuously at a constant stress level (Figure 7.27d). Figure 7.27e represents the stress–strain diagram for *rigid-linearly plastic* behavior. The stress–strain diagram of a *linearly elastic-linearly plastic* material has two distinct regions with two different slopes (bilinear), in which stresses and strains are linearly proportional (Figure 7.27f).

7.16 Mechanical Properties of Materials

Table 7.2 lists properties of selected materials in terms of their tensile yield strengths (σ_y), tensile ultimate strengths (σ_u), elastic moduli (E), shear moduli (G), and Poisson's ratios (ν). The significance of the Poisson's ratio will be discussed in the following chapter. Note that mechanical properties of a material can vary depending on many factors including its content (for example, in the case of steel, its carbon content) and the processes used to manufacture the material (for example, hard rolled, strain hardened, or heat treated). As will be discussed in later chapters, biological materials exhibit time-dependent properties. That is, their response to external forces depends on the rate at which the forces are applied. Bone, for example, is an anisotropic material. Its response to tensile loading in different directions is different, and different elastic moduli are established to account for its response in different directions. Therefore, the values listed in Table 7.2 are some averages and ranges, and are aimed to provide a sense of the orders of magnitude of numbers involved.

Table 7.2 *Average mechanical properties of selected materials.*
(1 MPa = 1,000,000 Pa, 1 GPa = 1,000,000,000 Pa)

MATERIAL	YIELD STRENGTH σ_y (MPa)	ULTIMATE STRENGTH σ_u (MPa)	ELASTIC MODULUS E (GPa)	SHEAR MODULUS G (MPa)	POISSON'S RATIO ν
Muscle	–	0.2	–	–	0.49
Tendon	–	70	0.4	–	0.40
Skin	–	8	0.5	–	0.49
Cortical Bone	80	130	17	3.3	0.40
Glass	35–70	–	70–80	–	–
Cast Iron	40–260	140–380	100–190	42–90	0.29
Aluminum	60–220	90–390	70	28	0.33
Steel	200–700	400–850	200	80	0.30
Titanium	400–800	500–900	100	45	0.34

Figure 7.28 illustrates the tensile stress–strain diagrams of a few materials. Steel and aluminum are relatively stiff and have high ultimate strengths. Glass and dry bone are brittle. Wet bone has the lowest ultimate strength and elastic modulus.

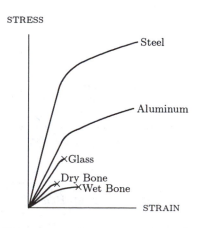

Figure 7.28 *Gross comparison of stress–strain diagrams of selected materials.*

7.17 Example Problems

The following examples will demonstrate some of the uses of the concepts introduced in this chapter.

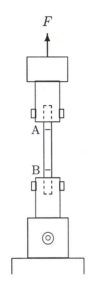

Figure 7.29 *Example 7.1.*

Example 7.1 A circular cylindrical rod with radius $r = 1.26$ cm is tested in a uniaxial tension test (Figure 7.29). Before applying a tensile force of $F = 1000$ N, two points A and B that are at a distance $\ell_0 = 30$ cm (gage length) are marked on the rod. After the force is applied, the distance between A and B is measured as $\ell_1 = 31.5$ cm.

Determine the tensile strain and average tensile stress generated in the rod.

Solution: By definition, the tensile strain is equal to the ratio of the amount of elongation to the original length. The amount of elongation, $\Delta\ell$, is the difference between the gage lengths before and after deformation:

$$\Delta\ell = \ell_1 - \ell_0 = 31.5 - 30.0 = 1.5 \text{ cm}$$

Therefore, the tensile strain ϵ generated in the rod is:

$$\epsilon = \frac{\Delta\ell}{\ell_0} = \frac{1.5}{30} = 0.05 \text{ cm/cm}$$

The bar has a circular cross-section with radius $r = 1.26$ cm or $r = 0.0126$ m. The cross-sectional area A of the bar is:

$$A = \pi r^2 = (3.1416)(0.0126)^2 = 5 \times 10^{-4} \text{ m}^2$$

The average tensile stress is equal to the applied force per unit area of the surface that cuts the line of action of the force at right angles. In this case, it is the cross-sectional area of the rod. Therefore:

$$\sigma = \frac{F}{A} = \frac{1000}{0.0005} = 2,000,000 = 2 \times 10^6 \text{ Pa} = 2 \text{ MPa}$$

SPECIMEN 1: ALUMINUM BAR

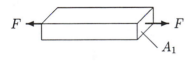

SPECIMEN 2: STEEL ROD

Figure 7.30 *Example 7.2.*

Example 7.2 Two specimens made of two different materials are tested in a uniaxial tension test by applying a force of $F = 20$ kN (20×10^3 N) on each specimen (Figure 7.30). Specimen 1 is an aluminum bar with elastic modulus $E_1 = 70$ GPa (70×10^9 Pa) and a rectangular cross-section (1 cm by 2 cm). Specimen 2 is a steel rod with elastic modulus $E_2 = 200$ GPa (200×10^9 Pa) and a circular cross-section (radius 1 cm).

Calculate tensile stresses developed in each specimen. Assuming that the tensile stress in each specimen is below the proportionality limit of the material, calculate the tensile strain for each specimen. Also, if the original length of each specimen was 30 cm, what are their lengths after deformation?

Solution: The tensile stress is equal to the ratio of the applied force and the cross-sectional area of the specimen. The cross-sectional areas of the specimens are:

$$A_1 = (1 \text{ cm})(2 \text{ cm}) = 2 \text{ cm}^2 = 2 \times 10^{-4} \text{ m}^2$$
$$A_2 = \pi(1 \text{ cm})^2 = 3.14 \text{ cm}^2 = 3.14 \times 10^{-4} \text{ m}^2$$

Therefore, the tensile stresses developed in each specimen are:

$$\sigma_1 = \frac{F}{A_1} = \frac{20 \times 10^3}{2 \times 10^{-4}} = 100 \times 10^6 \text{ Pa} = 100 \text{ MPa}$$

$$\sigma_2 = \frac{F}{A_2} = \frac{20 \times 10^3}{3.14 \times 10^{-4}} = 63.7 \times 10^6 \text{ Pa} = 63.7 \text{ MPa}$$

That is, the aluminum bar (specimen 1) is stressed more than the steel rod (specimen 2).

To calculate the tensile strains corresponding to tensile stresses σ_1 and σ_2, we can assume that the deformations are elastic and that the stresses σ_1 and σ_2 are below the proportionality limits for aluminum and steel. In other words, the stresses are linearly proportional to strains and the elastic moduli E_1 and E_2 are the constants of proportionality. Hence:

$$\epsilon_1 = \frac{\sigma_1}{E_1} = \frac{100 \times 10^6}{70 \times 10^9} = 1.43 \times 10^{-3}$$

$$\epsilon_2 = \frac{\sigma_2}{E_2} = \frac{63.7 \times 10^6}{200 \times 10^9} = 0.32 \times 10^{-3}$$

These results suggest that the aluminum bar is stretched more than the steel rod.

The original length of each specimen was $\ell_0 = 30$ cm. After deformation, assume that the aluminum bar is elongated by $\Delta\ell_1$ to length ℓ_1, and the steel rod is elongated by $\Delta\ell_2$ to length ℓ_2. By definition, tensile strain is $\epsilon = \Delta\ell/\ell_0$, or the amount of elongation is $\Delta\ell = \epsilon\ell_0$. On the other hand, the length of the specimen after deformation is equal to the original length plus the amount of elongation. For example, for the aluminum bar, $\ell_1 = \ell_0 + \Delta\ell$, or $\ell_1 = \ell_0(1 + \epsilon_1)$. Therefore:

$$\ell_1 = \ell_0(1 + \epsilon_1) = 30(1 + 0.00143) = 30.0429 \text{ cm}$$

$$\ell_2 = \ell_0(1 + \epsilon_2) = 30(1 + 0.00320) = 30.0960 \text{ cm}$$

In other words, the increase in length of the aluminum bar and the steel rod is less than 1 mm.

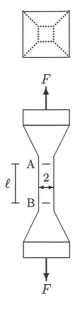

Figure 7.31 *The top and side views of the specimen.*

Example 7.3 An experiment was designed to determine the elastic modulus of the human bone (cortical) tissue. Three almost identical bone specimens were prepared. The specimen size and shape used is shown in Figure 7.31, which has a square (2 × 2 mm) cross-section. Two sections, A and B, are marked on each specimen at a fixed distance apart. Each specimen was then subjected to tensile loading of varying magnitudes, and the lengths between the marked sections were again measured electronically. The following data were obtained:

Applied Force, F (N)	Measured Gage Length, ℓ (mm)
0	5.000
240	5.017
480	5.033
720	5.050

Determine the tensile stresses and strains developed in each specimen, plot a stress–strain diagram for the bone, and determine the elastic modulus (E) for the bone.

Solution: The cross-sectional area of each specimen is $A = 4$ mm² or 4×10^{-6} m². When the applied load is zero, the gage length is 5 mm, which is the original (undeformed) gage length, ℓ_0. Therefore, the stress and strain developed in each specimen can be calculated using:

$$\sigma = \frac{F}{A} \qquad \epsilon = \frac{\ell - \ell_0}{\ell_0}$$

The following table lists stresses and strains calculated using the above formulas:

F (N)	$\sigma \times 10^6$ (Pa)	ℓ (mm)	ϵ (mm/mm)
0	0	5.000	0.0
240	60	5.017	0.0034
480	120	5.033	0.0066
720	180	5.050	0.0100

In Figure 7.32, the stress and strain values computed are plotted to obtain a σ-ϵ graph for the bone. Notice that the relationship between the stress and strain is almost linear, which is indicated in Figure 7.32 by a straight line.

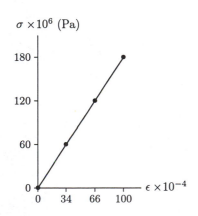

Figure 7.32 *Stress–strain diagram for the bone.*

Recall that the elastic modulus of a linearly elastic material is equal to the slope of the straight line representing the σ-ϵ relationship for that material. Therefore:

$$E = \frac{\sigma}{\epsilon} = \frac{180 \times 10^6}{0.0100} = 18 \times 10^9 \text{ Pa} = 18 \text{ GPa}$$

Example 7.4 Figure 7.33 illustrates a fixation device consisting of a plate and two screws, which can be used to stabilize fractured bones. During a single leg stance, a person can apply his/her entire weight to the ground via a single foot. In such situations, the total weight of the person is applied back on the person through the same foot, which has a compressive effect on the leg, its bones, and joints. In the case of a patient with a fractured leg bone (in this case, the femur), this force is transferred from below to above (distal to proximal) the fracture through the screws of the fixation device.

If the diameter of the screws is $D = 5$ mm and the weight of the patient is $W = 700$ N, determine the shear stress exerted on the screws of a two-screw fixation device during a single leg stance on the leg with a fractured bone.

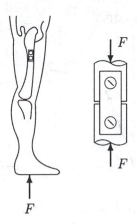

Figure 7.33 *Example 7.4.*

Solution: Free-body diagrams of the fixation device and the screws are shown in Figure 7.34. Note that the screw above the fracture is pushing the plate downward, whereas the screw below the fracture is pushing the plate upward. Each screw is applying a force on the plate equal to the weight of the person. The same magnitude force is also acting on the screws but in the opposite directions. For example, for the screw above the fracture, the plate is exerting an upward force on the head of the screw and the bone is applying a downward force. The effects of the forces applied on the screws are such that they are trying to shear the screws in a plane perpendicular to the centerline of the screws. With respect to the cross-sectional areas of the screws, these are shearing forces.

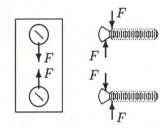

Figure 7.34 *Forces applied and the plate and screws.*

The shear stress τ generated in the screws can be calculated by using the following relationship between the shear force F and area A over which the shear stress is to be determined:

$$\tau = \frac{F}{A}$$

In this case, F is equal to the weight ($W = 700$ N) of the patient. Since the diameters of the screws are given, the cross-sectional area of each screw can be calculated as $A = \pi D^2 / 4 = 19.6$ mm^2 or $A = 19.6 \times 10^{-6}$ m^2. Substituting the numerical value of A and $F = W = 700$ N into the above formula and carrying out the computation will yield $\tau = 35.7 \times 10^6$ Pa.

Note that if we had a four-screw rather than a two-screw fixation device as shown in Figure 7.35, then each screw would be subjected to a shearing force equal to half of the total weight of the patient.

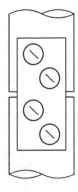

Figure 7.35 *A four-screw fixation device.*

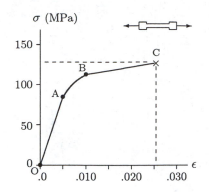

Figure 7.36 *Tensile stress–strain diagram for cortical bone* $(1 \text{ MPa} = 10^6 \text{ Pa})$.

Example 7.5 Specimens of human cortical bone tissue were subjected to a simple tension test until fracture. The test results revealed a stress–strain diagram shown in Figure 7.36, which has three distinct regions. These regions are an initial linearly elastic region (between O and A), an intermediate nonlinear elasto-plastic region (between A and B), and a final linearly plastic region (between B and C). The average stresses and corresponding strains at points O, A, B, and C are measured as:

Point	Stress σ (MPa)	Strain ϵ (mm/mm)
O	0	0.0
A	85	0.005
B	114	0.010
C	128	0.026

Using this information, determine the elastic and strain hardening moduli of the bone tissue in the linear regions of its σ-ϵ diagram. Note that the *strain hardening modulus* is the slope of the σ-ϵ curve in the plastic region. Also, find mathematical expressions relating stresses to strains in the linearly elastic and linearly plastic regions.

Solution: The elastic modulus E is the slope of the σ-ϵ curve in the elastic region. Between O and A, the bone exhibits linearly elastic material behavior, and the σ-ϵ curve is a straight line. The slope of this line is:

$$E = \frac{\sigma_A - \sigma_O}{\epsilon_A - \epsilon_O} = \frac{85 \times 10^6 - 0}{0.005 - 0.0} = 17 \times 10^9 \text{ Pa} = 17 \text{ GPa}$$

The strain hardening modulus E' is equal to the slope of the σ-ϵ curve in the plastic region. Between B and C, the bone exhibits a linearly plastic material behavior, and its σ-ϵ curve is a straight line. Therefore:

$$E' = \frac{\sigma_C - \sigma_B}{\epsilon_C - \epsilon_B} = \frac{128 \times 10^6 - 114 \times 10^6}{0.026 - 0.010} = 0.875 \times 10^9 \text{ Pa}$$

The stress–strain relationship between O and A is:

$$\sigma = E \epsilon \quad \text{or} \quad \epsilon = \frac{\sigma}{E}$$

The relationship between σ and ϵ in the linearly plastic region between B and C can be expressed as:

$$\sigma = \sigma_B + E'(\epsilon - \epsilon_B) \quad \text{or} \quad \epsilon = \epsilon_B + \frac{1}{E'}(\sigma - \sigma_B)$$

For example, the strain corresponding to a tensile stress of $\sigma = 120$ MPa can be calculated as:

$$\epsilon = 0.010 + \frac{120 \times 10^6 - 114 \times 10^6}{0.875 \times 10^9} = 0.017$$

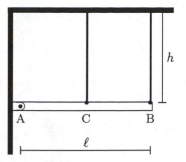

Example 7.6 Consider the structure shown in Figure 7.37. The horizontal beam AB has a length $\ell = 4$ m, weight $W = 500$ N, and is hinged to the wall at A. The beam is supported by two identical steel bars of length $h = 3$ m, cross-sectional area $A = 2$ cm^2, and elastic modulus $E = 200$ GPa. The steel bars are attached to the beam at B and C, where C is equidistant from both ends of the beam.

Determine the forces applied by the beam on the steel bars, the reaction force at A, the amount of elongation in each steel bar, and the stresses generated in each bar. Assume that the beam material is much stiffer than the steel bars.

Figure 7.37 *A statically indeterminate system.*

Solution: This problem will be analyzed in three stages.

<u>Static analysis.</u> The free-body diagram of the beam is shown in Figure 7.38. The total weight, W, of the beam is assumed to act at its geometric center located at C. R_A is the magnitude of the ground reaction force applied on the beam through the hinge joint at A, and T_1 and T_2 are the forces applied by the steel bars on the beam. Note that only a vertical reaction force is considered at A since there is no horizontal force acting on the beam.

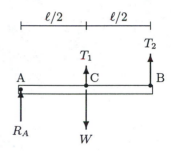

We have a coplanar force system with three unknowns: R_A, T_1, and T_2. There are three equations of equilibrium available from statics. Since there is no horizontal force, the horizontal equilibrium of the beam is automatically satisfied. Therefore, we have two equations but three unknowns. In other words, we have a statically indeterminate system and the equations of equilibrium are not sufficient to fully analyze this problem. We need an additional equation that will be derived by taking into consideration the deformability of the parts constituting the system.

Figure 7.38 *The free-body diagram of the beam.*

The vertical equilibrium of the beam requires that:

$$\sum F_y = 0: \qquad T_1 + T_2 + R_A = W \qquad (i)$$

For the rotational equilibrium of the beam about point A:

$$\sum M_A = 0: \qquad \frac{1}{2}T_1 + T_2 = \frac{1}{2}W \qquad (ii)$$

<u>Geometric compatibility.</u> The horizontal beam is hinged to the wall at A, and the weight of the beam tends to rotate the beam about A in the clockwise direction. Because of the weight of the beam, which is applied as tensile forces T_1 and T_2 on the bars, the steel bars will deflect (elongate) and the beam will slightly swing about A (Figure 7.39). Because of the forces acting on the beam, the beam may bend a little as well. Since it is stated that the beam material is much stiffer than the steel bars, we can ignore the deformability of the beam and assume that it maintains its straight configuration. If δ_1 and δ_2 refer to the amount

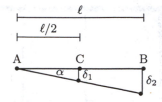

Figure 7.39 *Deflection of the beam.*

of deflections in the steel bars, then from Figure 7.39:

$$\tan \alpha \simeq \frac{\delta_1}{\ell/2} \simeq \frac{\delta_2}{\ell} \qquad (iii)$$

Note that this relationship is correct when deflections (δ_1 and δ_2) are small for which angle α is small.

Next we need take into consideration the relationship between applied forces and corresponding deformations.

Stress–strain (force-deformation) analyses. The steel bars with length h elongate by δ_1 and δ_2. Therefore, the tensile strains in the bars are:

$$\epsilon_1 = \frac{\delta_1}{h} \qquad \epsilon_2 = \frac{\delta_2}{h} \qquad (iv)$$

The bars are subjected to tensile forces T_1 and T_2. The cross-sectional area of each bar is given as A. Therefore, the tensile stresses exerted by the beam on the steel rods are:

$$\sigma_1 = \frac{T_1}{A} \qquad \sigma_2 = \frac{T_2}{A} \qquad (v)$$

Assuming that the stresses involved are within the proportionality limit for steel, we can apply the Hooke's law to relate stresses to strains:

$$\sigma_1 = E \, \epsilon_1 \qquad \sigma_2 = E \, \epsilon_2 \qquad (vi)$$

Now, we can subtitute Eqs. (*iv*) and (*v*) into Eqs. (*vi*) so as to eliminate stresses and strains. This will yield:

$$\delta_1 = \frac{T_1 \, h}{E \, A} \qquad \delta_2 = \frac{T_2 \, h}{E \, A} \qquad (vii)$$

Note that from Eq. (*iii*):

$$\delta_2 = 2 \, \delta_1 \qquad (viii)$$

Substituting Eqs. (*vii*) into Eq. (*viii*) will yield:

$$T_2 = 2 \, T_1 \qquad (ix)$$

Now, we have a total of three equations, Eqs. (*i*), (*ii*), and (*ix*), with three unknowns, R_A, T_1, and R_2. Solving these equations simultaneously will yield:

$$T_1 = \frac{1}{5} W = 100 \, \text{N}$$

$$T_2 = \frac{2}{5} W = 200 \, \text{N}$$

$$R_A = \frac{2}{5} W = 200 \, \text{N}$$

Once tensile forces T_1 and T_2 are determined, Eqs. (*vii*) can be used to calculate the amount of elongations, Eqs. (*v*) can be used to calculate the tensile stresses, and Eqs. (*iv*) can be used to calculate the tensile strains developed in the steel bars:

$$\sigma_1 = \frac{T_1}{A} = 0.5 \times 10^6 \, \text{Pa} = 0.5 \, \text{MPa}$$

$$\sigma_2 = \frac{T_2}{A} = 1.0 \times 10^6 \, \text{Pa} = 1.0 \, \text{MPa}$$

$$\epsilon_1 = \frac{\sigma_1}{E} = 2.5 \times 10^{-6}$$

$$\epsilon_2 = \frac{\sigma_2}{E} = 5.0 \times 10^{-6}$$

Note that the calculated strains are very small. Correspondingly, the deformations are very small as predicted earlier while deriving the relationship in Eq. (*iii*). Also note that the stresses developed in the steel bars are much lower than the proportionality limit for steel. Therefore, the assumption made to relate stresses and strains in Eqs. (*vi*) was correct as well.

7.18 Exercise Problems

Problem 7.1 Complete the following definitions with appropriate expressions.

(a) Unit deformation of a material as a result of an applied load is called _____.

(b) The internal resistance of a material to deformation due to externally applied forces is called _____.

(c) _____ is a measure of the intensity of internal forces acting parallel or tangent to a plane of cut, while _____ are associated with the intensity of internal forces that are perpendicular to the plane of cut.

(d) On the stress–strain diagram, the stress corresponding to the _____ is the highest stress that can be applied to the material without causing permanent deformation.

(e) On the stress–strain diagram, the highest stress level corresponds to the _____ of the material.

(f) For some materials, it may not be easy to distinguish the yield point. The yield strength of such materials may be determined by the _____.

(g) _____ is defined as the ability of a material to resume its original (stress-free) size and shape upon removal of applied loads.

(h) For linearly elastic materials, stress is linearly proportional to strain and the constant of proportionality is called the _____ of the material.

(i) The distinguishing factor in linearly elastic materials is their _____.

(j) Materials for which the stress–strain curve in the elastic region is not a straight line are known as _____ materials.

(k) _____ is the constant of proportionality between shear stress and shear strain for linearly elastic materials.

(l) A mathematical equation that relates stresses to strains is called a _____.

(m) The analogy between elastic materials and springs is known as _____.

(n) _____ implies permanent (unrecoverable) deformations.

(o) The area under the stress–strain diagram in the elastic region corresponds to the _____ energy stored in the material while deforming the material.

(p) _____ energy is dissipated as heat while deforming the material.

(q) The area enclosed by the _____ signifies the total strain energy dissipated as heat while loading and unloading a material.

(r) The technique of changing the yield point of a material by loading the material beyond its yield point is called _____.

(s) The elastic modulus of a material is a relative measure of the _____ of one material with respect to another.

(t) A _____ material is one that exhibits a large plastic deformation prior to failure.

(u) A _____ material is one that shows a sudden failure (rupture) without undergoing a considerable plastic deformation.

(v) _____ is a measure of the capacity of a material to sustain permanent deformation. The toughness of a material is measured by considering the total area under its stress–strain diagram.

(w) The ability of a material to store or absorb energy without permanent deformation is called the _____ of the material.

(x) If the mechanical properties of a material do not vary from location to location within the material, then the material is called _____.

(y) If a material has constant density, then the material is called _____.

(z) If the mechanical properties of a material are independent of direction or orientation, then the material is called _____.

Answers: Given at the end of the chapter.

σ

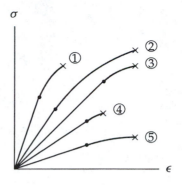

Figure 7.40 *Problem 7.2.*

Problem 7.2 Curves in Figure 7.40 represent the relationship between tensile stress and tensile strain for five different materials. The "dot" on each curve indicates the yield point and the "cross" represents the rupture point. Fill in the blank spaces below with the correct number referring to a material.

Material _____ has the highest elastic modulus.

Material _____ is the most ductile.

Material _____ is the most brittle.

Material _____ has the lowest yield strength.

Material _____ has the highest strength.

Material _____ is the toughest.

Material _____ is the most resilient.

Material _____ is the most stiff.

Answers: Given at the end of the chapter.

Problem 7.3 Consider two bars, 1 and 2, made of two different materials. Assume that these bars were tested in a uniaxial tension test. Let F_1 and F_2 be the magnitudes of tensile forces applied on bars 1 and 2, respectively. E_1 and E_2 are the elastic moduli and A_1 and A_2 are the cross-sectional areas perpendicular to the applied forces for bar 1 and 2, respectively. For the conditions indicated below, determine the correct symbol relating tensile stresses σ_1 and σ_2 and tensile strains ϵ_1 and ϵ_2. Note that ">" indicates greater than, "<" indicates less than, "=" indicates equal to, and "?" indicates that the information provided is not sufficient to make a judgement.

(a) If $A_1 > A_2$ and $F_1 = F_2$, then $\sigma_1 >=? < \sigma_2$ and $\epsilon_1 >=? < \epsilon_2$
(b) If $E_1 > E_2$, $A_1 = A_2$, and $F_1 = F_2$, then $\sigma_1 >=? < \sigma_2$ and $\epsilon_1 >=? < \epsilon_2$

Answers: Given at the end of the chapter.

Problem 7.4 Figure 7.41 illustrates a bone specimen with a circular cross-section. Two sections, A and B, that are $\ell_0 = 6$ mm distance apart are marked on the specimen. The radius of the specimen in the region between A and B is $r_0 = 1$ mm.

This specimen was subjected to a series of uniaxial tension tests until fracture by gradually increasing the magnitude of the applied force and measuring corresponding deformations. As a

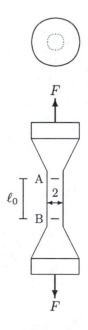

Figure 7.41 *Problem 7.4.*

result of these tests, the following data were recorded:

Record #	Force, F (N)	Deformation, $\Delta\ell$ (mm)
1	94	0.009
2	190	0.018
3	284	0.027
4	376	0.050
5	440	0.094

If record 3 corresponds to the end of the linearly elastic region and record 5 corresponds to fracture point, carry out the following:

(a) Calculate average tensile stresses and strains for each record.
(b) Draw the tensile stress–strain diagram for the bone specimen.
(c) Calculate the elastic modulus, E, of the bone specimen.
(d) What is the ultimate strength of the bone specimen?
(e) What is the yield strength of the bone specimen? (Use the offset method.)

Answers (approximately):

(c) $E = 20$ GPa
(d) $\sigma_u = 140$ MPa
(e) $\sigma_y = 118$ MPa

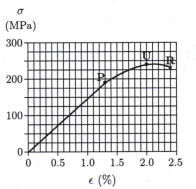

Figure 7.42 Problem 7.5.

Problem 7.5 Human femur bone was subjected to a uniaxial tension test. As a result of a series of experiments, the stress–strain curve shown in Figure 7.42 was obtained. Based on the graph in Figure 7.42, determine the following parameters:

(a) Elastic modulus, E.
(b) Apparent yield strength, σ_y (use the offset method).
(c) Ultimate strength, σ_u.
(d) Strain, ϵ_1, corresponding to yield stress.
(e) Strain, ϵ_2, corresponding to ultimate stress.
(f) Strain energy when stress is at the proportionality limit.

Answers (approximately):

(a) $E = 14.6$ GPa
(b) $\sigma_y = 235$ MPa
(c) $\sigma_u = 240$ MPa
(d) $\epsilon_1 = 0.018$
(e) $\epsilon_2 = 0.020$
(f) 1.235 MPa

Problem 7.6 Consider the uniform, horizontal beam shown in Figure 7.43a. The beam is hinged to a wall at A and a weight W_2 is attached on the beam at B. Point C represents the center of gravity of the beam, which is equidistant from A and B. The beam has a weight $W_1 = 100$ N and length $\ell = 4$ m. The beam is supported by two vertical rods, 1 and 2, attached to the beam at D and E. Rod 1 is made of steel with elastic modulus $E_1 = 200$ GPa and a cross-sectional area $A_1 = 500$ mm², and rod 2 is bronze with elastic modulus $E_2 = 80$ GPa and $A_2 = 400$ mm². The original (undeformed) lengths of both rods is $h = 2$ m. The distance between A and D is $d_1 = 1$ m and the distance between A and E is $d_2 = 3$ m.

The free-body diagram of the beam and its deflected orientation is shown in Figure 7.43b, where T_1 and T_2 represent the forces exerted by the rods on the beam. Symbols δ_1 and δ_2 represent the amount of deflection the steel and bronze rods undergo, respectively. Note that the beam material is assumed to be very stiff (almost rigid) as compared to the rods so that it maintains its straight shape.

(a) Calculate tensions T_1 and T_2, and the reactive force R_A on the beam at A.
(b) Calculate the average tensile stresses σ_1 and σ_2 generated in the rods.

Answers:

(a) $T_1 = 464$ N, $T_2 = 445$ N, $R_A = 409$ N
(b) $\sigma_1 = 0.928$ GPa, $\sigma_2 = 1.113$ GPa

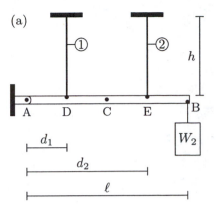

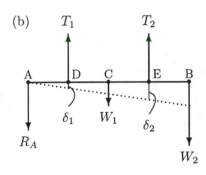

Figure 7.43 *Problem 7.6.*

Answers to Problem 7.1:

(a) strain
(b) stress
(c) Shear stress, normal stress
(d) elastic limit
(e) ultimate strength
(f) offset method
(g) Elasticity
(h) elastic (Young's) modulus
(i) elastic (Young's) modulus
(j) nonlinear elastic
(k) Shear modulus
(l) material function
(m) Hooke's Law

(n) Plasticity
(o) elastic strain
(p) Plastic strain
(q) hysteresis loop
(r) strain hardening
(s) stiffness
(t) ductile
(u) brittle
(v) Toughness
(w) resilience
(x) homogeneous
(y) incompressible
(z) isotropic

Answers to Problem 7.2:   1, 2, 4, 5, 2, 2, 3, 1

Answers to Problem 7.3:   (a) $\sigma_1 < \sigma_2$, $\epsilon_1 ? \epsilon_2$ (b) $\sigma_1 = \sigma_2$, $\epsilon_1 < \epsilon_2$

Chapter 8

Multiaxial Deformations and Stress Analyses

8.1 Poisson's Ratio

When a structure is subjected to uniaxial tension, the transverse dimensions decrease (the structure undergoes lateral contractions) while simultaneously elongating in the direction of the applied load. This was illustrated in the previous chapter through the phenomenon called necking. For stresses within the proportionality limit, the results of uniaxial tension and compression experiments suggest that the ratio of deformations occurring in the axial and lateral directions is constant. For a given material, this constant is called the *Poisson's ratio* and is commonly denoted by the symbol v (nu):

$$v = -\frac{\text{lateral strain}}{\text{axial strain}}$$

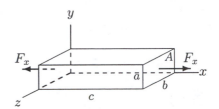

Figure 8.1 *A rectangular bar subjected to uniaxial tension.*

Consider the rectangular bar with dimensions a, b, and c shown in Figure 8.1. To be able to differentiate strains involved in different directions, a rectangular coordinate system is adopted. The bar is subjected to tensile forces of magnitude F_x in the x direction that induces a tensile stress σ_x (Figure 8.2). Assuming that this stress is uniformly distributed over the cross-sectional area ($A = ab$) of the bar, its magnitude can be determined using:

$$\sigma_x = \frac{F_x}{A} \qquad (8.1)$$

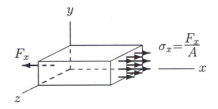

Figure 8.2 *Stress distribution is uniform over the cross-sectional area $A = ab$ of the bar.*

Under the effect of F_x, the bar elongates in the x direction, and contracts in the y and z directions. If the elastic modulus E of the bar material is known and the deformations involved are within the proportionality limit, then the stress and strain in the x direction are related through Hooke's law:

$$\epsilon_x = \frac{\sigma_x}{E} \qquad (8.2)$$

Equation (8.2) yields the unit deformation of the bar in the direction of the applied forces. Strains in the lateral directions can now be determined by utilizing the definition of the Poisson's ratio. If ϵ_y and ϵ_z are unit contractions in the y and z directions due to the uniaxial loading in the x direction, then:

$$v = -\frac{\epsilon_y}{\epsilon_x} = -\frac{\epsilon_z}{\epsilon_x} \qquad (8.3)$$

In other words, if the Poisson's ratio of the bar material is known, then the strains in the lateral directions can be determined:

$$\epsilon_y = \epsilon_z = -v\,\epsilon_x = -v\,\frac{\sigma_x}{E} \qquad (8.4)$$

The minus signs in Eqs. (8.3) and (8.4) indicate a decrease in the lateral dimensions when there is an increase in the axial dimension. Strains ϵ_y and ϵ_z are negative when ϵ_x is positive,

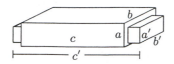

Figure 8.3 *The bar elongates and undergoes lateral contractions simultaneously.*

which is the case for tensile loading. These equations are also valid for compressive loading in the x direction for which σ_x and ϵ_x are negative, and ϵ_y and ϵ_z are positive.

Once the strains in all three directions are determined, then the deformed dimensions a', b', and c' (Figure 8.3) of the bar can also be calculated. By definition, strain is equal to the ratio of the change in length and the original length. Therefore:

$$\epsilon_x = \frac{c' - c}{c}$$

Solving this equation for the deformed length c' of the object in the x direction will yield $c' = (1 + \epsilon_x)c$. Similarly, $a' = (1 + \epsilon_y)a$ and $b' = (1 + \epsilon_z)b$.

Note that the stress–strain relationships provided here are valid only for linearly elastic materials, or within the proportionality limits of any elastic–plastic material.

For a given elastic material, the elastic modulus, shear modulus, and Poisson's ratio are related through the expression:

$$G = \frac{E}{2(1 + v)} \qquad \text{or} \qquad v = \frac{E}{2G} - 1 \qquad (8.5)$$

This formula can be used to calculate the Poisson's ratio of a material if the elastic and shear moduli are known.

8.2 Biaxial and Triaxial Stresses

As discussed in the previous section, when an object is subjected to uniaxial loading, strains can occur in all three directions. The strains in the lateral directions can be calculated by utilizing the definition of Poisson's ratio. Poisson's ratio also makes it possible to analyze situations in which there is more than one normal stress acting in more than one direction.

Consider the rectangular bar shown in Figure 8.4. The bar is subjected to biaxial loading in the xy-plane. Let P be a point in the bar. Stresses induced at point P can be analyzed by constructing a cubical material element around the point. A cubical material element with sides parallel to the sides of the bar itself is shown in Figure 8.4, along with the stresses acting on it. σ_x and σ_y are the normal stresses due to the tensile forces applied on the bar in the x and y directions, respectively. If $A_x = ab$ and $A_y = bc$ are the areas of the rectangular bar with normals in the x and y directions, respectively, then σ_x and σ_y can be calculated as:

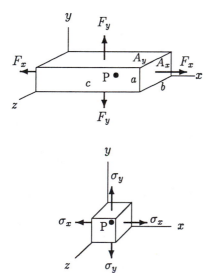

Figure 8.4 *A rectangular bar subjected to tensile forces in the x and y directions and a material element under biaxial stresses.*

$$\sigma_x = \frac{F_x}{A_x} = \frac{F_x}{a\,b}$$

$$\sigma_y = \frac{F_y}{A_y} = \frac{F_y}{b\,c}$$

The effects of these biaxial stresses are illustrated graphically in Figure 8.5. Stress σ_x elongates the material in the x direction and causes a contraction in the y (also z) direction. Strains due to σ_x in the x and y directions are:

$$\epsilon_{x1} = \frac{\sigma_x}{E}$$

$$\epsilon_{y1} = -\nu\,\epsilon_{x1} = -\nu\,\frac{\sigma_x}{E}$$

Similarly, σ_y elongates the material in the y direction and causes a contraction in the x direction (Figure 8.5b). Therefore, strains in the x and y directions due to σ_y are:

$$\epsilon_{y2} = \frac{\sigma_y}{E}$$

$$\epsilon_{x2} = -\nu\,\epsilon_{y2} = -\nu\,\frac{\sigma_y}{E}$$

The combined effect of σ_x and σ_y on the plane material element is shown in Figure 8.5c. The same effect can be represented mathematically by adding the individual effects of σ_x and σ_y. The resultant strains in the x and y directions can be determined as:

$$\epsilon_x = \epsilon_{x1} + \epsilon_{x2} = \frac{\sigma_x}{E} - \nu\,\frac{\sigma_y}{E}$$

$$\epsilon_y = \epsilon_{y1} + \epsilon_{y2} = \frac{\sigma_y}{E} - \nu\,\frac{\sigma_x}{E}$$

(8.6)

If required, these equations can be solved simultaneously to express stresses in terms of strains:

$$\sigma_x = \frac{(\epsilon_x + \nu\,\epsilon_y)\,E}{1 - \nu^2}$$

$$\sigma_y = \frac{(\epsilon_y + \nu\,\epsilon_x)\,E}{1 - \nu^2}$$

(8.7)

This discussion can be extended to derive the following stress–strain relationships for the case of triaxial loading (Figure 8.6):

$$\epsilon_x = \frac{1}{E}\left[\sigma_x - \nu(\sigma_y + \sigma_z)\right]$$

$$\epsilon_y = \frac{1}{E}\left[\sigma_y - \nu(\sigma_z + \sigma_x)\right]$$

(8.8)

$$\epsilon_z = \frac{1}{E}\left[\sigma_z - \nu(\sigma_x + \sigma_y)\right]$$

These formulas are valid for linearly elastic materials, and they can be used when the stresses induced are tensile or compressive, by adapting the convention that tensile stresses are positive and compressive stresses are negative.

Further generalization of stress–strain relationships for linearly elastic materials should take into consideration the relationships

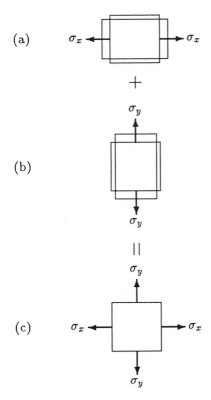

(a)

(b)

(c)

Figure 8.5 *Method of superposition.*

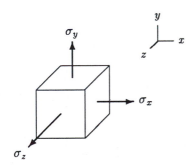

Figure 8.6 *Triaxial normal stresses.*

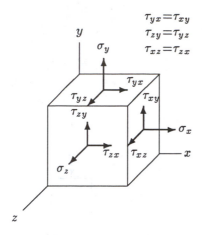

Figure 8.7 *Normal and shear components of the stress tensor.*

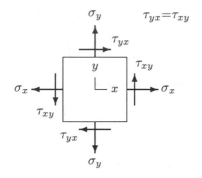

Figure 8.8 *Plane (biaxial) stress components.*

between shear stresses and shear strains. As illustrated in Figure 8.7, the most general case of material loading occurs when an object is subjected to normal and shear stresses in three mutually perpendicular directions. Corresponding to these stresses there are normal and shear strains. In Figure 8.7, shear stresses are identified by double subscripted symbols. The first subscript indicates the direction normal to the surface over which the shear stress is acting and the second subscript indicates the direction in which the stress is acting. A total of six shear stresses (τ_{xy}, τ_{yx}, τ_{yz}, τ_{zy}, τ_{zx}, τ_{xz}) are defined. However, the condition of static equilibrium requires that $\tau_{xy} = \tau_{yx}$, $\tau_{yz} = \tau_{zy}$, and $\tau_{zx} = \tau_{xz}$. Therefore, there are only three independent shear stresses (τ_{xy}, τ_{yz}, τ_{zx}), and three corresponding shear strains (γ_{xy}, γ_{yz}, γ_{zx}). For linearly elastic materials, shear stresses are linearly proportional to shear strains and the shear modulus, G, is the constant of proportionality:

$$\tau_{xy} = G\,\gamma_{xy}$$
$$\tau_{yz} = G\,\gamma_{yz} \qquad (8.9)$$
$$\tau_{zx} = G\,\gamma_{zx}$$

As illustrated in Figure 8.8, for two-dimensional problems in the xy-plane, only one shear stress (τ_{xy}) and two normal stresses (σ_x, σ_y) need to be considered.

The above discussion indicates that stress and strain have a total of nine components, only six of which are independent. Stress and strain are known as *second-order tensors*. Recall that a scalar quantity has a magnitude only. A vector quantity has both a magnitude and a direction, and can be represented in terms of its three components. Scalar quantities are also known as *zero-order tensors*, while vectors are *first-order tensors*. Second-order tensor components not only have magnitudes and directions associated with them, but they are also dependent upon the plane over which they are determined. For example, in the case of the stress tensor, magnitudes and directions of the stress tensor components at a material point may vary depending upon the orientation of the cubical material element constructed around it. However, if the components of the stress tensor with respect to one material element are known, then the components of the stress tensor with respect to another material element can be determined through appropriate coordinate transformations, which will be discussed in the following section. It is also important to note that although the magnitude and direction of the stress tensor components at a material may vary with the orientation of the material element constructed around it, under a given load condition, the overall state of stress at the material point is always the same.

Example 8.1 Consider the cube with sides $a = 10$ cm shown in Figure 8.9. This block is tested under biaxial forces that are applied in the x and y directions. Assume that the forces applied have equal magnitudes of $F_x = F_y = F = 2 \times 10^6$ N, and that the elastic modulus and Poisson's ratio of the block material are given as $E = 2 \times 10^{11}$ Pa and $\nu = 0.3$.

Determine the strains in the x, y, and z directions, and the deformed dimensions of the block if:

(a) both F_x and F_y are tensile (Figure 8.10a),
(b) F_x is tensile and F_y is compressive (Figure 8.10b), and
(c) both F_x and F_y are compressive (Figure 8.10c).

———

Solution: To be able to calculate the stresses involved, we need to know the areas to which forces F_x and F_y are applied. Let A_x and A_y be the areas of the sides of the cube with normals in the x and y directions, respectively. Since the object is a cube, these areas are equal:

$$A_x = A_y = A = a^2 = 100 \text{ cm}^2 = 1 \times 10^{-2} \text{ m}^2$$

We can now calculate the normal stresses in the x and y directions. Note that the magnitudes of the applied forces in the x and y directions and the areas upon which they are applied are equal. Therefore, the magnitudes of the stresses in the x and y directions are equal as well:

$$\sigma_x = \sigma_y = \sigma = \frac{F}{A} = \frac{2 \times 10^6 \text{ N}}{1 \times 10^{-2} \text{ m}^2} = 2 \times 10^8 \text{ Pa}$$

The normal stress component in the z direction is zero, since there is no force applied on the block in that direction.

Case (a)   Both forces are tensile, and therefore, both σ_x and σ_y are tensile and positive (Figure 8.10a). We can utilize Eqs. (8.8) to calculate the strains involved. Since $\sigma_x = \sigma_y = \sigma$ and $\sigma_z = 0$, these equations can be simplified as:

$$\epsilon_x = \frac{1}{E}(\sigma_x - \nu\,\sigma_y) = \frac{1-\nu}{E}\sigma$$

$$\epsilon_y = \frac{1}{E}(\sigma_y - \nu\,\sigma_x) = \frac{1-\nu}{E}\sigma$$

$$\epsilon_z = -\frac{\nu}{E}(\sigma_x + \sigma_y) = -\frac{2\nu}{E}\sigma$$

Substituting the numerical values of E, ν, and σ into these

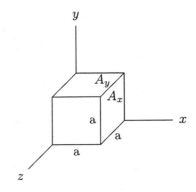

Figure 8.9 *Example 8.1.*

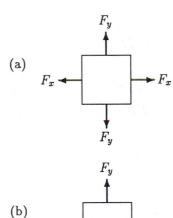

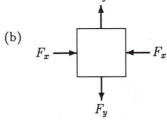

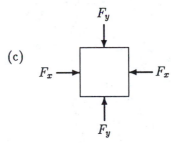

Figure 8.10 *Various modes of biaxial loading of the block.*

equations, and carrying out the computations will yield:

$$\epsilon_x = 0.7 \times 10^{-3}$$
$$\epsilon_y = 0.7 \times 10^{-3}$$
$$\epsilon_z = -0.6 \times 10^{-3}$$

Note that both ϵ_x and ϵ_y are positive, while ϵ_z is negative. As a result of the applied forces, the dimensions of the block in the x and y directions increase, while its dimension in the z direction decreases. If a_x, a_y, and a_z are the new (deformed) dimensions of the block, then:

$$a_x = (1 + \epsilon_x)\, a = (1 + 0.0007)(10 \text{ cm}) = 10.007 \text{ cm}$$
$$a_y = (1 + \epsilon_y)\, a = (1 + 0.0007)(10 \text{ cm}) = 10.007 \text{ cm}$$
$$a_z = (1 + \epsilon_z)\, a = (1 - 0.0006)(10 \text{ cm}) = 9.994 \text{ cm}$$

Case (b)   In this case, σ_x is tensile and positive while σ_y is compressive and negative (Figure 8.10b). Therefore, the stress–strain relationships take the following forms:

$$\epsilon_x = \frac{1}{E}(\sigma_x + v\,\sigma_y) = \frac{1+v}{E}\,\sigma$$
$$\epsilon_y = \frac{1}{E}(-\sigma_y - v\,\sigma_x) = -\frac{1+v}{E}\,\sigma$$
$$\epsilon_z = -\frac{v}{E}(\sigma_x - \sigma_y) = 0$$

Substituting the numerical values of the parameters involved into these equations, and carrying out the computations will yield:

$$\epsilon_x = 1.3 \times 10^{-3}$$
$$\epsilon_y = -1.3 \times 10^{-3}$$
$$\epsilon_z = 0$$

The tensile force applied in the x direction and the compressive force applied in the y direction elongates the block in the x direction and reduces its dimension in the y direction. In the z direction, the contraction caused by the tensile stress σ_x is counterbalanced by the expansion caused by the compressive stress σ_y.

Case (c)   In this case, both forces are compressive. Therefore, both σ_x and σ_y are compressive and negative (Figure 8.10c). The simplified equations relating normal strains to normal stresses for this case are:

$$\epsilon_x = \frac{1}{E}(-\sigma_x + v\,\sigma_y) = -\frac{1-v}{E}\,\sigma$$
$$\epsilon_y = \frac{1}{E}(-\sigma_y + v\,\sigma_x) = -\frac{1-v}{E}\,\sigma$$
$$\epsilon_z = -\frac{v}{E}(-\sigma_x - \sigma_y) = \frac{2\,v}{E}\,\sigma$$

Substituting the numerical values and carrying out the computations will yield:

$$\epsilon_x = -0.7 \times 10^{-3}$$
$$\epsilon_y = -0.7 \times 10^{-3}$$
$$\epsilon_z = 0.6 \times 10^{-3}$$

These results indicate that the dimensions of the block in the x and y directions decrease, while its dimension in the z direction increases. The deformed dimensions of the block can be determined in a similar manner as employed for Case (a).

8.3 Stress Transformation

Consider the rectangular bar shown in Figure 8.11. The bar is subjected to externally applied forces that cause various modes of deformation within the bar. Let P be a point within the structure. Assume that a small cubical material element at point P with sides parallel to the sides of the bar is cut out and analyzed. As illustrated in Figure 8.12, this material element is subjected to a combination of normal (σ_x and σ_y) and shear (τ_{xy}) stresses in the xy-plane. Now, consider a second element at the same material point but with a different orientation than the first element (Figure 8.13). One can assume that the second material element is obtained simply by rotating the first in the counterclockwise direction through an angle θ. Let x' and y' be two mutually perpendicular directions representing the normals to the surfaces of the transformed material element. The stress distribution on the transformed material element would be different than that of the first. In general, the second element may be subjected to normal stresses ($\sigma_{x'}$ and $\sigma_{y'}$) and shear stress ($\tau_{x'y'}$) as well. If stresses σ_x, σ_y, and τ_{xy}, and the angle of rotation θ are given, then stresses $\sigma_{x'}$, $\sigma_{y'}$, and $\tau_{x'y'}$ can be calculated using the following formulas:

$$\sigma_{x'} = \frac{\sigma_x + \sigma_y}{2} + \frac{\sigma_x - \sigma_y}{2} \cos(2\theta) + \tau_{xy} \sin(2\theta) \quad (8.10)$$

$$\sigma_{y'} = \frac{\sigma_y + \sigma_x}{2} - \frac{\sigma_x - \sigma_y}{2} \cos(2\theta) - \tau_{xy} \sin(2\theta) \quad (8.11)$$

$$\tau_{x'y'} = -\frac{\sigma_x - \sigma_y}{2} \sin(2\theta) + \tau_{xy} \cos(2\theta) \quad (8.12)$$

These equations can be used for transforming stresses from one set of coordinates (x, y) to another (x', y').

8.4 Principal Stresses

There are infinitely many possibilities of constructing elements around a given point within a structure. Among these possibilities, there may be one element for which the normal stresses

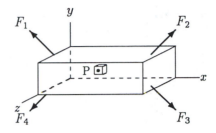

Figure 8.11 *A bar subjected to external forces applied in the xy-plane.*

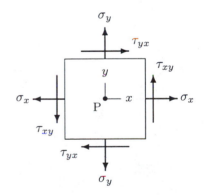

Figure 8.12 *Stress tensor components in the xy-plane.*

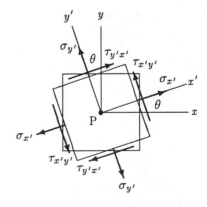

Figure 8.13 *Transformation of stress tensor components.*

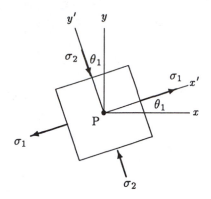

Figure 8.14 *Principal stresses.*

are maximum and minimum. These maximum and minimum normal stresses are called the *principal stresses*, and the planes with normals collinear with the directions of the maximum and minimum stresses are called the *principal planes*. On a principal plane, the shear stress is zero (Figure 8.14). This condition of zero shear stress and Eqs. (8.10) through (8.12) can be utilized to determine the principal normal stresses and the orientation of the principal planes. This can be achieved by setting Eq. (8.12) equal to zero and solving it for angle θ. If the angle thus determined is denoted as θ_1, then:

$$\theta_1 = \frac{1}{2}\tan^{-1}\left(\frac{2\,\tau_{xy}}{\sigma_x - \sigma_y}\right) \tag{8.13}$$

Here, θ_1 represents the angle of orientation of the principal planes relative to the x and y axes. If θ in Eqs. (8.10) and (8.11) is replaced by θ_1, then the following expressions can be derived for the principal stresses σ_1 and σ_2:

$$\sigma_1 = \frac{\sigma_x + \sigma_y}{2} + \sqrt{\left(\frac{\sigma_x - \sigma_y}{2}\right)^2 + \tau_{xy}{}^2} \tag{8.14}$$

$$\sigma_2 = \frac{\sigma_x + \sigma_y}{2} - \sqrt{\left(\frac{\sigma_x - \sigma_y}{2}\right)^2 + \tau_{xy}{}^2} \tag{8.15}$$

The concept of principal stresses is important in stress analyses. It is known that fracture or material failure occurs along the planes of maximum stresses, and structures must be designed by taking into consideration the maximum stresses involved.

Remember that the response of a material to different modes of loading are different, and different physical properties of a given material must be considered while analyzing its behavior under shear, tension, and compression. Note that Eqs. (8.14) and (8.15) are useful for calculating the maximum and minimum normal stresses. For a given structure and loading conditions, the maximum normal stress computed using Eq. (8.14) may be well within the limits of operational safety. However, the structure must also be checked for a critical shearing stress, called the *maximum shear stress*. The maximum shear stress, τ_{max}, occurs on a material element for which the normal stresses are equal (Figure 8.15). Therefore, Eqs. (8.10) and (8.11) can be set equal, and the resulting equation can be solved for the angle of orientation, θ_2, of the element on which the shear stress is maximum. This will yield:

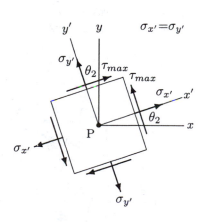

Figure 8.15 *Maximum shear stress.*

$$\theta_2 = \frac{1}{2}\tan^{-1}\left(\frac{\sigma_y - \sigma_x}{2\,\tau_{xy}}\right) \tag{8.16}$$

An expression for the maximum shear stress, τ_{max}, can then be derived by replacing θ in Eq. (8.12) with θ_2:

$$\tau_{max} = \sqrt{\left(\frac{\sigma_x - \sigma_y}{2}\right)^2 + \tau_{xy}^2} \qquad (8.17)$$

A graphical method of finding principal stresses will be discussed next.

8.5 Mohr's Circle

An effective way of visualizing the state of stress at a material point and calculating the principal stresses can be achieved by means of *Mohr's circle*. A typical Mohr's circle is illustrated in Figure 8.17 for the plane stress element shown in Figure 8.16. The procedure for constructing such a diagram and finding the maximum and minimum stresses is outlined below.

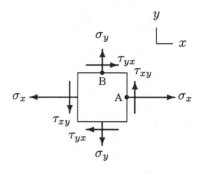

Figure 8.16 *Positive stresses.*

- As in Figure 8.16, make a sketch of the element for which the stresses are known and indicate on this element the proper directions of the stresses involved. The stresses shown in Figure 8.16 are all positive. The sign convention for positive and negative stresses is such that a tensile stress is positive while a compressive stress is negative. A shear stress on the right-hand surface that tends to rotate the material element in the counterclockwise direction and a shear stress on the upper surface that tends to rotate the element in the clockwise direction, are positive.

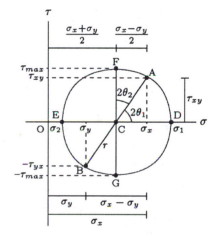

Figure 8.17 *Mohr's circle.*

- Set up a rectangular coordinate system in which the horizontal axis represents the normal stresses and the vertical axis represents the shear stresses. On the τ versus σ diagram, positive normal stresses are plotted to the right of the origin, O, whereas negative normal stresses are plotted to the left of O.

- Let A be a point on the τ-σ diagram with coordinates equal to the normal and shear stresses acting on the right-hand surface of the material element. That is, A has the coordinates σ_x and τ_{xy}. Similarly, B is a point with coordinates equal to the stress components σ_y and $-\tau_{yx}$ acting on the upper surface of the element.

- Connect points A and B with a straight line. Label the point of intersection of line AB and the horizontal axis as C. Point C is the center of the Mohr's circle, and is located at a distance $(\sigma_x + \sigma_y)/2$ from O. Therefore, the stress at C is:

$$\sigma_c = \frac{\sigma_x + \sigma_y}{2}$$

- The distance between A and C (or B and C) is the radius, r, of the Mohr's circle, which can be calculated as:

$$r = \sqrt{\left(\frac{\sigma_x - \sigma_y}{2}\right)^2 + \tau_{xy}^2}$$

• Draw a circle with radius r and center at C. Intersections of this circle with the horizontal axis (where $\tau = 0$) correspond to the maximum and minimum (principal) normal stresses, which can be calculated as:

$$\sigma_1 = \sigma_c + r = \frac{\sigma_x + \sigma_y}{2} + \sqrt{\left(\frac{\sigma_x - \sigma_y}{2}\right)^2 + \tau_{xy}^2}$$

$$\sigma_2 = \sigma_c - r = \frac{\sigma_x + \sigma_y}{2} - \sqrt{\left(\frac{\sigma_x - \sigma_y}{2}\right)^2 + \tau_{xy}^2}$$

If the Mohr's circle is drawn carefully, then σ_1 and σ_2 can also be measured directly from the τ-σ diagram. Note that the above equations are essentially Eqs. (8.14) and (8.15).

• On the τ-σ diagram, F and G are the points of intersection of the Mohr's circle and a vertical line passing through C. At F and G, normal stresses are both equal to σ_c and the magnitude of the shear stress is maximum. Therefore, points F and G on the Mohr's circle correspond to the state of maximum shear stress. The maximum shear stress is simply equal to the radius of the Mohr's circle:

$$\tau_{max} = \sqrt{\left(\frac{\sigma_x - \sigma_y}{2}\right)^2 + \tau_{xy}^2}$$

Again, this is the same equation provided in the previous section for τ_{max}.

• Mohr's circle has its own way of interpreting angles. On the plane stress element shown in Figure 8.16, the normals of surfaces A and B are at right angles. On the τ-σ diagram, A and B make an angle of 180°. Therefore, a rotation of θ degrees corresponds to an angle of 2θ on the Mohr's circle. Point D on the τ-σ diagram is related to the maximum normal stress, and therefore, to one of the principal directions. On the τ-σ diagram, point D is located at an angle of $2\theta_1$ clockwise from point A. The direction normal to the principal plane (direction of σ_1) can be obtained by rotating the x axis through an angle of θ_1 counterclockwise. A similar procedure is valid for finding the orientation of the element for which the shear stress is maximum.

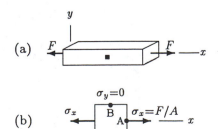

Figure 8.18 *Example 8.2.*

Example 8.2 Consider the bar shown in Figure 8.18a, which is subjected to tension in the x direction. Let F be the magnitude of the applied force and A the cross-sectional area of the bar. The state of uniaxial stress is shown in Figure 8.18b on a plane material element. The sides of this element have normals in the x and y directions.

Using Mohr's circle, determine the maximum shear stress induced and the plane of maximum shear stress.

Solution: For the given magnitude of the externally applied force and the cross-sectional area of the bar, the normal stress induced in the bar in the x direction can be determined as $\sigma_x = F/A$. As illustrated in Figure 8.18 b, σ_x is the only component of the stress tensor on a material element with sides parallel to the x and y directions.

Based on the plane stress element of Figure 8.18b, Mohr's circle is drawn in Figure 8.19a. Note that there is only a tensile stress of magnitude σ_x on surface A, and there is no stress on surface B. Therefore, on the τ-σ diagram, point A is located along the σ-axis at a distance σ_x from the origin, and point B is essentially the origin of the τ-σ diagram. The center C of the Mohr's circle lies along the σ-axis between B and A, at a distance $\sigma_x/2$ from both A and B. Therefore, the radius of the Mohr's circle is $\sigma_x/2$.

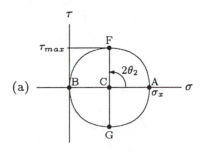

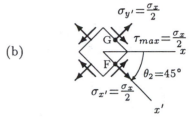

Point F on the Mohr's circle represents the orientation of the material element for which the shear stress is maximum. The magnitude of the maximum shear stress is equal to the radius of the Mohr's circle:

$$\tau_{max} = \frac{\sigma_x}{2} = \frac{F}{2A}$$

On the Mohr's circle, point F is located $90°$ counterclockwise from A. Therefore, as illustrated in Figure 8.19b, the material element for which the shear stress is maximum can be obtained by rotating the material element given in Figure 8.18b in the clockwise direction through an angle $\theta_2 = 45°$.

Note that σ_x is the maximum normal stress on the Mohr's circle. Therefore, point A on the Mohr's circle represents the orientation of the material element for which the normal stresses are maximum and minimum, and the material element in Figure 8.18b represents the state of principal stresses.

Figure 8.19 *Analyses of the material element in Figure 8.18.*

Example 8.3 Consider the material element shown in Figure 8.20. In the xy-plane, the element is subjected to compressive stress σ_x and shear stress τ_{xy}.

Using Mohr's circle, determine the principal stresses, maximum shear stress, and the principal planes.

Solution: Mohr's circle in Figure 8.21a is drawn by using the plane stress element in Figure 8.20. There is a negative (compressive) normal stress with magnitude σ_x and a negative shear stress with magnitude τ_{xy} on surface A of the material element in Figure 8.20. Therefore, point A on Mohr's circle has coordinates $-\sigma_x$ and $-\tau_{xy}$. Surface B on the stress element has only a negative shear stress with magnitude τ_{xy}. Therefore, point B on the τ-σ diagram lies along the τ-axis at a τ_{xy} distance above the origin

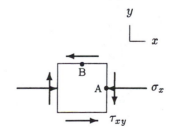

Figure 8.20 *Example 8.3.*

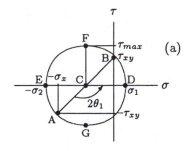

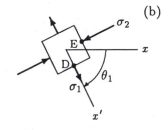

Figure 8.21 *Analyses of the material element in Figure 8.20.*

(positive). The center C of Mohr's circle can be determined as the point of intersection of the line connecting A and B with the σ-axis. The radius of the Mohr's circle can also be determined by utilizing the properties of right triangles. In this case:

$$r = \sqrt{\left(\frac{\sigma_x}{2}\right)^2 + \tau_{xy}^2}$$

Once the radius of the Mohr's circle is established, it is easy to find the principal stresses σ_1 and σ_2 and the maximum shear stress τ_{max}:

$$\sigma_1 = r - \frac{\sigma_x}{2} \quad \text{(tensile)}$$

$$\sigma_2 = r + \frac{\sigma_x}{2} \quad \text{(compressive)}$$

$$\tau_{max} = r$$

To determine the angle of rotation, θ_1, of the plane for which the stresses are maximum and minimum, we need to read the angle between lines CA and CD in Figure 8.21a, which is equal to $2\theta_1$. From the geometry of the problem:

$$2\theta_1 = 180° - \tan^{-1}\left(\frac{\tau_{xy}}{\sigma_x}\right)$$

As illustrated in Figure 8.21b, to construct the material element for which normal stresses are maximum and minimum, calculate angle θ_1 and rotate the material element in Figure 8.20 through an angle θ_1 in the clockwise direction.

8.6 Failure Theories

To ensure both safety and reliability, a structure must be designed and a proper material must be selected so that the strength of the structure is considerably greater than the stresses to which it will be subjected when it is put in service. For example, if a material is subjected to tensile loads only, then the strength of the material must be judged by its ultimate strength (or yield strength). As discussed in the previous chapter, there are well-established normal stress–strain and shear stress–strain diagrams for most of the common materials. Therefore, it is a relatively straightforward task to predict the response of a material subjected to uniaxial stress or pure shearing stress. However, such a direct approach is not available for a complex state of stress occurring under a combined loading.

Several criteria have been established to predict the conditions under which material failure may occur when it is subjected to combined loading. By material or structural failure, it is meant that the material either ruptures so that it can no longer support any load or undergoes excessive permanent deformation (yielding). Unfortunately, there is no single complete failure criterion that can be used to predict the material response under any type of loading. The purpose of these theories is to relate the stresses to the strength of the material. However, the available data are usually expressed in terms of the yield and ultimate strengths of the material as established in simple tension and pure shear experiments. Therefore, the idea is to utilize this relatively limited information to analyze complex situations in which there may be more than one stress component. A few of these failure criteria will be reviewed next.

The *maximum shear stress theory* is used to predict yielding and therefore is applicable to ductile materials. It is also known as *Coulomb theory* or *Tresca theory*. This theory assumes that yielding occurs when the maximum shear stress in a material element reaches the value of the maximum shear stress that would be observed at the instant when yielding occurred if the material were subjected to uniaxial tension.

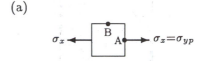
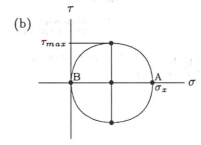

Figure 8.22 *Explaining the maximum shear stress theory.*

Assume that a material is subjected to a simple (uniaxial) tension test until yielding. The stress level at yielding is recorded as σ_{yp}. The maximum shear stress to which this material is subjected can be determined by constructing a Mohr's circle (see Example 8.17) which is illustrated in Figure 8.22. It is clear that the maximum shear stress in simple tension is equal to half of the normal stress. In this case, the normal stress is also the yield strength of the material in tension, and therefore:

$$\tau_{max} = \frac{\sigma_x}{2} = \frac{\sigma_{yp}}{2} \qquad (8.18)$$

The maximum stress theory states that if the same material is subjected to any combination of normal and shear stresses and the maximum shear stress is calculated, the yielding will start when the maximum shear stress is equal to τ_{max}. For example, consider that the material is subjected to a combination of normal and shear stresses in the xy plane as illustrated in Figure 8.23a. For this state of stress, the maximum shear stress τ_{max} can be determined by constructing a Mohr's circle (Figure 8.23b) or using Eq. (8.17). Yielding will occur if τ_{max} is equal to or greater than $\sigma_{yp}/2$.

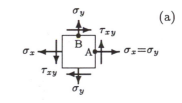
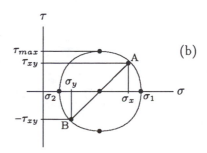

Figure 8.23 *Combined state of stress and its Mohr's circle.*

The *maximum distortion energy theory* is a widely accepted criterion that states the conditions for yielding of ductile materials. It is also known as the *von Mises yield theory* or the *Mises–Hencky theory*. This theory assumes that yielding can occur when the root mean square of the differences between the principal stresses is equal to the yield strength of the material established in a simple tension test.

Let σ_1 and σ_2 be the principal stresses in a plane state of stress, and σ_{yp} be the yield strength of the material. According to the distortion energy theory, the failure by yielding is predicted by the condition:

$$\sqrt{\sigma_1^2 - \sigma_1\sigma_2 + \sigma_2^2} = \sigma_{yp} \tag{8.19}$$

The *maximum normal stress theory* is based on the assumption that failure by yielding occurs whenever the largest principal stress is equal to the yield strength σ_{yp}, or that failure by rupture occurs whenever the largest principal stress is equal to the ultimate strength σ_u of the material. This is a relatively simple theory to implement, and it applies both for ductile and brittle materials.

The failure theories reviewed above can be compared by representing them on a common σ_1 versus σ_2 graph, as illustrated in Figure 8.24. For a given theory, failure occurs if the stress level falls on or outside the closed boundary representing that theory. Among these three failure theories, the maximum distortion energy theory predicts yielding with highest accuracy and shows the best agreement with the experimental results when it is applied to ductile materials. For brittle materials, the maximum normal stress theory is more suitable. If the distortion energy theory is accepted as the basis of comparison, then the maximum shear stress theory is always more conservative and safer, whereas the maximum normal stress theory is conservative whenever the signs of the principal stresses are alike.

The failure theories reviewed here are valid for static loading conditions. These theories must be modified to account for dynamic or repeatedly applied loads that can cause fatigue.

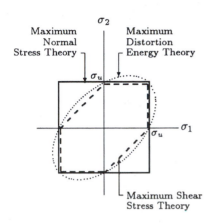

Figure 8.24 *Comparison of the failure theories.*

8.7 Allowable Stress and Factor of Safety

Stress and failure analyses constitute the fundamental components in the design of structures. A structure must be designed to withstand the maximum possible stress level, called *working stress*, to which it will be subjected when it is put in service. However, the exact magnitudes of the loads that will be acting upon the structure may not be known. The structure may be subjected to unexpectedly high loads, dynamic loading conditions, or a corrosive environment that can alter the physical properties of the structural material. To account for the effects of uncertainties, a stress level called the *allowable stress* must be set considerably lower than the ultimate strength of the material. The allowable stress must be low enough to provide a margin of safety. It must also allow for an efficient use of the material.

Safety against unpredictable conditions can be achieved by considering a *factor of safety*. The factor of safety n is usually determined as the ratio of the ultimate strength of the material to the allowable stress. The factor of safety is a number greater than one, and may vary depending on whether the material is loaded under tension, compression, or shear. Instead of the ultimate strength, the factor of safety can also be based on the yield strength of the material. This is particularly important for operational conditions in which excessive yielding or plastic deformations cannot be tolerated. In the case of fatigue loading, endurance limit or the fatigue strength of the material must be used. Once the factor of safety is established, the allowable stress σ_{all} can be determined. For example, based on the ultimate strength σ_u criteria:

$$\sigma_{all} = \frac{\sigma_u}{n} \tag{8.20}$$

8.8 Factors Affecting Strength of Materials

As discussed in the previous chapter, elastic and shear moduli, yield strength, ultimate strength, stress at failure, and the area under the stress–strain diagram for a material are indications of the strength of the material. There are many physical and enviromental factors that may influence the properties of a material. For example, temperature can alter the strength of a material by altering its physical properties. Common materials expand when heated and contract when cooled. An increase in temperature will lower the ultimate strength of a material. The stresses in structures caused by temperature changes can be quite considerable.

Friction occurs when two surfaces roll or slide over one another. Friction dissipates energy primarily as heat. Another consequence of the sliding action of two surfaces is the removal of material from the surfaces, which is called *wear*. Wear can alter the surface quality of structures, expose them to corrosive environments, and, consequently, reduce their mechanical strength. *Corrosion* is one of the primary causes of mechanical failure. Failure by corrosion can be accelerated by the presence of stresses. Corrosion can cause the development of minute cracks in the material, which can propagate in a stressed environment. In general, friction, wear, corrosion, and the presence of discontinuities in a material can lower its strength.

8.9 Fatigue and Endurance

The failure theories reviewed in Section 8.6 are based on the material properties established under static loading configurations. They attempt to predict the response of a material to a loading configuration that is applied once in a specific direction. Many structures, including machine parts and muscles and bones in the human body, are subjected to repeated loading and unloading. Loads that may not cause the failure of a structure in a single application may cause fracture when applied repeatedly. Failure may occur after a few cycles of loading and unloading, or after millions of cycles, depending on the amplitude of the applied load, the physical properties of the material, the size of the structure, the surface quality of the structure, and the operational conditions. Fracture due to repeated loading is called *fatigue*, and in mechanics, fatigue implies a condition of complete structural failure.

Fatigue analysis of structures is quite complicated. There are several experimental techniques developed to understand the fatigue behavior of materials. The fatigue behavior of a material can be determined in a fatigue test using tensile, compressive, bending, or torsional forces. Here, fatigue due to a combination of tension and compression will be explained.

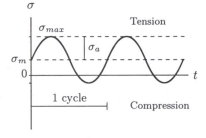

Figure 8.25 *Uniaxial fatigue test.*

Consider the bar shown in Figure 8.25. Assume that the bar is made of a material whose ultimate strength is σ_u. This bar is first stressed to a level σ_m (a mean stress) that is considerably lower than σ_u. The bar is then subjected to a stress fluctuating over time, sometimes tensile and other times compressive. The amplitude σ_a of the stress is such that the bar is subjected to a maximum tensile stress of $\sigma_{max} = \sigma_m + \sigma_a$, which is less than the ultimate strength σ_u of the material. This reversible and periodic stress is applied until the bar fractures and the number of cycles (N) to fracture is recorded. This experiment is repeated on specimens having the same geometric and material properties by applying sinusoidal stresses of varying amplitude. The results show that the number of cycles to failure depends on the stress amplitude σ_a. The higher the σ_a, the lower the N.

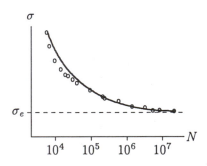

Figure 8.26 *Stress amplitude versus number of cycles to failure.*

A typical result of a fatigue test is plotted in Figure 8.26 on a diagram showing stress amplitude versus number of cycles to failure (σ-N). For a given N, the corresponding stress value is called the *fatigue strength* of the material at that number of cycles. For a given stress level, N represents the fatigue life of the material, which increases rapidly with decreasing stress amplitude. In Figure 8.26, the experimental data are represented by a single curve. For some materials, the σ-N curve levels off and becomes essentially a horizontal line. The stress at which the fatigue curve levels off is called the *endurance limit* of the material, which is denoted by σ_e in Figure 8.26. Below the endurance limit, the material has a high probability of not failing in fatigue no matter how many cycles of stress are imposed upon the material.

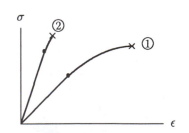

Figure 8.27 *Comparing ductile (1) and brittle (2) fractures.*

A brittle material such as glass or ceramic will undergo elastic deformation when it is subjected to a gradually increasing load. *Brittle fracture* occurs suddenly without exhibiting considerable plastic deformation (Figure 8.27). *Ductile fracture*, on the other hand, is characterized by failure accompanied by considerable elastic and plastic deformations. When a ductile material is subjected to fatigue loading, the failure occurs suddenly without showing considerable plastic deformation (yielding). The fatigue failure of a ductile material occurs in a manner similar to the static failure of a brittle material.

The fatigue behavior of a material depends upon several factors. The higher the temperature, the lower the fatigue strength. The fatigue behavior is very sensitive to surface imperfections and presence of discontinuities within the material that cause stress concentrations. The fatigue failure starts with the creation of a small crack on the surface of the material, which propagates under the effect of repeated loads, resulting in the rupture of the material. A good surface finish can improve the fatigue life of a material.

Orthopedic devices undergo repeated loading and unloading due to the activities of the patients and the actions of their muscles. Over a period of years, a weight-bearing prosthetic device or a fixation device can be subjected to a considerable number of cycles of stress reversals due to normal daily activity. This cyclic loading and unloading can cause fatigue failure of the components of a prosthetic device.

8.10 Stress Concentration

Consider the rectangular bar shown in Figure 8.28. The bar has a cross-sectional area A and is subjected to a tensile force F. The internal reaction force per unit cross-sectional area is defined as the stress, and in this case:

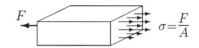

$$\sigma = \frac{F}{A} \qquad (8.21)$$

Figure 8.28 *Uniform stress distribution.*

The classic definition of stress is based on the assumption that the external force F is applied over a relatively large area rather than at a single point, and that the cross-sectional area of the bar is constant throughout the length of the bar. Consequently, as illustrated in Figure 8.28, the stress distribution is uniform over the cross-sectional area of the bar throughout the length of the bar. If the uniformity of the cross-sectional area of the bar is disturbed by the presence of holes, cracks, fillets, scratches, or notches, or if the force is applied over a very small area, then the stress distribution will no longer be uniform at the section where the discontinuity is present, or around the region where the force is applied.

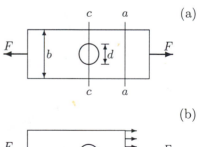

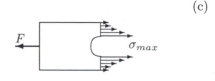

Consider the plate with a circular hole of diameter d shown in Figure 8.29a. The plate is subjected to a tensile load of F. The equilibrium considerations of the plate require that the resultant internal reaction force is equal to F at any section of the plate. As shown in Figure 8.29b, at a section away from the hole (for example, at section aa), the stress distribution is assumed to be uniform. If the cross-sectional area of the plate is A, the magnitude of this uniform stress can be calculated as $\sigma = F/A$. Now, consider the section cc passing through the center of the hole. The magnitude of the average stress at section cc is $\bar{\sigma} = F/(A - A_h)$, where A_h is the hollow area of the cross-section of the plate at section cc.

Figure 8.29 *Effects of stress concentration.*

Since $A - A_h$ is always less than A, the magnitude of the average stress at section cc is greater than the magnitude of the uniform stress at section aa. Futhermore, the distribution of stress at section cc is not uniform, but the stress is maximum along the edges of the hole (Figure 8.29c). That is, the stress is concentrated around the hole. This phenomena is known as the *stress concentration*.

Based on experimental observations, there are empirical formulas established to calculate the maximum stresses developed due to the presence of stress concentrators. The general relationship between the maximum stress σ_{max} (or τ_{max}) and the average stress $\bar{\sigma}$ (or $\bar{\tau}$) is such that:

$$\sigma_{max} = k\,\bar{\sigma} \qquad (8.22)$$

In Eq. (8.22), k is known as the *stress concentration factor*. The value of the stress concentration factor is greater than one and varies depending on many factors such as the size of the stress concentrator relative to the size of the structure (for example, the ratio of the diameter of the hole and the width of the structure), the type of applied load (tension, compression, shear, bending, torsion, or combined), and the physical properties of the material (ductility, brittleness, hardness).

Although the stress levels measured by considering the uniform cross-sectional area of the structure may be below the fracture strength of the material, the structure may fail unexpectedly due to stress concentration effects. The fracture or ultimate strength of a material may be exceeded locally due to the presence of a stress concentrator. Note here that the fatigue failure of structures is explained by the localized stress theory due to the stress concentration effects. There may be very small imperfections or discontinuities inside or on the surface of a structure. These small holes or notches may not cause any serious problem when the structure is subjected to static loading configurations. However, repeatedly applied loads can start minute cracks in the material at the locations of discontinuities. With each application of the load these cracks may propagate, and eventually cause the material to rupture.

The effect of stress concentrations on the lives of bones and orthopedic devices is very important. It is noted that after the removal of orthopedic screws from a bone, the screw holes remain in the bone for many months. During the first few months after the removal of screws, the bones may fracture through the sections of one of the screw holes. A screw hole in the bone causes stress concentration effects and makes the bone weaker, particularly in bending and torsion.

The effects of stress concentrations can be reduced by good surface finish and by avoiding unnecessary holes or any other sudden shape changes in the structure.

8.11 Torsion

Torsion is one of the fundamental modes of loading resulting from the twisting action of applied forces. Here, torsional analyses will be limited to circular shafts. The analyses of structures with noncircular cross-sections subjected to torsional loading are complex and beyond the scope of this text.

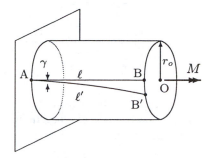

Figure 8.30 *A circular shaft subjected to torsion.*

Consider the solid circular shaft shown in Figure 8.30. The shaft has a length ℓ and a radius r_0. AB represents a straight line on the outer surface of the shaft that is parallel to its centerline. Note that a plane passing through the centerline and cutting the shaft into two semicylinders is called a *longitudinal plane*, and a plane perpendicular to the longitudinal planes is called a *transverse plane* (plane abcd in Figure 8.32). In this case, line AB lies along one of the longitudinal planes. The shaft is mounted to the wall at one end, and a twisting torque with magnitude M is applied at the other end (shown in Figure 8.30 with a double-headed arrow). Owing to the externally applied torque, the shaft deforms in such a way that the straight line AB is twisted into a helix AB'. The deformation at A is zero because the shaft is fixed at that end. The extent of deformation increases in the direction from the fixed end toward the free end. Angle γ in Figure 8.30 is a measure of the deformation of the shaft, and it represents the shear strain due to the shear stresses induced in the transverse planes. The tangent of angle γ is approximately equal to the ratio of the arc length BB' and the length ℓ of the shaft. For small deformations, the tangent of this angle is approximately equal to the angle itself measured in radians. Therefore:

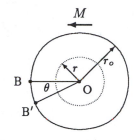

Figure 8.31 θ *is the angle of twist.*

$$\gamma = \frac{\text{arc length BB}'}{\ell} \qquad (8.23)$$

As illustrated in Figure 8.31, the amount of deformation within the shaft also varies with respect to the radial distance r measured from the centerline of the shaft. This variation is such that the deformation is zero at the center, increases toward the rim, and reaches a maximum on the outer surface. Angle θ in Figure 8.31 is called the *angle of twist* and it is a measure of the extent of the twisting action that the shaft suffers. From the geometry of the problem:

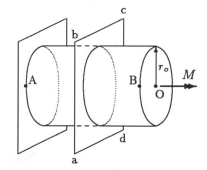

Figure 8.32 *A plane perpendicular to the centerline cuts the shaft into two.*

$$\theta = \frac{\text{arc length BB}'}{r_0} \qquad (8.24)$$

Equations (8.23) and (8.24) can be combined by eliminating the arc length BB'. Solving the resulting equation for the angle of twist will yield:

$$\theta = \frac{\ell}{r_0} \gamma \qquad (8.25)$$

Now consider a plane perpendicular to the centerline of the shaft (plane abcd in Figure 8.32) that cuts the shaft into two segments.

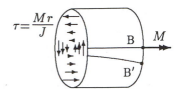

$$\tau = \frac{Mr}{J}$$

Figure 8.33 *Shear stress, τ, distribution over the cross-sectional area of the shaft.*

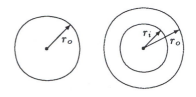

(a) Solid shaft   (b) Hollow shaft

$$J = \frac{\pi r_o^4}{2} \qquad J = \frac{\pi(r_o^4 - r_i^4)}{2}$$

Figure 8.34 *Polar moments of inertia for circular cross-sections.*

Since the shaft as a whole is in static equilibrium, its individual parts have to be in static equilibrium as well. This condition requires the presence of internal shearing forces distributed over the cross-sectional area of the shaft (Figure 8.33). The intensity of these internal forces (force per unit area) is the shear stress τ. The magnitude of the shear stress is related to the magnitude M of the applied torque, the cross-sectional area of the shaft, and the radial distance r between the centerline and the point at which the shear stress is to be determined. This relationship can be determined by satisfying the rotational equilibrium of either the right-hand or the left-hand segment of the shaft, which will yield:

$$\tau = \frac{Mr}{J} \tag{8.26}$$

This is known as the *torsion formula*. In Eq. (8.26), J is the *polar moment of inertia* of the cross-sectional area about the centerline of the shaft. The polar moment of inertia for a solid circular shaft with radius r_o (Figure 8.34a) about its centerline is:

$$J = \frac{\pi r_o^4}{2}$$

The polar moment of inertia for a hollow circular shaft (Figure 8.34b) with inner radius r_i and outer radius r_o is:

$$J = \frac{\pi}{2}\left(r_o^4 - r_i^4\right)$$

The polar moment of inertia has the dimension of length to the power four, and therefore, has a unit m^4 in SI.

If the shaft material is linearly elastic or the deformations are within the proportionality limit, then the stress and strain must be linearly proportional. In the case of shear loading, the constant of proportionality is the shear modulus, G, of the material:

$$\tau = G\gamma \tag{8.27}$$

Solving Eq. (8.27) for γ and substituting Eq. (8.26) into the resulting equation will yield:

$$\gamma = \frac{\tau}{G} = \frac{Mr}{GJ} \tag{8.28}$$

On the circumference of the shaft, r in Eq. (8.28) is equal to r_o. Substituting Eq. (8.28) into Eq. (8.25), a more useful expression for the angle of twist can be obtained:

$$\theta = \frac{M\ell}{GJ} \tag{8.29}$$

The following are some important remarks about torsion, torsion formula, and torsional stresses.

- To derive the torsion formula several assumptions and idealizations were made. For example, it is assumed that the material is isotropic, homogeneous, and linearly elastic.

- For a given shaft and applied torque, the torsional shear stress τ is a linear function of the radial distance r measured from the center of the shaft. The shear stress is distributed nonuniformly over the cross-sectional area of the shaft. At the center of the shaft, $r = 0$ and $\tau = 0$. The stress-free centerline of the solid circular shaft is called the *neutral axis*. The magnitude of the torsional shear stress increases in the direction from the center toward the rim, and reaches a maximum on the circumference of the shaft where $r = r_0$ and $\tau = Mr_0/J$ (Figure 8.35).

- Torsion formula takes a special form, $\tau = 2M/\pi r_0^3$, at the rim of a solid circular shaft for which $J = \pi r_0^4/2$. This equation indicates that the larger the radius of the shaft, the harder it is to deform it in torsion.

- Since $\gamma = \tau/G = Mr/GJ$, the greater the magnitude of the applied torque, the larger the shear stress and shear deformation. The greater the shear modulus of the shaft material, the stiffer the material and the more difficult to deform it in torsion.

- The shear stress discussed herein is that induced in the transverse planes. For a shaft subjected to torsional loading, shear stresses are also developed along the longitudinal planes. This is illustrated in Figure 8.36 on a material element that is obtained by cutting the shaft with two transverse and two longitudinal planes. The transverse and longitudinal stresses are denoted with τ_t and τ_ℓ, respectively, and the equilibrium of the material element requires that τ_t and τ_ℓ are numerically equal.

- A shaft subjected to torsion not only deforms in shear but is also subjected to normal stresses. This can be explained by the fact that the straight line AB deforms into a helix AB′, as illustrated in Figure 8.30. The length ℓ before deformation is increased to length ℓ' after the deformation, and an increase in length indicates the presence of tensile stresses along the direction of elongation.

- Consider the material element in Figure 8.37. The normals of the sides of this material element make an angle 45° with the centerline of the shaft. It can be illustrated by proper coordinate transformations that the only stresses induced on the sides of such an element are normal stresses (tensile stress σ_1 and compressive stress σ_2). The absence of shear stresses on a material element indicates that the normal stresses present are the principal (maximum and minimum) stresses, and that the planes on which these stresses act are the principal planes. (These concepts will be discussed in the following chapter.) For structures subjected to pure torsion, material failure occurs along one of the principal planes. This can be demonstrated by twisting a piece

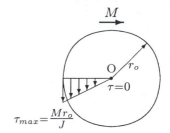

Figure 8.35 *Shear stress varies linearly with radial distance.*

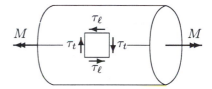

Figure 8.36 *Transverse (τ_t) and longitudinal (τ_ℓ) stresses.*

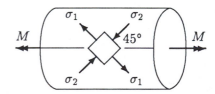

Figure 8.37 *Under pure torsion, principal normal stresses(τ_1, τ_2) occur on planes whose normals are at 45° with the centerline.*

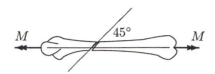

Figure 8.38 *Spiral fracture pattern for a bone subjected to pure torsion.*

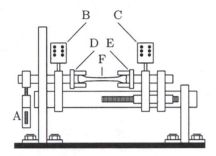

Figure 8.39 *Standard torsion testing machine.*

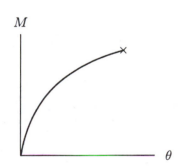

Figure 8.40 *Applied torque versus angular displacement.*

of chalk until it breaks into two pieces. A careful examination of the chalk will reveal the occurrence of the fracture along a spiral line normal to the direction of maximum tension. For circular shafts, the spiral lines make an angle of 45° with the neutral axis (centerline). The same fracture pattern has been observed for bones subjected to pure torsion (Figure 8.38).

There are various experimental methods to analyze the behavior of structures under torsion. Figure 8.39 illustrates a simplified, schematic drawing of a standard torsion testing machine. The important components of this machine are: an adjustable pendulum (A), an angular displacement transducer (B), a torque transducer (C), a rotating grip (D), and a stationary grip (E). This machine can be used to determine the torsional characteristics of specimens, in this case, of bone (F).

The pendulum generates a twisting torque about its shaft, which is also connected to the rotating grip of the machine, and forces an angular deformation in the specimen. The magnitude of the torque applied on the specimen can be controlled by adjusting the position of the mass of the pendulum relative to its center of rotation. The closer the mass to the center of rotation of the pendulum, the smaller the length of the moment arm as measured from the center and therefore the smaller the torque generated. Conversely, the more distal the mass of the pendulum from the center, the larger the length of the moment arm and the larger the twisting action (torque) of the weight of the pendulum. The torque generated by the weight of the pendulum is transmitted through a shaft that is connected to the rotating grip, thereby applying the same torque to the specimen firmly placed between the two grips. The torque and angular displacement transducers measure the amount of torque applied on the specimen and the corresponding angular deformation of the specimen. The fracture occurs when the torque applied is sufficiently high so that the stresses generated in the specimen are beyond the ultimate strength of the material.

The data collected by the transducers of the torsion test machine can be plotted on a torque (M) versus angular deformation (angle of twist, θ) graph. A typical M-θ graph is shown in Figure 8.40. This graph can be analyzed to gather information about the material properties of the specimen.

Example 8.4 A human femur is mounted in the grips of the torsion testing machine (Figure 8.41). The length of the bone at sections between the rotating (D) and stationary (E) grips is measured as $L = 37$ cm. The femur is subjected to a torsional loading until fracture, and the applied torque versus angular

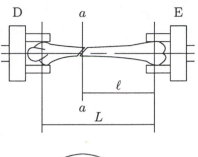

displacement (deflection) graph shown in Figure 8.42 is obtained. The femur is fractured at a section (section aa in Figure 8.41) that is $\ell = 25$ cm distance away from the stationary grip. The geometry of the bony tissue at the fractured section is observed to be a circular ring with inner radius $r_i = 7$ mm and outer radius $r_o = 13$ mm (Figure 8.41).

Calculate the maximum shear strain and shear stress at the fractured section of the femur, and determine the shear modulus of elasticity of the femur.

Section aa

Figure 8.41 *Fractured bone and its cross-sectional geometry.*

Solution: In Figure 8.41, $\theta = 20°$ is the maximum angle of deformation (angle of twist) measured at the rotating grip at the instant when fracture occurred. The total length of the bone between the rotating and stationary grips is measured as $L = 37$ cm. Therefore, the angular deformation is $(20°)/(37\text{cm}) = 0.54$ degrees per unit centimeter of bone length as measured from the stationary grip. The fracture occurred at section aa which is $\ell = 25$ cm away from the stationary grip. Therefore, the angular deflection θ_{aa} at section aa just before fracture is $(0.54°/\text{cm})(25 \text{ cm}) = 13.5°$ or $(13.5°)(\pi/180°) = 0.236$ radian.

The shear strain γ on the surface of the bone at section aa can be determined from Eq. (8.25). In this case, $\ell = 0.25$ m is the distance between section aa and the section of the femur where the stationary grip holds the bone, $r_o = 0.013$ m is the outer radius of the cross-section of the bone at the fracture, and $\theta = \theta_{aa} = 0.236$ is the angle of twist measured in radians. Therefore:

$$\gamma = \frac{r_o}{\ell} \theta_{aa} = \left(\frac{0.013}{0.25}\right)(0.236) = 0.0123 \text{ rad}$$

The torsion formula [Eq. (8.26)] relates the applied torque M, radial distance r, polar moment of inertia J of the cross-section, and the shear stress τ. The cross-sectional geometry of the bony structure of the femur at section aa is a ring with inner radius $r_i = 0.007$ m and outer radius $r_o = 0.013$ m. Therefore, the polar moment of inertia of the cross-section of the femur at section aa is:

$$J = \frac{\pi}{2}\left(r_o^4 - r_i^4\right) = \frac{3.14}{2}[(0.013)^4 - (0.007)^4] = 41.1 \times 10^{-9} \text{ m}^4$$

In Figure 8.42, the magnitude of the applied torque at the instant when the fracture occurred (maximum torque) is $M = 180$ N-m. Hence, the maximum shear stress on the surface of the bone at section aa can be determined using the torsion formula:

$$\tau = \frac{Mr_o}{J} = \frac{(180)(0.013)}{41.1 \times 10^{-9}} = 56.9 \times 10^6 \text{ Pa} = 56.9 \text{ MPa}$$

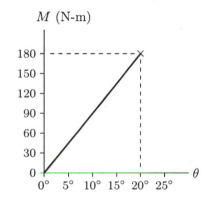

M (N-m)

Figure 8.42 *Torque versus angle of twist diagram.*

Assuming that the deformations are elastic and the relationship between the shear stress and shear strain is linear, the shear modulus G of the bone can be determined from Eq. (8.27):

$$G = \frac{\tau}{\gamma} = \frac{56.9 \times 10^6}{0.0123} = 4.6 \times 10^9 \text{ Pa} = 4.6 \text{ GPa}$$

Remarks:

• For a specimen subjected to torsion, the maximum shear stress before fracture is the *torsional strength* of that specimen. In this case, the torsional strength of the femur is 56.9 MPa.

• The *torsional stiffness* is the ratio of the applied torque and the resultant angular deformation. In this case, the torsional stiffness of the femur is (180 N-m)/(20° × π/180°) = 515.7 N-m/rad.

• The *torsional rigidity* is the torsional stiffness multiplied by the length of the specimen. In this case, the torsional rigidity of the femur is (509.9 N-m/rad)(0.37 m) = 190.8 N-m²/rad.

• The maximum amount of torque applied to a specimen before fracture is defined as the *torsional load capacity* of the specimen. In this case, the torsional load capacity of the femur is 180 N-m.

• The total area under the torque versus angular displacement diagram represents the *torsional energy storage capacity* of the specimen or the *torsional energy absorbed* by the specimen. In this case, the torsional energy storage capacity of the femur is $\frac{1}{2}$(180 N-m)(20° × π/180°) = 31.4 N-m-rad.

• Torsional fractures are usually initiated at regions of the bones where the cross-sections are the smallest. Some particularly weak sections of human bones are the upper and lower thirds of the humerus, femur, and fibula; the upper third of the radius; and the lower fourth of the ulna and tibia.

Example 8.5 Consider the solid circular cylinder shown in Figure 8.43a. The cylinder is subjected to pure torsion by an externally applied torque, M. As illustrated in Figure 8.43b, the state of stress on a material element with sides parallel to the longitudinal and transverse planes of the cylinder is pure shear. For given M and the parameters defining the geometry of the cylinder, the magnitude τ_{xy} of the torsional shear stress can be calculated using the torsion formula provided in Eq. (8.26).

Using Mohr's circle, investigate the state of stress in the cylinder.

Solution: Mohr's circle in Figure 8.44a is drawn by using the stress element of Figure 8.43b. On surfaces A and B of the stress element shown in Figure 8.43b, there is only a negative shear

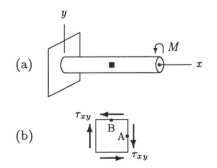

Figure 8.43 *Example 8.5.*

stress with magnitude τ_{xy}. Therefore, both A and B on the τ-σ diagram lie along the τ axis where $\sigma = 0$. Furthermore, the origin of the τ-σ diagram constitutes the midpoint between A and B, and hence, the center, C, of the Mohr's circle. The distance between C and A is equal to τ_{xy}, which is the radius of the Mohr's circle. Mohr's circle cuts the horizontal axis at two locations, both at a distance τ_{xy} from the origin. Therefore, the principal stresses are such that $\sigma_1 = \tau_{xy}$ (tensile) and $\sigma_2 = \tau_{xy}$ (compressive). Furthermore, τ_{xy} is also the maximum shear stress.

The point where $\sigma = \sigma_1$ on the τ-σ diagram in Figure 8.44a is labeled as D. The angle between lines CA and CD is $90°$, and it is equal to half of the angle of orientation of the plane with normal in one of the principal directions. Therefore, the planes of maximum and minimum normal stresses can be obtained by rotating the element in Figure 8.43b $45°$ (clockwise). This is illustrated in Figure 8.44b.

The lines that follow the directions of principal stresses are called the *stress trajectories*. As illustrated in Figure 8.45 for a circular cylinder subjected to pure torsion, the stress trajectories are in the form of helices making an angle $45°$ (clockwise and counterclockwise) with the longitudinal axis of the cylinder. As discussed previously, the significance of these stress trajectories is such that if the material is weakest in tension, the failure occurs along a helix such as *bb* in Figure 8.45, where the tensile stresses are at a maximum.

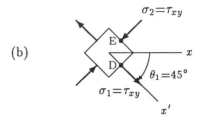

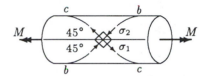

Figure 8.44 *Analyses of the material element in Figure 8.43.*

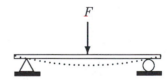

Figure 8.45 *Stress trajectories.*

8.12 Bending

Consider the simply supported, originally straight beam shown in Figure 8.46. A force with magnitude F applied downward bends the beam, subjecting parts of the beam to shear, tension, and compression. There are several ways to subject structures to bending. The free-body diagram of the beam in Figure 8.46 is shown in Figure 8.47a. The number of parallel forces acting on the beam are three. F is the applied load, and R_1 and R_2 are the reaction forces. The type of bending to which this beam is subjected is called a *three-point bending*. The beam in Figure 8.47b, on the other hand, is subjected to a *four-point bending*.

The stress analyses of structures subjected to bending start with static analyses that can be employed to determine both external reactions and internal resistances. For two-dimensional problems in the xy-plane, the number of equations available from statics is three. Two of these equations are translational equilibrium conditions ($\sum F_x = 0$ and $\sum F_y = 0$) and the third equation guarantees rotational equilibrium ($\sum M_z = 0$). The internal resistance of structures to externally applied loads can be determined by applying the *method of sections*, which is based on the

Figure 8.46 *Bending.*

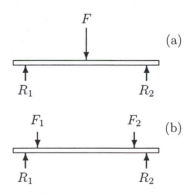

Figure 8.47 *Three- and four-point bending.*

fact that the individual parts of a structure that is itself in static equilibrium must also be in equilibrium. This concept makes it possible to utilize the equations of equilibrium for calculating internal forces and moments, which can then be used to determine stresses. Stresses in a structure subjected to bending may vary from one section to another and from one point to another over a given section. For design and failure analyses, maximum normal and shear stresses must be considered. These critical stresses can be determined by the repeated application of the method of sections throughout the structure.

Consider the simply supported beam shown in Figure 8.48a. Assume that the weight of the beam is negligible. The length of the beam is ℓ, and the two ends of the beam are labeled as A and B. The beam is subjected to a concentrated load with magnitude F, which is applied vertically downward at point C. The distance between A and C is ℓ_1. The free-body diagram of the beam is shown in Figure 8.48b. By applying the equations of equilibrium, the reaction forces at A and B can be determined as $R_A = (1 - \ell_1/\ell)F$ and $R_B = (\ell_1/\ell)F$.

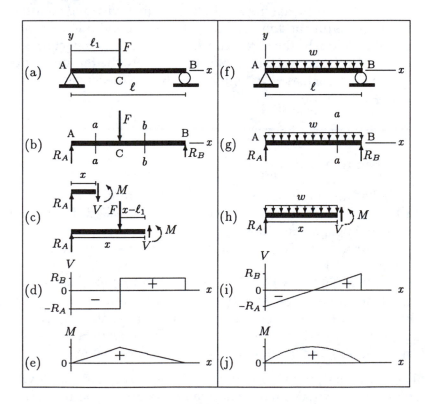

Figure 8.48 *Free-body, shear, and moment diagrams for simply supported beams subjected to concentrated and distributed loads.*

To determine the internal reactions, the method of sections can be applied. As shown in Figure 8.48c, this method is applied at sections aa and bb because the nature of the internal reactions on the left-hand and right-hand sides of the concentrated load are different. Once a hypothetical cut is made, either the left-hand or the right-hand segment of the beam can be analyzed for the internal reactions. In Figure 8.48c, the free-body diagrams of the left-hand segments of the beam are shown. For the vertical equilibrium of each of these segments, there must be an internal shear force at the cut. The magnitude of this force at section aa can be determined as $V = R_A$ by applying the equilibrium condition $\sum F_y = 0$. This force acts vertically downward and is constant between A and C. The magnitude of the shear force at section bb is $V = F - R_A = R_B$, and since F is greater than R_A, it acts vertically upward. The sign convention adopted in this text for the shear force is illustrated in Figures 8.49 and 8.50. An upward internal force on the left-hand segment (or the downward internal force on the right-hand segment) of a cut is positive. Otherwise the shear force is negative. Therefore, as illustrated in Figure 8.48d, the shear force is negative between A and C, and positive between C and B.

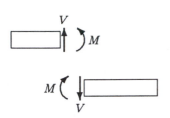

Figure 8.49 *Positive shear forces and bending moments.*

In addition to the vertical equilibrium, the segments must also be checked for rotational equilibrium. As illustrated in Figure 8.48c, this condition is satisfied if there are internal resisting moments at sections aa and bb. By utilizing the equilibrium condition $\sum M = 0$, the magnitudes of these counterclockwise moments can be determined for sections aa and bb as $M = xR_A$ and $M = xR_A - (x - \ell_1)F$, respectively. Note that M is a function of the axial distance x measured from A. The function relating M and x between A and C is different than that between C and B. These functions are plotted on an M versus x graph in Figure 8.48e. The moment is maximum at C where the load is applied. As illustrated in Figures 8.49 and 8.50, the sign convention adopted for the moment is such that a counterclockwise moment on the left-hand segment (or the clockwise moment on the right-hand segment) of a cut is positive.

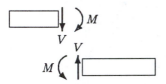

Figure 8.50 *Negative shear forces and bending moments.*

It can also be demonstrated that shear force and bending moment are related through the following equation:

$$V = -\frac{dM}{dx} \qquad (8.30)$$

If the variation of M along the length of the structure is known, then the shear force at a given section of the structure can also be determined using Eq. (8.30).

The procedure outlined above is also applied to analyze a simply supported beam subjected to a distributed load of w (force per unit length of the beam), which is illustrated in Figure 8.48f through j. The procedure for determining the internal shear forces and resisting moments in cantilever beams subjected to concentrated and distributed loads is outlined graphically in Figure 8.51.

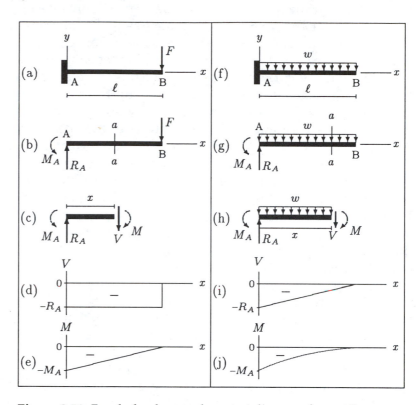

Figure 8.51 *Free-body, shear, and moment diagrams for cantilever beams subjected to concentrated and distributed loads.*

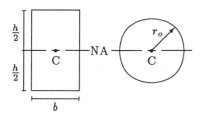

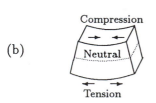

Figure 8.52 *Tension and compression in bending.*

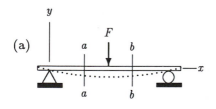

Figure 8.53 *Centroids (C) and neutral axes (NA) for a rectangle and a circle.*

When a structure is subjected to bending, both normal and shear stresses are generated in the structure. For example, consider the beam shown in Figure 8.52. The beam is bent by a downward force. If the beam is assumed to consist of layers of material, then the upper layers of the beam are compressed while the layers on the lower portion of the beam are subjected to tension. The extent of compression or the amount of contraction is maximum at the uppermost layer, and the amount of elongation is maximum on the bottom layer. There is a layer somewhere in the middle of the beam where a transition from tension to compression occurs. For such a layer, there is no tension or compression, and therefore, no deformation in the longitudinal direction. This stress-free layer that separates the zones of tension and compression is called the *neutral plane* of the beam. The line of intersection of the neutral plane with a plane (transverse) cutting the longitudinal axis of the beam at right angles is called the *neutral axis*. The neutral axis passes through the centroid. Centroids of symmetrical cross-sections are located at their geometric centers (Figure 8.53).

The above discussion indicates that when a beam is bent, it is subjected to stresses occurring in the longitudinal direction or in a direction normal to the cross-section of the beam. Furthermore, for the loading configuration shown in Figure 8.52, the distribution of these normal stresses over the cross-section of the beam is such that it is zero on the neutral axis, negative (compressive) above the neutral axis, and positive (tensile) below the neutral axis. For a beam subjected to pure bending, the following equation can be derived from the equilibrium considerations of a beam segment:

$$\sigma_x = -\frac{My}{I} \qquad (8.31)$$

This equation is known as the *flexure formula*, and the stress σ_x is called the *flexural stress*. In Eq. (8.31), M is the bending moment, y is the vertical distance between the neutral axis and the point at which the stress is sought, and I is the *area moment of inertia* of the cross-section of the beam about the neutral axis. The area moments of inertia for a number of simple geometries are listed in Table 8.1.

Table 8.1 *Area (A), area moment of inertia (I) about the neutral axis (NA), first moment of the area (Q) about the neutral axis, and maximum normal(σ_{max}) and shear (τ_{max}) stresses for beams subjected to bending and with cross-section as shown.*

Cross-section	A, I	Q, σ_{max}, τ_{max}
Rectangle, h, b, NA	$A = bh$ $I = \dfrac{bh^3}{12}$	$Q = \dfrac{bh^2}{8}$ $\sigma_{max} = \dfrac{Mh}{2I}$ $\tau_{max} = \dfrac{3V}{2A}$
Circle, r_o, NA	$A = \pi r_o^2$ $I = \dfrac{\pi r_o^4}{4}$	$Q = \dfrac{2r_o^3}{3}$ $\sigma_{max} = \dfrac{Mr_o}{I}$ $\tau_{max} = \dfrac{4V}{3A}$
Annulus, r_o, r_i, NA	$A = \pi(r_o^2 - r_i^2)$ $I = \dfrac{\pi(r_o^4 - r_i^4)}{4}$	$Q = r_o(r_o^2 + r_i^2)$ $\sigma_{max} = \dfrac{Mr_o}{I}$ $\tau_{max} = \dfrac{2V}{A}$

The stress distribution at a section of a beam subjected to pure bending is shown in Figure 8.54. At a given section of the beam, both the bending moment and the area moment of inertia of the cross-section are constant. According to the flexure formula, the flexural stress σ_x is a linear function of the vertical distance y measured from the neutral axis, which can take both positive and negative values. At the neutral axis, $y = 0$ and σ_x is zero. For points above the neutral axis, y is positive and σ_x is negative,

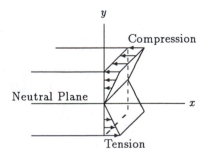

Figure 8.54 *Normal (flexural) stress distribution over the cross-section of the beam.*

indicating compression. For points below the neutral axis, y is negative and σ_x is positive and tensile. At a given section, the stress reaches its absolute maximum value either at the top or the bottom of the beam where y is maximum. It is a common practice to indicate the maximum value of y with c, eliminate the negative sign indicating direction (which can be found by inspection), and write the flexure formula as:

$$\sigma_{max} = \frac{Mc}{I} \qquad (8.32)$$

For example, for a beam with a rectangular cross-section, width b, height h, and bending moment M, $c = h/2$ and the magnitude of the maximum flexural stress is:

$$\sigma_{max} = \frac{Mh}{I\,2} \qquad \text{with} \qquad I = \frac{bh^3}{12}$$

Note that while deriving the flexure formula, a number of assumptions and idealizations are made. For example, it is assumed that the beam is subjected to pure bending. That is, it is assumed that shear, torsional, or axial forces are not present. The beam is initially straight with a uniform, symmetric cross-section. The beam material is isotropic and homogeneous, and linearly elastic. Therefore, Hooke's law ($\sigma_x = E\epsilon_x$) can be used to determine the strains due to flexural stresses. Furthermore, Poisson's ratio of the beam material can be used to calculate lateral contractions and/or elongations.

When the internal shear force in a beam subjected to bending is not zero, a shear stress is also developed in the beam. The distribution of this shear stress on the cross-section of the beam is such that it is maximum on the neutral axis and zero on the top and bottom surfaces of the beam. The following formula is established to calculate shear stresses in bending:

$$\tau_{xy} = \frac{VQ}{Ib} \qquad (8.33)$$

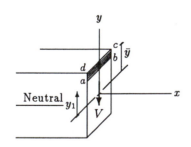

Figure 8.55 *Definitions of the parameters involved.*

In Eq. (8.33), V is the shear force at a section where the shear stress τ_{xy} is sought, I is the moment of inertia of the cross-sectional area about the neutral axis, and b is the width of the cross-section. As shown in Figure 8.55, let y_1 be the vertical distance between the neutral axis and the point at which τ_{xy} is to be determined. Then Q is the *first moment* of the area $abcd$ about the neutral axis. The first moment of area $abcd$ can be calculated as $Q = A\bar{y}$, where A is the area enclosed by $abcd$ and $\bar{y}$ is the distance between the neutral axis and the centroid of area $abcd$. Note that both A and $\bar{y}$ are maximum at the neutral axis, and A is zero at the top and bottom surfaces. Therefore, Q is maximum at the neutral axis, and zero at the top and bottom surfaces. Maximum Q's for different cross-sections are listed in Table 8.1.

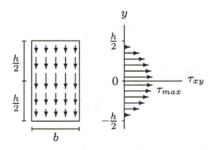

Figure 8.56 *Shear stress distribution over a section of the beam.*

The shear stress distribution over the cross-section of a beam is shown in Figure 8.56. The shear stress is constant along lines parallel to the neutral axis. The shear stress is maximum at the neutral axis, where $y_1 = 0$ and Q is maximum. Maximum shear stresses for different cross-sections can be obtained by substituting the values of I and Q into Eq. (8.19), which are listed in Table 8.1. For example, for a rectangular cross-section:

$$\tau_{max} = \frac{3\,V}{2\,b\,h}$$

Consider a cubical material element in a beam subjected to shear force and bending moment as illustrated in Figure 8.57. On this material element, the effect of bending moment M is represented by a normal (flexural) stress σ_x, and the effect of shear force V is represented by the shear stress τ_{xy} acting on the surfaces with normals in the positive and negative x (longitudinal) directions. As shown in Figure 8.57b, for the rotational equilibrium of this material element, there have to be additional shear stresses (τ_{yx}) on the upper and lower surfaces of the cube (with normals in the positive and negative y directions) such that numerically $\tau_{yx} = \tau_{xy}$. The occurrence of τ_{yx} can be understood by assuming that the beam is made of layers of material, and that these layers tend to slide over one another when the beam is subjected to bending (Figure 8.58).

There are various experiments that may be conducted to analyze the behavior of specimens subjected to bending forces. Some of these experiments will be introduced within the context of the following examples.

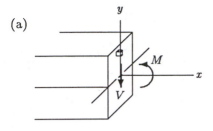

(a)

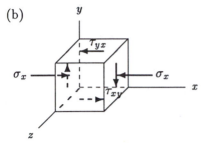

(b)

Figure 8.57 *Both normal and shear stresses occur on a material element subjected to bending.*

Figure 8.58 *Shear stresses in the longitudinal direction.*

Example 8.6 Figure 8.59 illustrates an apparatus that may be used to conduct three-point bending experiments. This apparatus consists of a stationary head (A) to which the specimen (B) is attached, two rings (C and D), and a mass (E) with weight W applied to the specimen through the rings.

For a weight $W = 1000$ N applied to the middle of the specimen and for a support length of $\ell = 16$ cm (the distance between the left and the right supports), determine the maximum flexural and shear stresses generated at section bb of a specimen. The specimen has a square ($a = 1$ cm) cross-section, and the distance between the left support and section bb is $d = 4$ cm (Figure 8.60a).

Solution: The free-body diagram of the specimen is shown in Figure 8.60a. The force (W) is applied to the middle of the specimen. The rotational and translational equilibrium of the specimen requires that the magnitude R of the reaction forces generated at the supports must be equal to half of W. That is, $R = 500$ N.

The specimen has a square cross-section, and its neutral axis is located at a vertical distance $a/2$ measured from both the top and bottom surfaces of the specimen. The normal (flexural) stresses generated at section bb of the specimen depend on

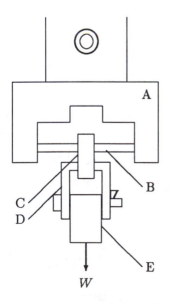

Figure 8.59 *Three-point bending apparatus.*

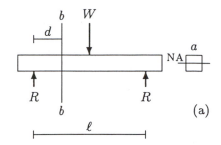

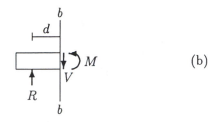

Figure 8.60 *The free-body diagrams.*

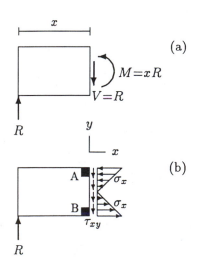

Figure 8.61 *Stress distribution at a section of the specimen in Figure 8.60a.*

the magnitude of the bending moment M at section bb and the area moment of inertia I of the specimen at section bb about the neutral axis. At section bb, the magnitude of the flexural stress is maximum (σ_{max}) at the top (compressive) and the bottom (tensile) surfaces of the specimen:

$$\sigma_{max} = \frac{M}{I} \frac{a}{2}$$

The internal resistances at section bb of the specimen are shown in Figure 8.60b. For the rotational equilibrium of the specimen:

$$M = d\,R = (0.04)(500) = 20 \text{ N-m}$$

The area moment of inertia of a square with sides a is:

$$I = \frac{a^4}{12} = \frac{(0.01)^4}{12} = 8.33 \times 10^{-10} \text{ m}^4$$

Substituting M and I into the flexure formula will yield:

$$\sigma_{max} = \left(\frac{20}{8.33 \times 10^{-10}}\right)\frac{0.01}{2} = 120 \times 10^6 \text{ Pa} = 120 \text{ MPa}$$

The shear stress generated at section bb of the specimen is a function of the shear force V at section bb, and the first moment Q and the area moment of inertia I of the cross-section of the specimen at section bb. The shear stress is maximum (τ_{max}) along the neutral axis, such that:

$$\tau_{max} = \frac{V}{I} \frac{Q}{a}$$

For the vertical equilibrium of the specimen (Figure 8.60b), the magnitude V of the internal shear force at section bb must be equal to the magnitude R of the reaction force; $V = R = 500$ N. The first moment of the cross-sectional area of the specimen about the neutral axis is:

$$Q = \frac{a^3}{8} = \frac{(0.01)^3}{8} = 0.125 \times 10^{-6} \text{ m}^3$$

Therefore, the maximum shear stress occurring at section bb along the neutral axis is:

$$\tau_{max} = \frac{(500)(0.125 \times 10^{-6})}{(8.33 \times 10^{-10})(0.01)} = 7.5 \times 10^6 \text{ Pa} = 7.5 \text{ MPa}$$

Remarks: Note that magnitudes and directions of internal shear force and bending moment are different at different sections of the specimen. At a given section, the internal shear force and bending moment are functions of the horizontal distance between the left support (or the right support) and the section. This is illustrated in Figure 8.61a for a section that lies somewhere between the left support and where the load W is applied on the specimen. If the distance between the left support and the section is x, then the magnitude of the internal bending moment at the section is:

$$M = xR$$

In other words, M is a function of x. The stress distribution at the same section is shown in Figure 8.61b. The upper layers of the specimen are under compression (negative σ_x), while the lower layers are in tension (positive σ_x). Also, a downward shear stress (τ_{xy}) acts throughout the section. Also indicated in Figure 8.61b are two material elements, A and B. The nonzero stress tensor components acting on the surfaces of these material elements are shown in Figure 8.62. It is clear that the state of stress at the upper layers of the specimen is different than the state of stress at the lower layers.

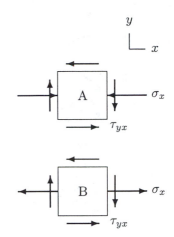

Figure 8.62 *Stress components on material elements A and B in Figure 8.61b.*

Example 8.7 Consider the beam shown in Figure 8.63a. The externally applied force and the reactions at the supports bend the beam (three-force bending), subjecting it to shear stresses. In addition to shear, the upper layers of the beam are subjected to compression and the lower layers to tension. Figure 8.63b shows the state of stress occurring at a material point along a section (section aa) on the left-hand side of the applied force F and above the neutral plane of the beam. Figure 8.63c illustrates the state of stress at the same section below the neutral plane.

Using Mohr's circle, determine the principal stresses and maximum shear stresses occurring in the beam for the states of stress shown in Figures 8.63b and 8.63c.

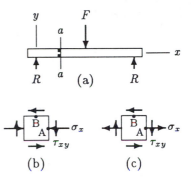

Figure 8.63 *Example 8.7.*

Solution: At a given section, magnitude F of the externally applied force and parameters defining the geometry of the beam, the normal (flexural) stress σ_x and the shear stress τ_{xy} distributions can be determined in a manner similar to that described in Example 8.6. Once the state of stress at a material point is known, Mohr's circle can be used for further analyses of the stresses involved.

Mohr's circle in Figure 8.64a is drawn by using the plane stress element of Figure 8.63b. There is a negative normal stress with magnitude σ_x and a negative shear stress with magnitude τ_{xy} on surface A of the stress element in Figure 8.63b. Therefore, point A on Mohr's circle has coordinates $-\sigma_x$ and $-\tau_{xy}$. Surface B on the stress element has only a negative shear stress with magnitude τ_{xy}. Therefore, point B on the τ-σ diagram lies along the τ-axis, τ_{xy} distance above the origin. The center C of Mohr's circle can be determined as the point of intersection of the line connecting A and B with the σ-axis. The radius of the Mohr's circle can also be determined by utilizing the properties of right triangles. In this case, $r = \sqrt{(\sigma_x/2)^2 + \tau_{xy}^2}$.

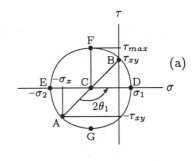

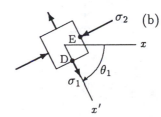

Figure 8.64 *Analyses of the element in Figure 8.63b.*

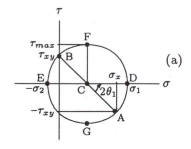

(a)

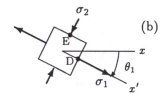

(b)

Figure 8.65 *Analyses of the element in Figure 8.63c.*

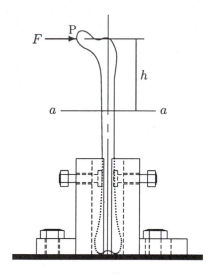

Figure 8.66 *A bench test.*

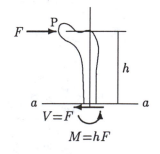

Figure 8.67 *The free-body diagram.*

Once the radius of the Mohr's circle is established, principal stresses and maximum shear stress can be determined as $\sigma_1 = r - \sigma_x/2$ (tensile), $\sigma_2 = r + \sigma_x/2$ (compressive), and $\tau_{max} = r$. To determine the angle of rotation, θ_1, of the plane for which the stresses are maximum and minimum, we need to read the angle between lines CA and CD (in this case, it is $2\theta_1$ counterclockwise), divide it by two, and rotate the stress element in Figure 8.63b in the clockwise direction through an angle θ_1. This is illustrated in Figure 8.64b.

Analyses of the material element in Figure 8.63c are illustrated graphically in Figure 8.65.

Example 8.8 Figure 8.66 illustrates a bench test that may be used to subject bones to bending. In the case shown, the distal end of a human femur is firmly clamped to the bench and a horizontal force with magnitude $F = 500$ N is applied to the head of the femur at point P.

Determine the maximum normal and shear stresses generated at section aa of the femur that is located at a vertical distance $h = 16$ cm measured from point P. Assume that the geometry of the femur at section aa is circular with inner radius $r_i = 6$ mm and outer radius $r_o = 13$ mm.

Solution: The femur is hypothetically cut into two parts by a plane passing through section aa, and the free-body diagram of the proximal part of the femur is shown in Figure 8.67. The internal resistances at section aa are the shear force V and bending moment M. The translational and rotational equilibrium of the proximal part of the femur require that:

$$V = F = 500 \text{ N}$$
$$M = h F = (0.16)(500) = 80 \text{ N-m}$$

Formulas to determine the cross-sectional area, A, the area moment of inertia, I, and the first moment of cross-sectional area about the neutral axis of a structure with a hollow circular cross-section, Q, are provided in Table 8.1. Accordingly:

$$A = \pi \left(r_o^2 - r_i^2 \right) = 4.18 \times 10^{-4} \text{ m}^2$$

$$I = \frac{\pi}{4} \left(r_o^4 - r_i^4 \right) = 2.14 \times 10^{-8} \text{ m}^4$$

$$Q = r_o \left(r_o^2 + r_i^2 \right) = 2.67 \times 10^{-6} \text{ m}^3$$

Formulas to determine the maximum normal stress σ_{max} and maximum shear stress τ_{max} for a structure with a hollow circular cross-section and those subjected to bending forces are also provided in Table 8.1:

$$\sigma_{max} = \frac{M r_o}{I} = \frac{(80)(0.013)}{2.14 \times 10^{-8}} = 48.6 \times 10^6 \text{ Pa} = 48.6 \text{ MPa}$$

$$\tau_{max} = \frac{2V}{A} = \frac{2(500)}{4.18 \times 10^{-4}} = 2.4 \times 10^6 \text{ Pa} = 2.4 \text{ MPa}$$

Owing to the direction of the applied force, the flexural stress is maximum on the medial and lateral sides of the femur. The flexural stress is tensile on the medial side and compressive on the lateral side. The shear stress is maximum along the inner surface of the bony structure of the femur on the ventral and dorsal sides. The loading configuration of the bone is such that it behaves like a cantilever beam.

8.13 Combined Loading

The stress analyses discussed so far were concerned with axial (tension or compression), pure shear, torsional, and flexural (bending) loading of structures based on the assumption that these loads were applied on a structure one at a time. The stresses due to these basic types of loading configurations can be calculated using the following formulas:

$$\text{Axial loading:} \quad \sigma_a = \frac{F_a}{A_a}$$

$$\text{Pure shear loading:} \quad \tau_s = \frac{F_s}{A_s}$$

$$\text{Torsional loading:} \quad \tau_t = \frac{M_t r}{J}$$

$$\text{Flexural loading:} \quad \sigma_b = \frac{M_b y}{I}$$

Here, σ_a is the normal stress due to axial load F_a applied on an area A_a, τ_s is the shear stress due to shear load F_s applied on an area A_s, τ_t is the shear stress due to applied twisting torque M_t at a point r from the centerline of a cylindrical shaft and at a section with polar moment of inertia J, and σ_b is the normal stress due to bending moment M_b at a distance y from the neutral axis of the structure at a section with area moment of inertia of I.

A structure may be subjected to two or more of these loads simultaneously. To analyze the overall effects of such combined loading configurations, first the stresses generated at a given section of the structure due to each load are determined individually. Next, the normal stresses are combined (added or subtracted) together, and the shear stresses are combined together. The following example is aimed to illustrate how combined stresses can be handled.

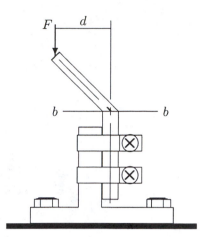

Figure 8.68 *A bench test.*

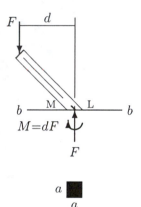

Figure 8.69 *The free-body diagram.*

Example 8.9 Figure 8.68 illustrates a bench test performed on an intertrochanteric nail. The nail is firmly clamped to the bench and a downward force with magnitude $F = 1000$ N is applied.

Determine the stresses generated at section bb of the nail which is located at a horizontal distance $d = 6$ cm measured from the point of application of the force on the nail. The geometry of the nail at section bb is a square with sides $a = 15$ mm.

Solution: The nail is hypothetically cut into two parts by a plane passing through section bb, and the free-body diagram of the proximal part of the nail is shown in Figure 8.69. The translational equilibrium of the nail in the vertical direction requires the presence of a compressive force at section bb with a magnitude equal to the magnitude $F = 1000$ N of the external force applied on the nail. The rotational equilibrium condition requires that there is a clockwise internal bending moment at section bb with magnitude:

$$M = d\,F = (0.06)(1000) = 60 \text{ N-m}$$

The compressive force at section bb gives rise to an axial stress σ_a. The nail has a square geometry at section bb, and its cross-sectional area is:

$$A = a^2 = (0.015)^2 = 2.25 \times 10^{-4} \text{ m}^2$$

Therefore, the magnitude of the axial stress at section bb due to the compressive force is:

$$\sigma_a = \frac{F}{A} = \frac{1000}{2.25 \times 10^{-4}} = 4.4 \times 10^6 \text{ Pa} = 4.4 \text{ MPa}$$

The area moment of inertia I of the nail at section bb is:

$$I = \frac{a^4}{12} = \frac{(0.015)^4}{12} = 4.2 \times 10^{-9} \text{ m}^4$$

The bending moment M at section bb gives rise to a flexural stress σ_b. The flexural stress is maximum on the medial and lateral sides of the nail, which are indicated as M and L in Figure 8.69. The maximum flexural stress is:

$$\sigma_{b_{max}} = \frac{M\,a}{2\,I} = \frac{(60)(0.015)}{2(4.2 \times 10^{-9})} = 107.1 \times 10^6 \text{ Pa} = 107.1 \text{ MPa}$$

The flexural stress σ_b varies linearly over section bb. It is compressive on the medial half of the nail, zero in the middle, and tensile on the lateral half of the nail. The distribution of normal stresses σ_a and σ_b due to the compressive force F and bending moment M at section bb are shown in Figures 8.70a and 8.70b, respectively. The combined effect of these stresses is shown in Figure 8.70c. Note from Figure 8.70c that the resultant normal stress generated at section bb is maximum at the medial (M) side of the nail. This maximum stress is compressive and has a magnitude:

$$\sigma_{max} = \sigma_a + \sigma_{b_{max}} = 111.5 \text{ MPa}$$

Since $\sigma_{b_{max}}$ (tensile) is greater than σ_a (compressive), the resultant stress σ_L on the lateral end of section bb is tensile, and its magnitude is equal to:

$$\sigma_L = \sigma_{b_{max}} - \sigma_a = 102.7 \text{ MPa}$$

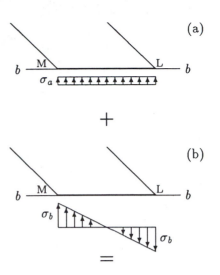

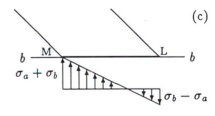

8.14 Exercise Problems

Problem 8.1 Complete the following definitions with appropriate expressions.

(a) For a given material and for stresses within the proportionality limit, the ratio of deformations occurring in the axial and lateral directions is constant, and this constant is called the _____ .

(b) Stress and strain are known as _____ , while force is a vector quantity.

(c) Maximum and minimum normal stresses at a material point are called the _____ .

(d) Planes with normals collinear with the directions of the maximum and minimum normal stresses are called the

_____ .

(e) The _____ occurs on a material element for which normal stresses are equal.

(f) _____ is an effective graphical method of investigating the state of stress at a material point.

(g) By material _____ , it is meant that the material either ruptures or undergoes excessive permanent deformation.

(h) Fracture due to repeated loading and unloading is called

_____ .

(i) The stress at which the fatigue curve levels off is called the _____ of the material.

(j) _____ fracture occurs suddenly without exhibiting considerable plastic deformation.

(k) _____ fracture is characterized by failure accompanied by considerable elastic and plastic deformations.

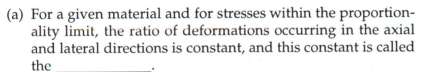

Figure 8.70 *Combined stresses.*

(l) Presence of holes, cracks, fillets, scratches, and notches in a structure can cause _____ that may lead to unexpected structural failure.

(m) For a circular cylindrical shaft, the _____ formula relates shear stresses to applied torque, polar moment of inertia, and the radial distance from the centerline of the shaft.

(n) The shear stress is zero at the center of a circular shaft subject to pure torsion. The stress-free centerline of the shaft is called the _____.

(o) For structures subjected to pure torsion, material failure occurs along one of the principal planes that make an angle _____ degrees with the centerline.

(p) The stress analyses of structures subjected to bending start with _____ analyses that can be employed to determine both external reactions and internal resistances.

(q) The internal resistance of structures to externally applied forces can be determined by applying the _____.

(r) For a structure subjected to bending, the _____ formula relates normal stresses to bending moment and the geometric parameters of the structure.

Answers: Given at the end of the chapter.

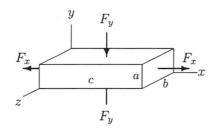

Figure 8.71 *Problem 8.2. (Dimensions of the bar are not drawn to scale.)*

Problem 8.2 Consider the rectangular bar shown in Figure 8.71, with original (undeformed) dimensions $a = b = 2$ cm and $c = 20$ cm. The elastic modulus of the bar material is $E = 100$ GPa and its Poisson's ratio is $\nu = 0.30$. The bar is subjected to biaxial forces in the x and y directions such that $F_x = 4 \times 10^6$ N and that $\underline{F}_x$ is tensile while $\underline{F}_y$ is compressive. Assuming that the bar material is linearly elastic, determine:

(a) average normal stresses σ_x, σ_y, and σ_z developed in the bar,
(b) average normal strains ϵ_x, ϵ_y, and ϵ_z, and
(c) dimension, c', of the bar in the x direction after deformation.

Answers:

(a) $\sigma_x = 10$ GPa, $\sigma_y = 1$ GPa, $\sigma_z = 0$
(b) $\epsilon_x = 0.1030$, $\epsilon_y = -0.0400$, $\epsilon_z = -0.0027$
(c) $c' = 22.06$ cm

Problem 8.3 Consider the solid circular cylinder shown in Figure 8.72. The cylinder has a length $\ell = 10$ cm and radius $r_0 = 2$ cm. The cylinder is made of a linearly elastic material with shear modulus $G = 10$ GPa. If the cylinder is subjected to a twisting torque with magnitude $M = 3000$ N-m, calculate:

(a) the polar moment of inertia, J, of the cross-section of the cylinder,
(b) the maximum angle of twist, θ, in degrees,
(c) the maximum shear stress, τ, in the transverse plane, and
(d) the maximum shear strain, γ, in the transverse plane.

Answers:

(a) $J = 25.13 \times 10^{-8}$ m^4
(b) $\theta = 6.84°$
(c) $\tau = 0.23876$ GPa
(d) $\gamma = 0.02388$

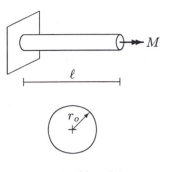

Figure 8.72 *Problem 8.3.*

Problem 8.4 Consider the uniform horizontal beam shown in Figure 8.73. Also shown is the cross-section of the beam. The beam has a length $\ell = 4$ m, width $b = 10$ cm, and height $h = 20$ cm. The beam is hinged to the ground at A, supported by a roller at B, and a downward force with magnitude $F = 400$ N is applied on the beam at C which is located at a distance $d = 1$ m from A.

Assuming that the weight of the beam is negligible, calculate:

(a) the reactive forces on the beam at A and B;
(b) the internal shearing force, V_{aa}, and bending moment, M_{aa}, at section aa of the beam which is 2 meters from A;
(c) the internal shearing force, V_{bb}, and bending moment, M_{bb}, at section bb of the beam which is 3 meters from A;
(d) the first moment, Q, of the cross-sectional area of the beam; and
(e) the maximum shear stress, τ_{aa}, generated in the beam at section aa.

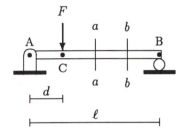

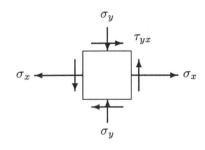

Figure 8.73 *Problem 8.4.*

Answers:

(a) $R_A = 300$ N ($\uparrow$), $R_B = 100$ N ($\uparrow$)
(b) $V_{aa} = 100$ N, $M_{aa} = 200$ N-m
(c) $V_{bb} = 100$ N, $M_{bb} = 100$ N-m
(d) $Q = 0.0005$ m^3
(e) $\tau_{aa} = 7500$ Pa

Problem 8.5 Consider the plane stress element shown in Figure 8.74 representing the state of stress at a material point. If the magnitudes of the stresses are such that $\sigma_x = 200$ Pa, $\sigma_y = 100$ Pa, and $\tau_{xy} = 50$ Pa, calculate the magnitudes of principal stresses

Figure 8.74 *Problem 8.5.*

σ_1 and σ_2, and the maximum shear stress, τ_{max}, generated at this material point.

Answers:

$\sigma_1 = 208$ Pa (tensile), $\sigma_2 = 108$ Pa (compressive),
$\tau_{max} = 158$ Pa

Answers to Problem 8.1:

(a) Poisson's ratio
(b) second-order tensor
(c) principal stresses
(d) principal planes
(e) maximum shear stress
(f) Mohr's circle
(g) failure
(h) fatigue
(i) endurance limit

(j) Brittle
(k) Ductile
(l) stress concentration effects
(m) torsion
(n) neutral axis
(o) 45
(p) static
(q) method of sections
(r) flexure

Chapter 9

Mechanical Properties
of Biological Tissues

9.1 Viscoelasticity

The material response discussed in the previous chapters was limited to the response of elastic materials, in particular to linearly elastic materials. Most metals, for example, exhibit linearly elastic behavior when they are subjected to relatively low stresses at room temperature. They undergo plastic deformations at high stress levels. For an elastic material, the relationship between stress and strain can be expressed in the following general form:

$$\sigma = \sigma(\epsilon) \tag{9.1}$$

Equation (9.1) states that the normal stress σ is a function of normal strain ϵ only. The relationship between the shear stress τ and shear strain γ can be expressed in a similar manner. For a linearly elastic material, stress is linearly proportional to strain, and in the case of normal stress and strain, the constant of proportionality is the elastic modulus E of the material (Figure 9.1):

$$\sigma = E\,\epsilon \tag{9.2}$$

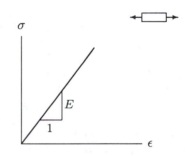

Figure 9.1 *Linearly elastic material behavior.*

While investigating the response of an elastic material, the concept of time does not enter into the discussions. Elastic materials show time-independent material behavior. Elastic materials deform instantaneously when they are subjected to externally applied loads. They resume their original (unstressed) shapes almost instantly when the applied loads are removed.

There is a different group of materials – such as polymer plastics, almost all biological materials, and metals at high temperatures – that exhibits gradual deformation and recovery when they are subjected to loading and unloading. The response of such materials is dependent upon how quickly the load is applied or removed, the extent of deformation being dependent upon the rate at which the deformation-causing loads are applied. This time-dependent material behavior is called *viscoelasticity*. Viscoelasticity is made up of two words: viscosity and elasticity. *Viscosity* is a fluid property and is a measure of resistance to flow. *Elasticity*, on the other hand, is a solid material property. Therefore, a viscoelastic material is one that possesses both fluid and solid properties.

For viscoelastic materials, the relationship between stress and strain can be expressed as:

$$\sigma = \sigma(\epsilon, \dot{\epsilon}) \tag{9.3}$$

Equation (9.3) states that stress, σ, is not only a function of strain, ϵ, but is also a function of the *strain rate*, $\dot{\epsilon} = d\epsilon/dt$, where t is time. A more general form of Eq. (9.3) can be obtained by including higher-order time derivatives of strain. Equation (9.3) indicates that the stress–strain diagram of a viscoelastic material is not

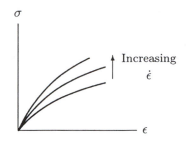

Figure 9.2 *Strain rate ($\dot{\epsilon}$) dependent viscoelastic behavior.*

$$\sigma = E\epsilon \qquad F = kx$$

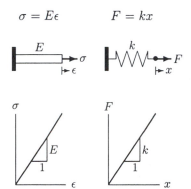

Figure 9.3 *Analogy between a linear spring and an elastic solid.*

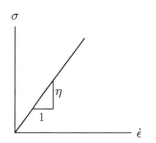

Figure 9.4 *Stress–strain rate diagram for a linearly viscous fluid.*

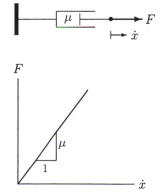

Figure 9.5 *A linear dashpot and its force–displacement rate diagram.*

unique but is dependent upon the rate at which the strain is developed in the material (Figure 9.2).

9.2 Analogies Based on Springs and Dashpots

In Section 7.8, while covering Hooke's law, an analogy was made between linearly elastic materials and linear springs. An elastic material deforms, stores potential energy, and recovers deformations in a manner similar to that of a spring. The elastic modulus E for a linearly elastic material relates stresses and strains, whereas the constant k for a linear spring relates applied forces and corresponding deformations (Figure 9.3). Both E and k are measures of stiffness. The similarities between elastic materials and springs suggest that springs can be used to represent elastic material behavior. Since these similarities were first noted by Robert Hooke, elastic materials are also known as *Hookean solids*.

When subjected to external loads, fluids deform as well. Fluids deform continuously, or *flow*. For fluids, stresses are not dependent upon the strains but on the strain rates. If the stresses and strain rates in a fluid are linearly proportional, then the fluid is called a *linearly viscous fluid* or a *Newtonian fluid*. Examples of linearly viscous fluids include water and blood plasma. For a linearly viscous fluid:

$$\sigma = \eta\,\dot{\epsilon} \qquad (9.4)$$

In Eq. (9.4), η (eta) is the constant of proportionality between the stress σ and the strain rate $\dot{\epsilon}$, and is called the *coefficient of viscosity* of the fluid. As illustrated in Figure 9.4, the coefficient of viscosity is the slope of the σ-$\dot{\epsilon}$ graph of a Newtonian fluid. The physical significance of this coefficient is similar to that of the coefficient of friction between the contact surfaces of solid bodies. The higher the coefficient of viscosity, the "thicker" the fluid and the more difficult it is to deform. The coefficient of viscosity for water is about 1 centipoise at room temperature, while it is about 1.2 centipoise for blood plasma.

The spring is one of the two basic mechanical elements used to simulate the mechanical behavior of materials. The second basic mechanical element is called the *dashpot*, which is used to simulate fluid behavior. As illustrated in Figure 9.5, a dashpot is a simple piston-cylinder or a syringe type of arrangement. A force applied on the piston will advance the piston in the direction of the applied force. The speed of the piston is dependent upon the magnitude of the applied force and the friction occurring between the contact surfaces of the piston and cylinder. For a linear dashpot, the applied force and speed (rate of displacement) are linearly proportional, the *coefficient of friction* μ (mu) being the constant of proportionality. If the applied force and the displacement are both in the x direction, then:

$$F = \mu \, \dot{x} \qquad (9.5)$$

In Eq. (9.5), $\dot{x} = dx/dt$ is the time rate of change of displacement or the speed.

By comparing Eqs. (9.4) and (9.5), an analogy can be made between linearly viscous fluids and linear dashpots. The stress and the strain rate for a linearly viscous fluid are respectively analogous to the force and the displacement rate for a dashpot, and the coefficient of viscosity is analogous to the coefficient of viscous friction for a dashpot. These analogies suggest that dashpots can be used to represent fluid behavior.

9.3 Empirical Models of Viscoelasticity

Springs and dashpots constitute the building blocks of model analyses in viscoelasticity. Springs and dashpots connected to one another in various forms are used to construct empirical viscoelastic models. Springs are used to account for the elastic solid behavior and dashpots are used to describe the viscous fluid behavior (Figure 9.6). It is assumed that a constantly applied force (stress) produces a constant deformation (strain) in a spring and a constant rate of deformation (strain rate) in a dashpot. The deformation in a spring is completely recoverable upon release of applied forces, whereas the deformation that the dashpot undergoes is permanent.

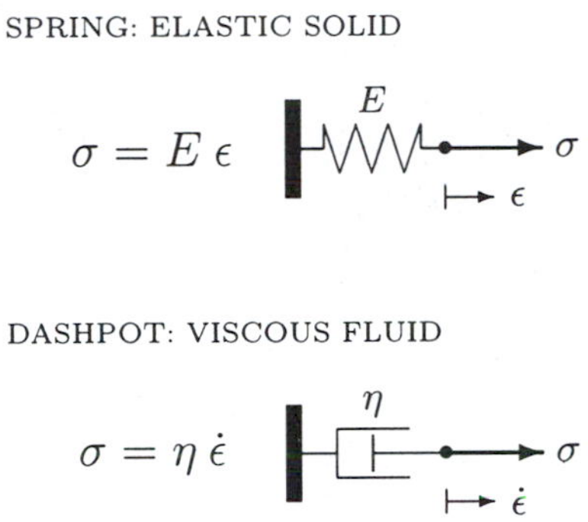

9.3.1 Kelvin–Voight model

The simplest forms of empirical models are obtained by connecting a spring and a dashpot together in parallel and in series configurations. As illustrated in Figure 9.7, the *Kelvin–Voight model* is a system consisting of a spring and a dashpot connected in a parallel arrangement. If subscripts "s" and "d" denote the spring and dashpot, respectively, then a stress σ applied to the entire system will produce stresses σ_s and σ_d in the spring and the dashpot. The total stress applied to the system will be shared by the spring and the dashpot such that:

$$\sigma = \sigma_s + \sigma_d \qquad (9.6)$$

As the stress σ is applied, the spring and dashpot will deform by an equal amount because of their parallel arrangement. Therefore, the strain ϵ of the system will be equal to the strains ϵ_s and ϵ_d occurring in the spring and the dashpot:

$$\epsilon = \epsilon_s = \epsilon_d \qquad (9.7)$$

The stress–strain relationship for the spring and the stress–strain rate relationship for the dashpot are:

$$\sigma_s = E \, \epsilon_s \qquad (9.8)$$
$$\sigma_d = \eta \, \dot{\epsilon}_d \qquad (9.9)$$

SPRING: ELASTIC SOLID

$$\sigma = E \, \epsilon$$

DASHPOT: VISCOUS FLUID

$$\sigma = \eta \, \dot{\epsilon}$$

Figure 9.6 *Spring represents elastic and dashpot represents viscous material behaviors.*

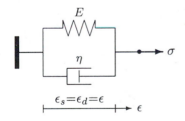

Figure 9.7 *Kelvin–Voight model.*

Substituting Eqs. (9.8) and (9.9) into Eq. (9.6) will yield:

$$\sigma = E\,\epsilon_s + \eta\,\dot{\epsilon}_d \qquad (9.10)$$

From Eq. (9.7), $\epsilon_s = \epsilon_d = \epsilon$. Therefore:

$$\sigma = E\,\epsilon + \eta\,\dot{\epsilon} \qquad (9.11)$$

Note that the strain rate $\dot{\epsilon}$ can alternatively be written as $d\epsilon/dt$. Consequently:

$$\sigma = E\,\epsilon + \eta\,\frac{d\epsilon}{dt} \qquad (9.12)$$

Equation (9.12) relates stress to strain and the strain rate for the Kelvin–Voight model, which is a two-parameter (E and η) viscoelastic model. Equation (9.12) is an *ordinary differential equation*. More specifically, it is a first-order, linear ordinary differential equation. For a given stress σ, Eq. (9.12) can be solved for the corresponding strain ϵ. For prescribed strain ϵ, it can be solved for stress σ.

Note that the review of how to handle ordinary differential equations is beyond the scope of this text. The interested reader is encouraged to review textbooks in "differential equations."

9.3.2 Maxwell model

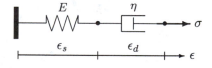

Figure 9.8 *Maxwell model.*

As shown in Figure 9.8, the *Maxwell model* is constructed by connecting a spring and a dashpot in a series. In this case, a stress σ applied to the entire system is applied equally on the spring and the dashpot ($\sigma = \sigma_s = \sigma_d$), and the resulting strain ϵ is the sum of the strains in the spring and the dashpot ($\epsilon = \epsilon_s + \epsilon_d$). Through stress–strain analyses similar to those carried out for the Kelvin–Voight model, a differential equation relating stresses and strains for the Maxwell model can be derived in the following form:

$$\eta\,\dot{\sigma} + E\,\sigma = E\,\eta\,\dot{\epsilon} \qquad (9.13)$$

This is also a first-order, linear ordinary differential equation representing a two-parameter (E and η) viscoelastic behavior. For a given stress (or strain), Eq. (9.13) can be solved for the corresponding strain (or stress).

Note that springs are used to represent the elastic solid behavior, and there is a limit to how much a spring can deform. On the other hand, dashpots are used to represent fluid behavior and are assumed to deform continuously (flow) as long as there is a force to deform them. For example, in the case of a Maxwell model, a force applied will cause both the spring and the dashpot to deform. The deformation of the spring will be finite. The dashpot will keep deforming as long as the force is maintained. Therefore, the overall behavior of the Maxwell model is more like a fluid than a solid, and is known to be a *viscoelastic fluid* model. The deformation of a dashpot connected in parallel to a spring, as in the Kelvin–Voight model, is restricted by the response of the spring to the applied loads. The dashpot in the Kelvin–Voight model cannot undergo continuous deformations. Therefore, the Kelvin–Voight model represents a *viscoelastic solid* behavior.

9.3.3 Standard solid model

The Kelvin–Voight solid and Maxwell fluid are the basic viscoelastic models constructed by connecting a spring and a dashpot together. They do not represent any known real material. However, in addition to springs and dashpots, they can be used to construct more complex viscoelastic models, such as the standard solid model. As illustrated in Figure 9.9, the *standard solid model* is composed of a spring and a Kelvin–Voight solid connected in a series. The standard solid model is a three-parameter (E_1, E_2, and η) model and is used to describe the viscoelastic behavior of a number of biological materials such as the cartilage and the white blood cell membrane. The material function relating the stress, strain, and their rates for this model is:

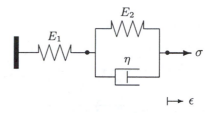

Figure 9.9 *Standard solid model.*

$$(E_1 + E_2)\,\sigma + \eta\,\dot{\sigma} = E_1 E_2\,\epsilon + E_1 \eta\,\dot{\epsilon} \qquad (9.14)$$

In Eq. (9.14), $\dot{\sigma} = d\sigma/dt$ is the stress rate and $\dot{\epsilon} = d\epsilon/dt$ is the strain rate. This equation can be derived as follows. As illustrated in Figure 9.10, the model can be represented by two units, A and B, connected in a series such that unit A is an elastic solid and unit B is a Kelvin–Voight solid. If σ_A and ϵ_A represent stress and strain in unit A, and σ_B and ϵ_B are stress and strain in unit B, then:

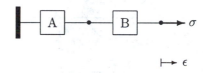

Figure 9.10 *Standard solid model is represented by units A and B.*

$$\sigma_A = E_1\,\epsilon_A \qquad (i)$$

$$\sigma_B = E_2\,\epsilon_B + \eta\,\frac{d\epsilon_B}{dt} = \left(E_2 + \eta\,\frac{d}{dt}\right)\epsilon_B \qquad (ii)$$

Since units A and B are connected in a series:

$$\epsilon_A + \epsilon_B = \epsilon \qquad (iii)$$

$$\sigma_A = \sigma_B = \sigma \qquad (iv)$$

Substitute Eq. (*iv*) into Eqs. (*i*) and (*ii*) and express them in terms of strains ϵ_A and ϵ_B:

$$\epsilon_A = \frac{\sigma}{E_1} \qquad (v)$$

$$\epsilon_B = \frac{\sigma}{E_2 + \eta\,\frac{d}{dt}} \qquad (vi)$$

Substitute Eqs. (*v*) and (*vi*) into Eq. (*iii*):

$$\frac{\sigma}{E_1} + \frac{\sigma}{E_2 + \eta\,\frac{d}{dt}} = \epsilon$$

Employ cross multiplication and rearrange the order of terms to obtain:

$$(E_1 + E_2)\,\sigma + \eta\,\frac{d\sigma}{dt} = E_1 E_2\,\epsilon + E_1\eta\,\frac{d\epsilon}{dt}$$

9.4 Time-Dependent Material Response

An empirical model for a given viscoelastic material can be established through a series of experiments. There are several experimental techniques designed to analyze the time-dependent aspects of material behavior. As illustrated in Figure 9.11a, a *creep and recovery (recoil)* test is conducted by applying a load (stress σ_o) on the material at time t_0, maintaining the load at a constant level until time t_1, suddenly removing the load at t_1, and observing the material response. As illustrated in Figure 9.11b, the *stress relaxation* experiment is done by straining the material to a level ϵ_0 and maintaining the constant strain while observing the stress response of the material. In an *oscillatory response* test, a harmonic stress is applied and the strain response of the material is measured (Figure 9.11c).

Consider a viscoelastic material. Assume that the material is subjected to a creep test. The results of the creep test can be represented by plotting the measured strain as a function of time. An empirical viscoelastic model for the material behavior can be established through a series of trials. For this purpose, an empirical model is constructed by connecting a number of springs and dashpots together. A differential equation relating stress, strain, and their rates is derived through the procedure outlined in Section 9.3 for the Kelvin–Voight model. The imposed condition in a creep test is $\sigma = \sigma_0$. This condition of constant stress is substituted into the differential equation, which is then solved (integrated) for strain ϵ. The result obtained is another equation relating strain to stress constant σ_0, the elastic moduli and coefficients of viscosity of the empirical model, and time. For a given σ_0 and assigned elastic and viscous moduli, this equation is reduced to a function relating strain to time. This function is then used to plot a strain versus time graph and is compared to the experimentally obtained graph. If the general characteristics of the two (experimental and analytical) curves match, the analyses are furthered to establish the elastic and viscous moduli (material constants) of the material. This is achieved by varying the values of the elastic and viscous moduli in the empirical model until the analytical curve matches the experimental curve as closely as possible. In general, this procedure is called *curve fitting*. If there is no general match between the two curves, the model is abandoned and a new model is constructed and checked.

The result of these mathematical model analyses is an empirical model and a differential equation relating stresses and strains.

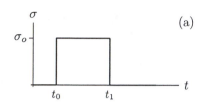

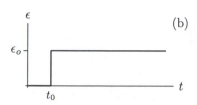

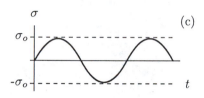

Figure 9.11 *(a) Creep and recovery, (b) stress relaxation, and (c) oscillatory response tests.*

The stress–strain relationship for the material can be used in conjunction with the fundamental laws of mechanics to analyze the response of the material to different loading conditions.

Note that the deformation processes occurring in viscoelastic materials are quite complex, and it is sometimes necessary to use an array of empirical models to describe the response of a viscoelastic material to different loading conditions. For example, the shear response of a viscoelastic material may be explained with one model and a different model may be needed to explain its response to normal loading. Different models may also be needed to describe the response of a viscoelastic material at low and high strain rates.

9.5 Comparison of Elasticity and Viscoelasticity

There are various criteria with which the elastic and viscoelastic behavior of materials can be compared. Some of these criteria will be discussed in this section.

An elastic material has a unique stress–strain relationship that is independent of the time or strain rate. For elastic materials, normal and shear stresses can be expressed as functions of normal and shear strains:

$$\sigma = \sigma(\epsilon) \quad \text{and} \quad \tau = \tau(\gamma)$$

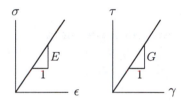

Figure 9.12 *An elastic material has unique normal and shear stress–strain diagrams.*

For example, the stress–strain relationships for a linearly elastic solid are $\sigma = E\epsilon$ and $\tau = G\gamma$, where E and G are constant elastic moduli of the material. As illustrated in Figure 9.12, a linearly elastic material has a unique normal stress–strain diagram and a unique shear stress–strain diagram.

Viscoelastic materials exhibit time-dependent material behavior. The response of a viscoelastic material to an applied stress not only depends upon the magnitude of the stress but also on how fast the stress is applied to or removed from the material. Therefore, the stress–strain relationship for a viscoelastic material is not unique but is a function of the time or the rate at which the stresses and strains are developed in the material:

$$\sigma = \sigma(\epsilon, \dot{\epsilon}, \ldots, t) \quad \text{and} \quad \tau = \tau(\gamma, \dot{\gamma}, \ldots, t)$$

Consequently, as illustrated in Figure 9.13, a viscoelastic material does not have a unique stress–strain diagram.

For an elastic body, the energy supplied to deform the body (strain energy) is stored in the body as potential energy. This energy is available to return the body to its original (unstressed) size and shape once the applied stress is removed. As illustrated in Figure 9.14, the loading and unloading paths for an elastic material coincide. This indicates that there is no loss of energy during loading and unloading.

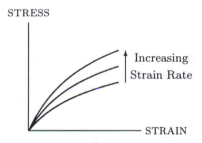

Figure 9.13 *Stress–strain diagram for a viscoelastic material may not be unique.*

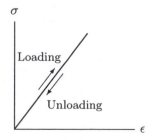

Figure 9.14 *For an elastic material, loading and unloading paths coincide.*

For a viscoelastic body, some of the strain energy is stored in the body as potential energy and some of it is dissipated as heat. For example, consider the Maxwell model. The energy provided to stretch the spring is stored in the spring while the energy supplied to deform the dashpot is dissipated as heat due to the friction between the moving parts of the dashpot. Once the applied load is removed, the potential energy stored in the spring is available to recover the deformation of the spring, but there is no energy available in the dashpot to regain its original configuration.

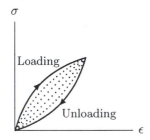

Figure 9.15 *Hysteresis loop.*

Consider the three-parameter standard solid model shown in Figure 9.9. A typical loading and unloading diagram for this model is shown in Figure 9.15. The area enclosed by the loading and unloading paths is called the *hysteresis loop*, which represents the energy dissipated as heat during the deformation and recovery phases. This area, and consequently the amount of energy dissipated as heat, is dependent upon the rate of strain employed to deform the body. The presence of the hysteresis loop in the stress–strain diagram for a viscoelastic material indicates that continuous loading and unloading would result in an increase in the temperature of the material.

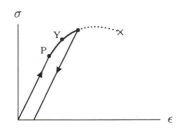

Figure 9.16 *Hysteresis loop for an elastic–plastic material.*

Note here that most of the elastic materials exhibit plastic behavior at stress levels beyond the yield point. For elastic–plastic materials, some of the strain energy is dissipated as heat during plastic deformations. This is indicated with the presence of a hysteresis loop in their loading and unloading diagrams (Figure 9.16). For such materials, energy is dissipated as heat only if the plastic region is entered. Viscoelastic materials dissipate energy regardless of whether the strains or stresses are small or large.

Since viscoelastic materials exhibit time-dependent material behavior, the differences between elastic and viscoelastic material responses are most evident under time-dependent loading conditions, such as during the creep and stress relaxation experiments.

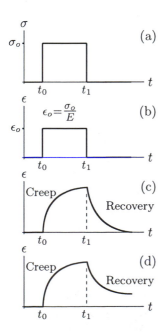

Figure 9.17 *Creep and recovery.*

As discussed earlier, a creep and recovery test is conducted by observing the response of a material to a constant stress σ_o applied at time t_0 and removed at a later time t_1 (Figure 9.17a). As illustrated in Figure 9.17b, such a load will cause a strain $\epsilon_o = \sigma_o/E$ in a linearly elastic material instantly at time t_0. This constant strain will remain in the material until time t_1. At time t_1, the material will instantly and completely recover the deformation. To the same constant loading condition, a viscoelastic material will respond with a strain gradually increasing between times t_0 and t_1. At time t_1, gradual recovery will start. For a viscoelastic solid material, the recovery will eventually be complete (Figure 9.17c). For a viscoelastic fluid, complete recovery will never be achieved and there will be a residue of deformation left in the material (Figure 9.17d).

As illustrated in Figure 9.18a, the stress relaxation test is performed by straining a material instantaneously, maintaining the constant strain level ϵ_0 in the material, and observing the response of the material. A linearly elastic material response is illustrated in Figure 9.18b. The constant stress $\sigma_0 = E\epsilon_0$ developed in the material will remain as long as the strain ϵ_0 is maintained. In other words, an elastic material will not exhibit a stress relaxation behavior. A viscoelastic material, on the other hand, will respond with an initial high stress that will decrease over time. If the material is a viscoelastic solid, the stress level will never reduce to zero (Figure 9.18c). As illustrated in Figure 9.18d, the stress will eventually reduce to zero for a viscoelastic fluid.

Because of their time-dependent material behavior, viscoelastic materials are said to have a "memory." In other words, viscoelastic materials remember the history of deformations they undergo and react accordingly.

Almost all biological materials exhibit viscoelastic properties, and the remainder of this chapter is devoted to the discussion and review of the mechanical properties of biological tissues including bone, tendons, ligaments, muscles, and articular cartilage.

9.6 Common Characteristics of Biological Tissues

One of the objectives of studies in the field of biomechanics is to establish the mechanical properties of biological tissues so as to develop mathematical models that help us describe and further investigate their behavior under various loading conditions. While conducting studies in biomechanics, it has been a common practice to utilize engineering methods and principles, and at the same time to treat biological tissues like engineering materials. However, living tissues have characteristics that are very different than engineering materials. For example, living tissues can be self-adapting and self-repairing. That is, they can adapt to changing mechanical demand by altering their mechanical properties, and they can repair themselves. The mechanical properties of living tissues tend to change with age. Most biological tissues are composite materials (consisting of materials with different properties) with nonhomogeneous and anisotropic properties. In other words, the mechanical properties of living tissues may vary from point to point within the tissue, and their response to forces applied in different directions may be different. For example, values for strength and stiffness of bone may vary between different bones and at different points within the same bone. Furthermore, almost all biological tissues are viscoelastic in nature. Therefore, the strain or loading rate at which a specific test is conducted must also be provided while reporting the results of the strength measurements. These considerations require that most of the mechanical

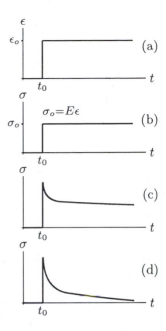

Figure 9.18 *Stress relaxation.*

properties reported for living tissues are only approximations, and a mathematical model aimed to describe the behavior of a living tissue is usually limited to describing its response under a specific loading configuration.

From a mechanical point of view, all tissues are composite materials. Among the common components of biological tissues, collagen and elastin fibers have the most important mechanical properties affecting the overall mechanical behavior of the tissues in which they appear. *Collagen* is a protein made of crimped fibrils that aggregate into fibers. The mechanical properties of collagen fibrils are such that each fibril can be considered a mechanical spring and each fiber as an assemblage of springs. The primary mechanical function of collagen fibers is to withstand axial tension. Because of their high length-to-diameter ratios (aspect ratio), collagen fibers are not effective under compressive loads. Whenever a fiber is pulled, its crimp straightens, and its length increases. Like a mechanical spring, the energy supplied to stretch the fiber is stored, and it is the release of this energy that returns the fiber to its unstretched configuration when the applied load is removed. The individual fibrils of the collagen fibers are surrounded by a gel-like *ground substance* that consists largely of water. Collagen fibers possess a two-phase, solid–fluid, or viscoelastic material behavior with a relatively high tensile strength and poor resistance to compression.

The geometric configuration of collagen fibers and their interaction with the noncollagenous tissue components form the basis of the mechanical properties of biological tissues. Among the noncollagenous tissue components, *elastin* is another fibrous protein with material properties that resemble the properties of rubber. Elastin and microfibrils form elastic fibers that are highly extensible, and their extension is reversible even at high strains. Elastin fibers behave elastically with low stiffness up to about 200% elongation followed by a short region where the stiffness increases sharply until failure (Figure 9.19). The elastin fibers do not exhibit considerable plastic deformation before failure, and their loading and unloading paths do not show significant hysteresis. In summary, elastin fibers possess a low-modulus elastic material property, while collagen fibers show a higher-modulus viscoelastic material behavior.

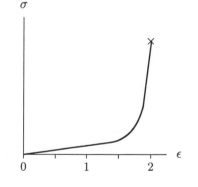

Figure 9.19 *Stress–strain diagram for elastin.*

9.7 Biomechanics of Bone

Bone is the primary structural element of the human body. Bones form the building blocks of the skeletal system that protects the internal organs, provides kinematic links, provides muscle attachment sites, and facilitates muscle actions and body movements. Bone has unique structural and mechanical properties that allow it to carry out these functions. As compared to other structural materials, bone is also unique in that it is self-repairing. Bone can also alter its shape, mechanical behavior, and mechanical properties to adapt to the changes in mechanical demand. The major factors that influence the mechanical behavior of bone are: the composition of bone; the mechanical properties of the tissues comprising the bone; the size and geometry of the bone; and the direction, magnitude, and rate of applied loads.

9.7.1 Composition of bone

In biological terms, bone is a *connective tissue* that binds together various structural elements of the body. In mechanical terms, bone is a *composite material* with various solid and fluid phases. Bone consists of cells and an organic mineral matrix of fibers and a ground substance surrounding collagen fibers. Bone also contains inorganic substances in the form of mineral salts. The inorganic component of bone makes it hard and relatively rigid, and its organic component provides flexibility and resilience. The composition of bone varies with species, age, sex, type of bone, type of bone tissue, and the presence of bone disease.

At the macroscopic level, all bones consist of two types of tissues (Figure 9.20). The *cortical* or *compact* bone tissue is a dense material forming the outer shell (cortex) of bones and the diaphysial region of long bones. The *cancellous, trabecular,* or *spongy* bone tissue consists of thin plates (trabeculae) in a loose mesh structure that is enclosed by the cortical bone. Bones are surrounded by a dense fibrous membrane called the *periosteum*. The periosteum covers the entire bone except for the joint surfaces that are covered with articular cartilage.

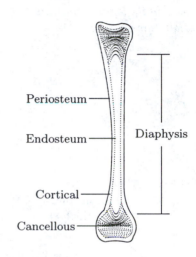

Figure 9.20 *Sectional view of a whole bone showing cortical and cancellous tissues.*

9.7.2 Mechanical properties of bone

Bone is a nonhomogeneous material because it consists of various cells and organic and inorganic substances with different material properties. Bone is an anisotropic material because its mechanical properties are different in different directions. That is, the mechanical response of bone is dependent upon the direction as well as the magnitude of the applied load. For example, the compressive strength of bone is greater than its tensile strength. Bone possesses viscoelastic (time-dependent) material properties. The mechanical response of bone is dependent on the rate at which the loads are applied. Bone can resist rapidly applied loads much better than slowly applied loads. In other words, bone is stiffer and stronger at higher strain rates.

Bone is a complex structural material. The mechanical response of bone can be observed by subjecting it to tension, compression, bending, and torsion. Various tests to implement these conditions were discussed in the previous chapters. These tests can be

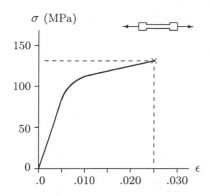

Figure 9.21 *Tensile stress–strain diagram for human cortical bone loaded in the longitudinal direction (strain rate $\dot{\epsilon} = 0.05\ s^{-1}$).*

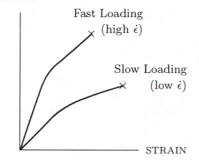

Figure 9.22 *The strain rate dependent stress–strain curves for cortical bone tissue.*

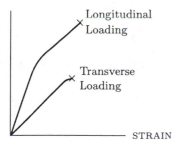

Figure 9.23 *The direction dependent stress–strain curves for bone tissue.*

performed using uniform bone specimens or whole bones. If the purpose is to investigate the mechanical response of a specific bone tissue (cortical or cancellous), then the tests are performed using bone specimens. Testing a whole bone, on the other hand, attempts to determine the "bulk" properties of that bone.

The tensile stress–strain diagram for the cortical bone is shown in Figure 9.21. This σ-ϵ curve is drawn using the averages of the elastic modulus, strain hardening modulus, ultimate stress, and ultimate strain values determined for the human femoral cortical bone tested under tensile and compressive loads applied in the longitudinal direction at a moderate strain rate ($\dot{\epsilon} = 0.05\ s^{-1}$). The σ-ϵ curve in Figure 9.21 has three distinct regions. In the initial linearly elastic region, the σ-ϵ curve is nearly a straight line and the slope of this line is equal to the elastic modulus (E) of the bone, which is about 17 GPa. In the intermediate region, the bone exhibits nonlinear elastoplastic material behavior. Material yielding also occurs in this region. By the offset method discussed in Chapter 7, the yield strength of the cortical bone for the σ-ϵ diagram shown in Figure 9.21 can be determined to be about 110 MPa. In the final region, the bone exhibits a linearly plastic material behavior and the σ-ϵ diagram is another straight line. The slope of this line is the strain hardening modulus (E') of bone tissue, which is about 0.9 GPa. The bone fractures when the tensile stress is about 128 MPa, for which the tensile strain is about 0.026.

The elastic moduli and strength values for bone are dependent upon many factors including the test conditions, such as the rate at which the loads are applied. This viscoelastic nature of bone tissue is demonstrated in Figure 9.22. The stress–strain diagrams in Figure 9.22 for different strain rates indicate that a specimen of bone tissue that is subjected to rapid loading (high $\dot{\epsilon}$) has a greater elastic modulus and ultimate strength than a specimen that is loaded more slowly (low $\dot{\epsilon}$). Figure 9.22 also demonstrates that the energy absorbed (which is proportional to the area under the σ-ϵ curve) by the bone tissue increases with an increasing strain rate. Note that during normal daily activities, bone tissues are subjected to a strain rate of about 0.01 s^{-1}.

The stress–strain behavior of bone is also dependent upon the orientation of bone with respect to the direction of loading. This anisotropic material behavior of bone is demonstrated in Figure 9.23. Note that the cortical bone has a larger ultimate strength (stronger) and a larger elastic modulus (stiffer) in the longitudinal direction than the transverse direction. Furthermore, bone specimens loaded in the transverse direction fail in a more brittle manner (without showing considerable yielding) as compared to bone specimens loaded in the longitudinal direction. The ultimate strength values for adult femoral cortical bone under various modes of loading, and its elastic and shear moduli are listed in Table 9.1. The ultimate strength values in Table 9.1 demonstrate that the bone strength is highest under compressive loading in the longitudinal direction (the direction of osteon orientation) and lowest under tensile loading in the transverse direction (the direction perpendicular to the longitudinal direction). The elastic modulus of cortical bone in the longitudinal direction is higher than its elastic modulus in the transverse direction. Therefore, cortical bone is stiffer in the longitudinal direction than in the transverse direction.

It should be noted that there is a wide range of variation in values reported for the mechanical properties of bone. It may be useful to remember that the tensile strength of bone is less than 10 percent of that of stainless steel. Also, the stiffness of bone is about 5 percent of the stiffness of steel. In other words, for specimens of the same dimension and under the same tensile load, a bone specimen will deform 20 times as much as the steel specimen.

The chemical compositions of cortical and cancellous bone tissues are similar. The distinguishing characteristic of the cancellous bone is its porosity. This physical difference between the two bone tissues is quantified in terms of the *apparent density* of bone, which is defined as the mass of bone tissue present in a unit volume of bone. To a certain degree, both cortical and cancellous bone tissues can be regarded as a single material of variable density. The material properties such as strength and stiffness, and the stress–strain characteristics of cancellous bone depend not only on the apparent density that may be different for different bone types or at different parts of a single bone, but also on the mode of loading. The compressive stress–strain curves (Figure 9.24) of cancellous bone contain an initial linearly elastic region up to a strain of about 0.05. The material yielding occurs as the trabeculae begin to fracture. This initial elastic region is followed by a plateau region of almost constant stress until fracture, exhibiting a ductile material behavior. By contrast to compact bone, cancellous bone fractures abruptly under tensile forces, showing a brittle material behavior. Cancellous bone is about 25 to 30 percent as dense, 5 to 10 percent as stiff, and 5 times as ductile as cortical bone. The energy absorption capacity of cancellous bone is considerably higher under compressive loads than under tensile loads.

9.7.3 Structural integrity of bone

There are several factors that may affect the structural integrity of bones. For example, the size and geometry of a bone determine the distribution of the internal forces throughout the bone, thereby influencing its response to externally applied loads. The larger the bone, the larger the area upon which the internal forces are distributed and the smaller the intensity (stress) of these

Table 9.1 *Ultimate strength, and elastic and shear moduli for human femoral cortical bone.* (1 GPa = 10^9Pa, 1 MPa = 10^6Pa)

LOADING MODE	ULTIMATE STRENGTH
LONGITUDINAL	
Tension	133 MPa
Compression	193 MPa
Shear	68 MPa
TRANSVERSE	
Tension	51 MPa
Compression	133 MPa
ELASTIC MODULI, E	
Longitudinal	17.0 GPa
Transverse	11.5 GPa
SHEAR MODULUS, G	3.3 GPa

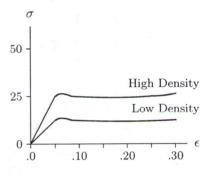

Figure 9.24 *Apparent density dependent stress–strain curves for cancellous bone tissue.*

forces. Consequently, the larger the bone, the more resistant the bone to applied loads.

A common characteristic of long bones is their tubular structure in the diaphysial region, which has considerable mechanical advantage over solid circular structures of the same mass. Recall from the previous chapter that the shear stresses in a structure subjected to torsion are inversely proportional with the polar moment of inertia (J) of the cross-sectional area of the structure, and the normal stresses in a structure subjected to bending are inversely proportional to the area moment of inertia (I) of the cross-section of the structure. The larger the polar and area moments of inertia of a structure, the lower the maximum normal stresses due to torsion and bending. Since tubular structures have larger polar and area moments of inertia as compared to solid cylindrical structures of the same volume, tubular structures are more resistant to torsional and bending loads as compared to solid cylindrical structures. Furthermore, a tubular structure can distribute the internal forces more evenly over its cross-section as compared to a solid cylindrical structure of the same cross-sectional area.

Certain skeletal conditions such as osteoporosis can reduce the structural integrity of bone by reducing its apparent density. Small decreases in bone density can generate large reductions in bone strength and stiffness. As compared to a normal bone with the same geometry, an osteoporotic bone will deform easier and fracture at lower loads. The density of bone can also change with aging, after periods of disuse, or after chronic exercise, thereby changing its overall strength. Certain surgical procedures that alter the normal bone geometry may also reduce the strength of bone. Bone defects such as screw holes reduce the load-bearing ability of bone by causing stress concentrations around the defects.

Bone becomes stiffer and less ductile with age. Also with age, the ability of bone to absorb energy and the maximum strain at failure are reduced, and the bone behaves more like dry bone. Although the properties of dry bone may not have any value in orthopaedics, it may be important to note that there are differences between bone in its wet and dry states. Dry bone is stiffer, has a higher ultimate strength, and is more brittle than wet bone (Figure 9.25).

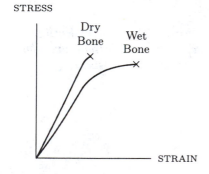

Figure 9.25 *Stress–strain curves for dry and wet bones.*

9.7.4 Bone fractures

When bones are subjected to moderate loading conditions, they respond by small deformations that are only present while the loads are applied. When the loads are removed, bones exhibit elastic material behavior by resuming their original (unstressed) shapes and positions. Large deformations occur when the applied loads are high. Bone fractures when the stresses generated in any region of bone are larger than the ultimate strength of bone.

Fractures caused by pure tensile forces are observed in bones with a large proportion of cancellous bone tissue. Fractures due to compressive loads are commonly encountered in the vertebrae of the elderly, whose bones are weakened as a result of aging. Bone fractures caused by compression occur in the diaphysial regions of long bones. Compressive fractures are identified by their oblique fracture pattern. Long bone fractures are usually caused by torsion and bending. Torsional fractures are identified by their spiral oblique pattern, whereas bending fractures are usually identified by the formation of "butterfly" fragments. Fatigue fracture of bone occurs when the damage caused by repeated mechanical stress outpaces the bone's ability to repair to prevent failure. Bone fractures caused by fatigue are common among professional athletes and dedicated joggers. Clinically, most bone fractures occur as a result of complex, combined loading situations rather than simple loading mechanisms.

9.8 Biomechanics of Tendons and Ligaments

Tendons and ligaments are fibrous connective tissues. Tendons help execute joint motion by transmitting mechanical forces (tensions) from muscles to bones. Ligaments join bones and provide stability to the joints. Unlike muscles, which are active tissues and can produce mechanical forces, tendons and ligaments are passive tissues and cannot actively contract to generate forces.

Around many joints of the human body, there is insufficient space to attach more than one or a few muscles. This requires that to accomplish a certain task, one or a few muscles must share the burden of generating and withstanding large loads with intensities (stress) even larger at regions closer to the bone attachments where the cross-sectional areas of the muscles are small. As compared to muscles, tendons are stiffer, have higher tensile strengths, and can endure larger stresses. Therefore, around the joints where the space is limited, muscle attachments to bones are made by tendons. Tendons are capable of supporting very large loads with very small deformations. This property of tendons enables the muscles to transmit forces to bones without wasting energy to stretch tendons.

The mechanical properties of tendons and ligaments depend upon their composition, which can vary considerably. The most common means of evaluating the mechanical response of tendons and ligaments is the uniaxial tension test. Figure 9.26 shows a typical tensile stress–strain diagram for tendons. The shape of this curve is the result of the interaction between elastic

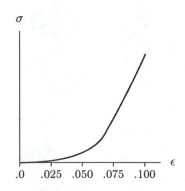

Figure 9.26 *Tensile stress–strain diagram for tendon.*

elastin fibers and the viscoelastic collagen fibers. At low strains (up to about 0.05), less stiff elastic fibers dominate and the crimp of the collagen fibers straightens, requiring very little force to stretch the tendon. The tendon becomes stiffer when the crimp is straightened. At the same time, the fluid-like ground substance in the collagen fibers tends to flow. At higher strains, therefore, the stiff and viscoelastic nature of the collagen fibers begins to take an increasing portion of the applied load. Tendons are believed to function in the body at strains of up to about 0.04, which is believed to be their yield strain (ϵ_y). Tendons rupture at strains of about 0.1 (ultimate strain, ϵ_u), or stresses of about 60 MPa (ultimate stress, σ_u).

Note that the shape of the stress–strain curve in Figure 9.26 is such that the area under the curve is considerably small. In other words, the energy stored in the tendon to stretch the tendon to a stress level is much smaller than the energy stored to stretch a linearly elastic material (with a stress–strain diagram that is a straight line) to the same stress level. Therefore, the tendon has higher resilience than linearly elastic materials.

The time-dependent viscoelastic nature of the tendon is illustrated in Figures 9.27 and 9.28. When the tendon is stretched rapidly, there is less chance for the ground substance to flow, and, consequently, the tendon becomes stiffer. The hysteresis loop shown in Figure 9.28 demonstrates the time-dependent loading and unloading behavior of the tendon. Note that more work is done in stretching the tendon than is recovered when the tendon is allowed to relax, and, therefore, some of the energy is dissipated in the process.

The mechanical role of ligaments is to transmit forces from one bone to another. Ligaments also have a stabilizing role for the skeletal joints. The composition and structure of ligaments depend upon their function and position within the body. Like tendons they are composite materials containing crimped collagen fibers surrounded by ground substance. As compared to tendons, they often contain a greater proportion of elastic fibers that accounts for their higher extensibility but lower strength and stiffness. The mechanical properties of ligaments are qualitatively similar to those of tendons. Like tendons, they are viscoelastic and exhibit hysteresis, but deform elastically up to strains of about $\epsilon_y = 0.25$ (about five times as much as the yield strain of tendons) and stresses of about $\sigma_y = 5$ MPa. They rupture at a stress of about 20 MPa.

Since tendons and ligaments are viscoelastic, some of the energy supplied to stretch them is dissipated by causing the flow of the fluid within the ground substance, and the rest of the energy is stored in the stretched tissue. Tendons and ligaments are tough materials and do not rupture easily. Most common damage to tendons and ligaments occurs at their junctions with bones.

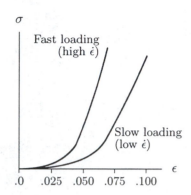

Figure 9.27 *The strain rate dependent stress–strain curves for tendon.*

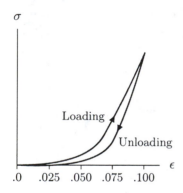

Figure 9.28 *The hysteresis loop of stretching and relaxing modes of the tendon.*

9.9 Biomechanics of Skeletal Muscles

There are three types of muscles: skeletal, smooth, and cardiac. Smooth muscles line the internal organs, and cardiac muscles form the heart. Here, we are concerned with the characteristics of the skeletal muscles, each of which is attached, via aponeuroses and/or tendons, to at least two bones causing and/or controlling the relative movement of one bone with respect to the other. When its fibers contract under the stimulation of a nerve, the muscle exerts a pulling effect on the bones to which it is attached. *Contraction* is a unique ability of the muscle tissue, which is defined as the development of tension in the muscle. Muscle contraction can occur as a result of muscle shortening (concentric contraction) or muscle lengthening (eccentric contraction), or it can occur without any change in the muscle length (static or isometric contraction).

The skeletal muscle is composed of muscle fibers and myofibrils. Myofibrils in turn are made of contractile elements: *actin* and *myosin* proteins. Actin and myosin appear in bands or filaments. Several relatively thick myosin filaments interact across cross-bridges with relatively thin actin filaments to form the basic structure of the contractile element of the muscle, called the *sarcomere* (Figure 9.29). Many sarcomere elements connected in a series arrangement form the contractile element (motor unit) of the muscle. It is within the sarcomere that the muscle force (tension) is generated, and where muscle shortening and lengthening take place. The active contractile elements of the muscle are contained within a fibrous passive connective tissue, called *fascia*. Fascia encloses the muscles, separates them into layers, and connects them to tendons.

The force and torque developed by a muscle is dependent on many factors, including the number of motor units within the muscle, the number of motor units recruited, the manner in which the muscle changes its length, the velocity of muscle contraction, and the length of the lever arm of the muscle force. For muscles, two different forces can be distinguished. *Active tension* is the force produced by the contractile elements of the muscle and is a result of voluntary muscle contraction. *Passive tension*, on the other hand, is the force developed within the connective muscle tissue when the muscle length surpasses its resting length. The net tensile force in a muscle is dependent on the force–length characteristics of both the active and passive components of the muscle. A typical tension versus muscle length diagram is shown in Figure 9.30. The number of cross-bridges between the filaments is maximum, and, therefore, the active tension (T_a) is maximum at the resting length (ℓ_o) of the muscle. As the muscle lengthens, the filaments are pulled apart, the number of cross-bridges is reduced, and the active tension is decreased. At full length, there are no

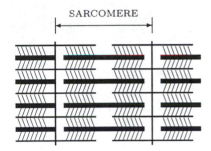

Figure 9.29 *Basic structure of the contractile element of muscle. (Thick lines represent myosin filaments, thin horizontal lines are actin filaments, and cross-hatched lines are cross-bridges.)*

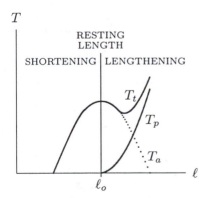

Figure 9.30 *Muscle force (T) versus muscle length (ℓ).*

cross-bridges and the active tension reduces to zero. As the muscle shortens, the cross-bridges overlap and the active tension is again reduced. When the muscle is at its resting length or less, the passive (connective) component of the muscle is in a loose state with no tension. As the muscle lengthens, a passive tensile force (T_p) starts building up in the connective tissues. The force–length characteristic of this passive component resembles that of a nonlinear spring. Passive tensile force increases at an increasing rate as the length of the muscle increases. The overall, total, or net muscle force (T_t) that is transmitted via tendons is the sum of the forces in the active and passive elements of the muscle. Note here that for a given muscle, the tension–length diagram is not unique but dependent on the number of motor units recruited. The magnitude of the active component of the muscle force can vary depending on how the muscle is excited, and is usually expressed as the percentage of the maximum voluntary contraction.

The force generated by a contracting muscle is usually transmitted to a bone through a tendon. There is a functional reason for tendons to make the transfer of forces from muscles to bones. As compared to tendons, muscles have lower tensile strengths. The relatively low ultimate strength requires muscles to have relatively large cross-sectional areas in order to transmit sufficiently high forces without tearing. Tendons are better designed to perform this function.

9.10 Biomechanics of Articular Cartilage

Cartilage covers the articulating surfaces of bones at the diarthrodial (synovial) joints. The primary function of cartilage is to facilitate the relative movement of articulating bones. Cartilage reduces stresses applied to bones by increasing the area of contact between the articulating surfaces and reduces bone wear by reducing the effects of friction.

Cartilage is a two-phase material consisting of about 75% water and 25% organic solid. A large portion of the solid phase of the cartilage material is made up of collagen fibers. The remaining ground substance is mainly proteoglycan (hydrophilic molecules). Collagen fibers are relatively strong and stiff in tension, while proteoglycans are strong in compression. The solid–fluid composition of cartilage makes it a viscoelastic material.

The mechanical properties of cartilage under various loading conditions have been investigated using a number of different techniques. For example, the response of the human patella to compressive loads has been investigated by using an *indentation test* in which a small cylindrical or hemispherical indenter is pressed into the articulating surface, and the resulting deformation is recorded (Figure 9.31a). A typical result of an indentation

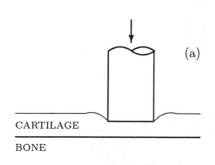

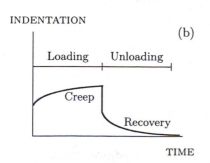

Figure 9.31 *Indentation test.*

test is shown in Figure 9.31b. When a constant magnitude load is applied, the material initially responds with a relatively large elastic deformation. The applied load causes pressure gradients to occur in the interstitial fluid, and the variations in pressure cause the fluid to flow through and out of the cartilage matrix. As the load is maintained, the amount of deformation increases at a decreasing rate. The deformation tends toward an equilibrium state as the pressure variations within the fluid are dissipated. When the applied load is removed (unloading phase), there is an instantaneous elastic recovery (recoil) that is followed by a more gradual recovery leading to complete recovery. This creep-recovery response of cartilage may be qualitatively represented by the three-parameter viscoelastic solid model (Figure 9.32), which consists of a linear spring and a Kelvin–Voight unit connected in series.

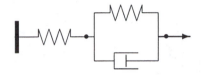

Figure 9.32 *The standard solid model has been used to represent the creep-recovery behavior of cartilage.*

Another experiment designed to investigate the response of cartilage to compressive loading conditions is the *confined compression test* illustrated in Figure 9.33. In this test, the specimen is confined in a rigid cylindrical die and loaded with a rigid permeable block. The compressive load causes pressure variations in the interstitial fluid and consequent fluid exudation. Eventually, the pressure variations dissipate and an equilibrium is reached. The state at which the equilibrium is reached is indicative of the compressive stiffness of the cartilage. The compressive stiffness and resistance of cartilage depend upon the water and proteoglycan content of the tissue. The higher the proteoglycan content, the higher the compressive resistance of the tissue.

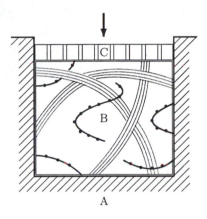

Figure 9.33 *Confined compression test. A is the rigid die, B is the specimen, and C is the permeable block.*

During daily activities, the articular cartilage is subjected to tensile and shear stresses as well as compressive stresses. Under tension, cartilage responds by realigning the collagen fibers that carry the tensile loads applied to the tissue. The tensile stiffness and strength of cartilage depend on the collagen content of the tissue. The higher the collagen content, the higher the tensile strength of cartilage. Shear stresses on the articular cartilage are due to the frictional forces between the relative movement of articulating surfaces. However, the coefficient of friction for synovial joints is so low (of the order 0.001–0.06) that friction has an insignificant effect on the stress resultants acting on the cartilage.

Both structural (such as intraarticular fracture) and anatomical abnormalities (such as rheumatoid arthritis and acetabular dysplasia) can cause cartilage damage, degeneration, wear, and failure. These abnormalities can change the load-bearing ability of the joint by altering its mechanical properties. The importance of the load-bearing ability of the cartilage and maintaining its mechanical integrity may become clear if we consider that the magnitude of the forces involved at the human hip joint is about five times body weight during ordinary walking (much higher during running or jumping). The hip contact area over which these forces are applied is about 15 cm^2 (0.0015 m^2). Therefore, the compressive stresses (pressures) involved are of the order 3 MPa for an 85 kg person.

9.11 Discussion

Here we have covered, very briefly, the mechanical properties of selected biological tissues. We believe that the knowledge of the mechanical properties and structural behavior of biological tissues is an essential prerequisite for any experimental or theoretical analysis of their physiological function in the body. We are aware of the fact that the proper coverage of each of these topics deserves at least a full chapter. Our purpose here was to provide a summary, to illustrate how biological phenomena can be described in terms of the mechanical concepts introduced earlier, and hope that the interested reader would refer to more complete sources of information to improve his or her knowledge of the subject matter.

9.12 Exercise Problems

Problem 9.1 Complete the following definitions with appropriate expressions.

(a) Elastic materials show time-independent material behavior. Elastic materials deform _____ when they are subjected to externally applied loads.

(b) Time-dependent material behavior is known as _____.

(c) Elasticity is a solid material behavior, whereas _____ is a fluid property and is a measure of resistance to flow.

(d) For a viscoelatic material, stress is not only a function of strain, but also a function of _____.

(e) _____ and _____ are basic mechanical elements that are used to simulate elastic solid and viscous fluid behaviors, respectively.

(f) The _____ is a viscoelastic model consisting of a spring and a dashpot connected in a parallel arrangement.

(g) The _____ is a viscoelastic model consisting of a spring and a dashpot connected in a series arrangement.

(h) The _____ is a viscoelastic model consisting of a spring and a Kelvin–Voight solid connected in a series.

(i) A _____ test is conducted by applying a load on the material, maintaining the load at a constant level for some time, suddenly removing the load, and observing the material response.

(j) A _____ test is conducted by straining the material at a level and maintaining the strain at a constant level while observing the stress response of the material.

(k) In a _____ test, a harmonic stress is applied on the material and the strain response of the material is observed.

(l) The area enclosed by the loading and unloading paths is called the _____, which represents the energy dissipated as heat.

(m) Because of their time-dependent behavior, viscoelastic materials are said to have a _____.

(n) Living tissues have characteristics that are very different than engineering materials. For example, they are _____ and _____.

(o) Among the common components of biological tissues, ____ and _____ fibers have the most important mechanical properties affecting the overall mechanical behavior of the tissues in which they appear.

(p) _____ is a protein made of crimped fibrils that aggregate into fibers.

(q) _____ is a fibrous protein with material properties that resemble the properties of rubber.

(r) In biological terms, bone is a _____ tissue that binds together various structural elements of the body. In mechanical terms, bone is a _____ material with various solid and fluid phases.

(s) The _____ bone tissue is a dense material forming the outer shell (cortex) of bones and the diaphysial region of long bones.

(t) The _____ bone tissue consists of thin plates (trabeculae) in a loose mesh structure that is enclosed by the cortical bone.

(u) Bone is stiffer and stronger at _____ strain rates.

(v) Cortical bone strength is highest under compressive loading in the _____ direction (direction of osteon orientation) and lowest under tensile loading in the _____ direction (direction perpendicular to the longitudinal direction).

(w) The tensile strength of bone is less than _____ percent of stainless steel, and the stiffness of bone is about _____ percent of the stiffness of steel.

(x) The chemical compositions of cortical and cancellous bone tissues are similar. The distinguishing characteristic of the cancellous bone is its _____.

(y) _____ is a unique ability of the muscle tissue, which is defined as the development of tension in the muscle.

(z) _____ tension is the force produced by the contractile elements of the muscle and is a result of voluntary muscle contraction, and _____ tension is the force developed within the connective muscle tissue when the muscle length surpasses its resting length.

Answer to Problem 9.1:

(a) instantaneously
(b) viscoelasticity
(c) viscosity
(d) strain rate or time
(e) Spring, dashpot
(f) Kelvin–Voight
(g) Maxwell
(h) standard solid
(i) creep and recovery
(j) stress relaxation
(k) oscillatory response
(l) hysteresis loop
(m) memory

(n) self-adapting, self-repairing
(o) collagen, elastin
(p) Collagen
(q) Elastin
(r) connective, composite
(s) cortical or compact
(t) cancellous or trabecular
(u) higher
(v) longitudinal, transverse
(w) 10, 5
(x) porosity
(y) Contraction
(z) Active, passive

Chapter 10

Introduction to Dynamics

10.1 Dynamics

Dynamics is the study of bodies in motion. Dynamics is concerned with describing motion and explaining its causes. The general field of dynamics consists of two major areas: kinematics and kinetics. Each of these areas can be further divided to describe and explain linear, angular, or general motion of bodies. The fundamental concepts in dynamics are space (relative position or displacement), time, mass, and force. Other important concepts include velocity, acceleration, torque, moment, work, energy, power, impulse, and momentum.

The broad definitions of basic terms and concepts in dynamics will be introduced in this chapter. The details of kinematic and kinetic characteristics of moving bodies will be covered in the following chapters.

10.2 Kinematics and Kinetics

The field of *kinematics* is concerned with the description of geometric and time-dependent aspects of motion without dealing with the forces causing the motion. Kinematic analyses are based on the relationships between position, velocity, and acceleration vectors. These relationships appear in the form of differential and integral equations.

The field of *kinetics* is based on kinematics, and it incorporates into the analyses the effects of forces and torques that cause the motion. Kinetic analyses utilize Newton's second law of motion that can take various mathematical forms. There are a number of different approaches to the solutions of problems in kinetics. These approaches are based on the equations of motion, work and energy methods, and impulse and momentum methods. Different methods may be applied to different situations, or depending on what is required to be determined. For example, the equations of motion are used for problems requiring the analysis of acceleration. Energy methods are suitable when a problem requires the analysis of forces related to changes in velocity. Momentum methods are applied if the forces involved are impulsive, which is the case during impact and collision.

There is also the field of *kinesiology* that is related to the study of human motion characteristics, joint and muscle forces, and neurological and other factors that may be important in studying human motion. The term "kinesiology" is not a mechanical but a medical term. It is commonly used to refer to the biomechanics of human motion.

10.3 Linear, Angular, and General Motions

To study both kinematics and kinetics in an organized manner, it is a common practice to divide them into branches according

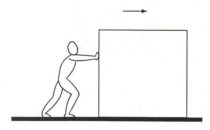

Figure 10.1 *Translational motion.*

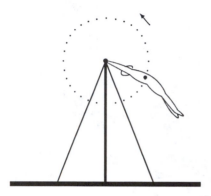

Figure 10.2 *Rotational motion.*

Figure 10.3 *General motion.*

to whether the motion is translational, rotational, or general. *Translational* or *linear motion* occurs if all parts of a body move the same distance at the same time and in the same direction. For example, if a block is pushed on a horizontal surface, the block will undergo translational motion only (Figure 10.1). Another typical example of translational motion is the vertical motion of an elevator in a shaft. It should be noted, however, that linear motion does not imply movement along a straight line. In a given time interval, an object may translate in one direction, and it may translate in a different direction during a different time interval.

Rotational or *angular motion* occurs when a body moves in a circular path such that all parts of the body move through the same angle at the same time. The angular motion occurs about a central line known as the *axis of rotation*, which lies perpendicular to the plane of motion. For example, for a gymnast doing giant circles, the center of gravity of the gymnast may undergo rotational motion with the centerline of the bar acting as the axis of rotation of the motion (Figure 10.2).

The third class of motion is called *general motion*, which occurs if a body undergoes translational and rotational motions simultaneously. It is more complex to analyze motions composed of both translation and rotation as compared to a pure translational or a pure rotational motion. The diver illustrated in Figure 10.3 is an example of a body undergoing general motion. Most human body segmental motions are of the general type. For example, while walking, the lower extremities both translate and rotate.

The branch of kinematics that deals with the description of translational motion is known as *linear kinematics* and the branch that deals with rotational motion is *angular kinematics*. Similarly, the field of kinetics can be divided into *linear* and *angular kinetics*.

Linear movements are direct consequences of applied forces. The linear motion of an object occurs in the direction of the net force acting on the object. On the other hand, angular movements are due to the rotational effects of applied forces, which are known as torques. There are linear and angular quantities defined to analyze linear and angular motions, respectively. For example, there are linear and angular displacements, linear and angular velocities, and linear and angular accelerations. It is important to note however that linear and angular quantities are not mutually independent. That is, if angular quantities are known, then linear quantities can also be determined, and vice versa.

10.4 Distance and Displacement

In mechanics, *distance* is defined as the total length of the path followed while moving from one point to another, and *displacement* is the length of the straight line joining the two points along with some indication of direction involved. Distance is a scalar quantity (has only a magnitude) and displacement is a vector quantity (has both a magnitude and a direction).

To understand the differences between distance and displacement, consider a person who lives in an apartment building located at the corner of Third Avenue and 18th Street, and walks to work in a building located at the corner of Second Avenue and 17th Street in New York City. In Figure 10.4, A represents the corner of Third Avenue and 18th Street, B represents the corner of Second Avenue and 18th Street, and C represents the corner of Second Avenue and 17th Street. Every morning this person walks toward the east from A to B, and then toward the south from B to C. Assume that the length of the straight line between A and B is 100 m, and between B and C is 50 m. Therefore, the total distance the person walks every morning is 150 m. On the other hand, the door-to-door southeasterly displacement of the person is equal to the length of the straight line joining A and C, which is $\sqrt{(100)^2 + (50)^2} = 112$ m.

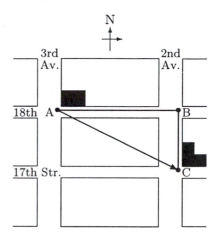

Figure 10.4 *Distance versus displacement.*

10.5 Speed and Velocity

While the terms speed and velocity are used interchangeably in ordinary language, they have distinctly different meanings in mechanics. *Velocity* is defined as the time rate of change of position. Velocity is a vector quantity having both a magnitude and a direction. *Speed* is a scalar quantity equal to the magnitude of the velocity vector.

10.6 Acceleration

Acceleration is defined as the time rate of change of velocity, and is a vector quantity. Although the term "acceleration" is more commonly used to describe situations where speed increases over time and the term "deceleration" is used to indicate decreasing speed over time, the mathematical definitions of the two are the same.

10.7 Inertia and Momentum

Inertia is the tendency of an object to maintain its state of motion. The more inertia an object has, the more difficult it is to start moving it from rest or to change its state of motion. The greater the mass of the object, the greater its inertia. For example, a truck has a greater inertia than a passenger car because of the difference in their mass. If both of them are traveling at the same speed, it is always more difficult to stop the truck as compared to the car.

Like inertia, *momentum* is a tendency to resist changes in the existing state of motion and is defined as the product of mass and velocity. Only moving objects have momentum, whereas every object – stationary or moving – has an inertia. If two moving objects having the same mass are considered, then the one with higher speed has the greater momentum. If two moving objects having the same speed are considered, then the one with higher mass has the greater momentum.

10.8   Degree of Freedom

Degree of freedom is an expression that describes the ability of an object to move in space. A completely unrestrained object, such as a ball, has six degrees of freedom (three related to translational motion along three mutually perpendicular axes and three related to rotational motion about the same axes). The human hip joint has three degrees of freedom because it enables the lower extremity to rotate about one axis and undergo angular movements in two planes. On the other hand, the elbow and forearm system has two degrees of freedom because it allows the lower arm to rotate about one axis and undergo angular movement in one plane.

10.9   Particle Concept

The "particle" concept in mechanics is rather a hypothetical one. It undermines the size and shape of the object under consideration, and assumes that the object is a particle with a mass equal to the total mass of the object and located at the center of gravity of the object. In some problems, the shape of the object under investigation may not be pertinent to the discussion of certain aspects of its motion. This is particularly true if the object is undergoing a translational motion only. For example, what is significant for a person pushing a wheelchair is the total mass of the wheelchair, not its size or shape. Therefore, the wheelchair may be treated as a particle with a mass equal to the total mass of the wheelchair, and proceed with relatively simple analyses. The size and shape of the object may become important if the object undergoes a rotational motion.

10.10   Reference Frames and Coordinate Systems

To be able to describe the motion of a body properly, a *reference frame* must be adopted. The *rectangular* or *Cartesian coordinate system* that is composed of three mutually perpendicular directions is the most suitable reference frame for describing linear movements. The axes of this system are commonly labeled with x, y, and z. For two-dimensional problems, the number of axes may be reduced to two by eliminating the z axis (Figure 10.5).

Another commonly used reference frame is the *polar coordinate system*, which is better suited for analyzing angular motions. As shown in Figure 10.5, the polar coordinates of a point P are defined by parameters r and θ. r is the distance between the origin O of the coordinate frame and point P, and θ is the angle line OP

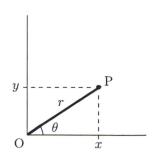

Figure 10.5 *Rectangular (x, y) and polar (r, θ) coordinates of a point.*

makes with the horizontal. The details of polar coordinates will be provided in later chapters.

10.11 Prerequisites for Dynamic Analysis

The prerequisites for dynamic analysis are vector algebra, differential calculus, and integral calculus. Vector algebra is reviewed in Appendix B. The principles of differential and integral calculus are provided in Appendix C, along with the definitions and properties of commonly encountered functions that form the basis of calculus. Appendices B and C must be reviewed before proceeding to the following chapters. Also important in dynamic analyses are the properties of force and torque vectors as covered in Chapters 2 and 3, respectively. It should be noted that the static analyses covered in Chapters 4 and 5 are specific cases of dynamic analyses for which acceleration is zero.

10.12 Topics to Be Covered

Chapters 10 through 15 constitute the last part of this textbook, which is devoted to the analyses of moving systems. In Chapter 11, mathematical definitions of displacement, velocity, and acceleration vectors are introduced, kinematic relationships between linear quantities are defined, uniaxial and biaxial motion analyses are discussed, and the concepts introduced are applied to problems of sports biomechanics. Linear kinetics is studied in Chapter 12. Solving problems in kinetics using the equations of motion and work and energy methods are discussed in Chapter 12. Angular kinematics and kinetics are covered in Chapters 13 and 14, respectively, and the concepts and procedures introduced are applied to investigate some of the problems of biomechanics. Topics such as impulse, momentum, impact, and collision are covered in Chapter 15.

Chapter 11

Linear Kinematics

11.1 Uniaxial Motion

Uniaxial motion is one in which the motion occurs only in one direction, and it is the simplest form of linear or translational motion. A car traveling on a straight highway, an elevator going up and down in a shaft, and a sprinter running a 100-m race are examples of uniaxial motion.

Kinematic analyses utilize the relationships between the position, velocity, and acceleration vectors. For uniaxial motion analyses, it is usually more practical to define a direction, such as x, to coincide with the direction of motion, define kinematic parameters in that direction, and carry out the analyses as if position, velocity, and acceleration are scalar quantities.

11.2 Position, Displacement, Velocity, and Acceleration

Consider the car illustrated in Figure 11.1. Assume that the car is initially stationary and located at 0. At time t_0, the car starts moving to the right on a straight horizontal path. At some time t_1, the car is observed to be at 1 and at a later time t_2 it is located at 2; 0, 1, and 2 represent *positions* of the car at different times, and 0 also represents the *initial position* of the car. It is a common practice to start measuring time beginning with the instant when the motion starts, in which case $t_0 = 0$.

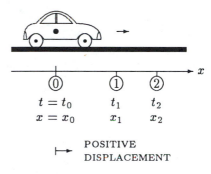

POSITIVE
DISPLACEMENT

Figure 11.1 *The car is located at positions 0, 1, and 2 at times t_0, t_1, and t_2.*

The position of the car at different times must be measured with respect to a point in space. Let x be a measure of horizontal distances relative to the initial position of the car. If x_0 represents the initial position of the car, then $x_0 = 0$. If 1 and 2 are located at x_1 and x_2 distances away from 0, then x_1 and x_2 define the relative positions of the object at times t_1 and t_2, respectively. Since the relative position of the car is changing with time, x is a function of time t, or $x = x(t)$. In the time interval between t_1 and t_2, the position of the car changed by an amount $\Delta x = x_2 - x_1$, where Δ (capital delta) implies change. This change in position is the *displacement* of the car in the time interval $\Delta t = t_2 - t_1$.

During a uniaxial horizontal motion, the car may be located on the right or the left of the origin 0 of the x axis. Assuming that the positive x axis is toward the right, the position of the car is positive if it is located on the right of 0 and negative if it is on the left of 0. Similarly, the displacement of the car is positive if it is moving toward the right, and it is negative if the car is moving toward the left.

Velocity is defined as the time rate of change of relative position. If the position of an object moving in the x direction is known as a function of time, then the *instantaneous velocity*, v, of the object can be determined by considering the derivative of x with respect to t:

$$v = \frac{dx}{dt} \tag{11.1}$$

If required, the *average velocity*, $\bar{v}$, of the object in any time interval can be determined by considering the ratio of change in position (displacement) of the object and the time it takes to make that change. For example, the average velocity of the car in Figure 11.1 in the time interval between t_1 and t_2 is:

$$\bar{v} = \frac{\Delta x}{\Delta t} = \frac{x_2 - x_1}{t_2 - t_1} \tag{11.2}$$

In Eq. (11.2), the "bar" over v indicates average, and x_1 and x_2 are the relative positions of the car at times t_1 and t_2, respectively.

Velocity is a vector quantity and may take positive and negative values, indicating the direction of motion. The velocity is positive if the object is moving away from the origin in the positive x direction, and it is negative if the object is moving in the negative x direction. The magnitude of the velocity vector is called *speed*, which is always a positive quantity.

The instantaneous velocity of an object may vary during a particular motion. In other words, velocity may be a function of time, or $v = v(t)$. *Acceleration* is defined as the time rate of change of velocity. If the velocity of an object is known as a function of time, then its *instantaneous acceleration*, a, can be determined by considering the derivative of v with respect to t:

$$a = \frac{dv}{dt} \tag{11.3}$$

In general, the acceleration of a moving object may vary with time. In other words, acceleration may be a function of time, or $a = a(t)$.

There is also *average acceleration*, $\bar{a}$, that can be determined by considering the ratio of the change in velocity of the object and the time elapsed during that change. For example, if the instantaneous velocities v_1 and v_2 of the car in Figure 11.2 at times t_1 and t_2 are known, then the average acceleration of the car in the time interval between t_1 and t_2 can be calculated:

$$\bar{a} = \frac{\Delta v}{\Delta t} = \frac{v_2 - v_1}{t_2 - t_1} \tag{11.4}$$

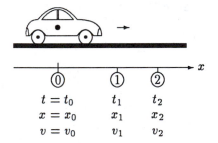

Figure 11.2 v_0 *is the initial velocity, and* v_1 *and* v_2 *are the velocities of the car at times* t_1 *and* t_2, *respectively.*

Acceleration is a vector quantity and may be positive or negative. Positive acceleration does not always mean that the object is speeding up and negative acceleration does not always imply that the object is slowing down. At a given instant, if the velocity and acceleration are both positive or negative, then the object is said to be speeding up or accelerating. For a uniaxial motion in the x direction, if both the velocity and acceleration are positive, then the object is moving in the positive x direction with an increasing speed. If both the velocity and acceleration are negative, then the object is moving in the negative x direction with an increasing speed. On the other hand, if the velocity and acceleration have opposite signs, then the object is slowing down or decelerating. For example, for a uniaxial motion in the x direction, if the velocity is positive and acceleration is negative, then the object is moving in the positive x direction with a decreasing speed. If the velocity is negative and acceleration is positive, then the object is moving in the negative x direction with a decreasing speed. Finally, if the acceleration is zero, then the object is said to have a constant or uniform velocity. All of these possibilities are summarized in Table 11.1.

Acceleration is derived from velocity, which is itself derived from position. Therefore, there must be a way to relate acceleration and position directly. This relationship can be derived by substituting Eq. (11.1) into Eq. (11.2):

$$v = \frac{dx}{dt} = \dot{x}$$

$$a = \frac{dv}{dt} = \frac{d}{dt}\left(\frac{dx}{dt}\right) = \frac{d^2x}{dt^2} = \ddot{x}$$

The "dots" over x in the above equations indicate differentiation with respect to time. One dot signifies the first derivative with respect to time and two dots imply the second derivative.

11.3 Dimensions and Units

The relative position is measured in units of length. By definition, displacement is equal to the change of position, velocity is the time rate of change of relative position, and acceleration is the time rate of change of velocity. Therefore, relative position and displacement have the dimension of length, velocity has the dimension of length divided by time, and acceleration has the dimension of velocity divided by time:

$$[\text{POSITION}] = L$$
$$[\text{DISPLACEMENT}] = L$$
$$[\text{VELOCITY}] = \frac{[\text{DISPLACEMENT}]}{[\text{TIME}]} = \frac{L}{T}$$
$$[\text{ACCELERATION}] = \frac{[\text{VELOCITY}]}{[\text{TIME}]} = \frac{L/T}{T} = \frac{L}{T^2}$$

Based on these dimensions, the units of displacement, velocity, and acceleration in different unit systems can be determined. Some of these units are listed in Table 11.2.

Table 11.1 *Acceleration, deceleration, and constant velocity conditions.*

	v	a
INCREASING SPEED OR ACCELERATION	$+$	$+$
	$-$	$-$
DECREASING SPEED OR DECELERATION	$+$	$-$
	$-$	$+$
CONSTANT SPEED	$\pm$	0

Table 11.2 *Units of displacement, velocity, and acceleration.*

UNIT SYSTEM	DISPLACEMENT	VELOCITY	ACCELERATION
SI	meter (m)	m/s	m/s^2
c-g-s	centimeter (cm)	cm/s	cm/s^2
British	foot (ft)	ft/s	ft/s^2

11.4 Measured and Derived Quantities

In practice, it is possible to measure position, velocity, and acceleration over time. From any one of the three, the other two quantities can be determined by employing proper differentiation and/or integration, or through the use of graphical and numerical techniques. If the position of an object undergoing uniaxial motion in the x direction is measured and recorded, then the position can be expressed as a function of time. Once the function representing the position of the object is established, the velocity and acceleration of the object at different times can be calculated using:

$$v = \frac{dx}{dt} \tag{11.5}$$

$$a = \frac{dv}{dt} = \frac{d^2x}{dt^2} \tag{11.6}$$

If the velocity of an object undergoing uniaxial motion in the x direction is measured and expressed as a function of time, then the position of the object relative to its initial position and instantaneous acceleration of the object can be calculated using:

$$x = x_0 + \int_{t_0}^{t} v\,dt \tag{11.7}$$

$$a = \frac{dv}{dt} \tag{11.8}$$

The lower limit of integration, t_0, in Eq. (11.7) corresponds to the time at which the first measurements are taken, and the upper limit corresponds to any time t. x_0 is the initial position of the object at time t_0. For practical purposes, t_0 can be taken to be zero. This would mean that all time measurements are made relative to the instant when the motion began. Also, $x_0 = 0$ if all position measurements are made relative to the initial position of the object.

If the acceleration of an object is measured and expressed as a function of time, then the instantaneous velocity and position of the object relative to its initial velocity and position can be calculated using:

$$v = v_0 + \int_{t_0}^{t} a\,dt \tag{11.9}$$

$$x = x_0 + \int_{t_0}^{t} v\,dt \tag{11.10}$$

In Eqs. (11.9) and (11.10), x_0 and v_0 correspond to the initial position and initial velocity of the object at time t_0. Note that these equations relate change of position and velocity relative to the initial position and velocity of the moving object. However, these equations are valid relative to the position and velocity of the object at any time. For example, if x_1 and v_1 represent the position and velocity of the object at time t_1, then Eqs. (11.9) and (11.10) can also be expressed as:

$$v = v_1 + \int_{t_1}^{t} a\, dt$$

$$x = x_1 + \int_{t_1}^{t} v\, dt$$

11.5 Uniaxial Motion with Constant Acceleration

A common type of uniaxial motion occurs when the acceleration is constant. If a_0 represents the constant acceleration of an object, v_0 is its initial velocity, and x_0 is its initial position at time $t_0 = 0$, then Eqs. (11.9) and (11.10) will yield:

$$v = v_0 + a_0\, t \tag{11.11}$$

$$x = x_0 + v_0\, t + \frac{1}{2}\, a_0\, t^2 \tag{11.12}$$

For a given initial position, initial velocity, and constant acceleration of an object undergoing uniaxial motion in the x direction, Eqs. (11.11) and (11.12) can be used to determine the velocity and position of the object as functions of time relative to its initial velocity and position. Note that Eqs. (11.11) and (11.12) can be expressed relative to any other time and position. For example, if x_1 and v_1 represent the known position and velocity of the object at time t_1, then:

$$v = v_1 + a_0\, (t - t_1)$$

$$x = x_1 + v_1\, (t - t_1) + \frac{1}{2}\, a_0\, (t - t_1)^2$$

Figure 11.3 shows an acceleration versus time graph for an object moving with constant acceleration, a_0. According to Eq. (11.11), velocity is a linear function of time. As illustrated in Figure 11.4, the velocity versus time graph is a straight line with constant slope that is equal to the magnitude of the constant acceleration. This is consistent with the fact that the slope of a function is equal to the derivative of that function, and that the derivative of velocity with respect to time is equal to acceleration. In Eq. (11.12), displacement is a quadratic function of time, and as illustrated in Figure 11.5, the graph of this function is a parabola. At any given time, the slope of this function is equal to the velocity of the object at that instant.

For a uniaxial motion with constant acceleration, it is also possible to derive an expression among velocity, displacement, and time by solving Eq. (11.11) for a_0 and substituting it into Eq. (11.12). This will yield:

$$x = x_0 + \frac{1}{2}\, (v + v_0)\, t \tag{11.13}$$

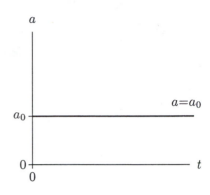

Figure 11.3 *Constant (uniform) acceleration.*

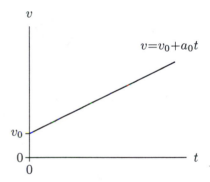

Figure 11.4 *When acceleration is constant, velocity is a linear function of time.*

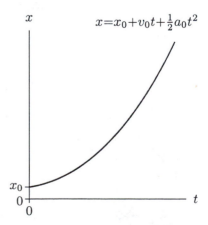

Figure 11.5 *When acceleration is constant, change of position is a quadratic function of time.*

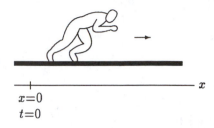

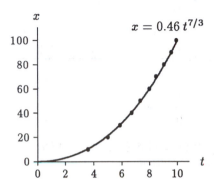

Figure 11.6 *Relative position, x, measured in meters versus time, t, measured in seconds.*

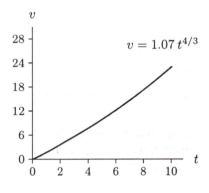

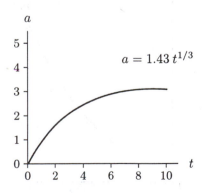

Figure 11.7 *Speed, v (m/s), and acceleration, a (m/s²), versus time, t (s), curves for the runner.*

Similarly, an expression among velocity, displacement, and acceleration can be derived by solving Eq. (11.11) for t and substituting it into Eq. (11.12):

$$v^2 = v_0{}^2 + 2\,a_0\,(x - x_0) \qquad (11.14)$$

Caution. Eqs. (11.11) through (11.14) are valid if the acceleration is constant. Furthermore, the direction of the parameters involved must be handled properly. For example, if the direction of acceleration is opposite to that of the positive x direction, then the "plus" sign in front of the terms carrying acceleration must be changed to a "minus" sign.

11.6 Examples of Uniaxial Motion

The following examples are aimed to demonstrate the use of the kinematic equations (11.5) through (11.10).

Example 11.1 The short distance runner illustrated in Figure 11.6 completed a 100-m race in 10 s. The time it took for the runner to reach the first 10 m and each successive 10 m mark were recorded by 10 observers using stopwatches. The data collected were then plotted to obtain the position versus time graph shown in Figure 11.6. It is suggested that the data may be represented with the following function:

$$x = 0.46\,t^{7/3}$$

Here, change of position x is measured in meters, and time t is measured in seconds.

Determine the velocity and acceleration of the runner as functions of time, and the instantaneous velocity and acceleration of the runner 5 s after the start.

Solution: Since the function representing the position of the runner is known, it can be differentiated with respect to time once to determine the velocity, and twice to determine the acceleration:

$$v = \frac{dx}{dt} = \frac{d}{dt}\left(0.46\,t^{7/3}\right) = 1.07\,t^{4/3}$$

$$a = \frac{dv}{dt} = \frac{d}{dt}\left(1.07\,t^{4/3}\right) = 1.43\,t^{1/3}$$

The graphs of these functions are shown in Figure 11.7.

To evaluate the velocity and acceleration of the runner 5 s after the start, substitute $t = 5$ s in the above equations and carry out the calculations. This will yield:

$$v = 9.15\ \text{m/s}$$

$$a = 2.45\ \text{m/s}^2$$

Example 11.2 The speedometer reading of a car driven on a straight highway is recorded for a total time interval of three minutes. The data collected are represented with a speed versus time diagram shown in Figure 11.8. The dotted curve in Figure 11.8 represents the actual measurements that are approximated by three straight lines (solid lines in Figure 11.8). According to the information presented in Figure 11.8, the speed of the car increases linearly from $v_0 = 0$ to $v_1 = 72$ km/h between times $t_0 = 0$ and $t_1 = 30$ s. Between times $t_1 = 30$ s and $t_2 = 120$ s, the speed of the car is constant at 72 km/h. Beginning at time $t_2 = 120$ s, the driver applies the brakes, decreases the speed of the car linearly with time, and brings the car to a stop in 60 seconds.

Determine expressions for the speed, displacement, and acceleration of the car as functions of time. Calculate the total distance traveled by the car in three minutes.

Solution: The speed measurements were made in kilometers per hour (km/h) that need to be converted to meters per second (m/s). This can be achieved by noting that 1 kilometer is equal to 1000 meters and that there are 3600 s in 1 h. Therefore, 72 km/h is equal to 20 m/s, which is calculated as:

$$72 \text{ km/h} = 72 \times \frac{1000}{3600} = 20 \text{ m/s}$$

The speed versus time graph in Figure 11.8 is redrawn in Figure 11.9, in which speed is expressed in meters per second and time in seconds.

Because of the approximations made, the speed versus time graph in Figure 11.9 has three distinct regions and there is not a single function that can represent the entire graph. Therefore, this problem should be analyzed in three phases.

Phase 1. Between $t_0 = 0$ and $t_1 = 30$ s, the speed of the car increased linearly with time from 0 to 20 m/s. As discussed in Appendix C, all linear functions can be represented as $X = A + BY$. In this expression, Y is the independent variable, X is the dependent variable, and A and B are some constant coefficients. In this case, we have time as the independent variable and speed is the dependent variable. Since the relationship between the speed of the car and time in phase 1 is linear, we can write:

$$v = A + Bt \qquad (i)$$

The function given in Eq. (*i*) is a general expression between v and t because coefficients A and B are not yet determined. We need two conditions to calculate A and B (two unknowns). These conditions can be obtained from Figure 11.9. When the

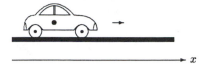

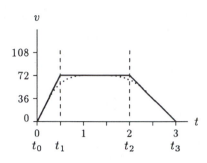

Figure 11.8 *Speed, v (km/h), versus time, t (min), diagram for the car.*

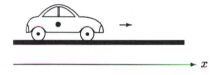

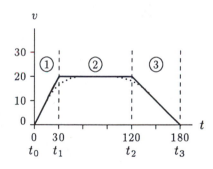

Figure 11.9 *Speed, v (m/s), versus time, t (s), diagram for the car.*

car first started to move, $t = 0$ and $v = 0$, and $v = 20$ m/s when $t = 30$ s. Substituting the initial condition ($v = 0$ when $t = 0$) into Eq. (*i*) will yield $A = 0$, and substituting the second condition ($v = 20$ m/s when $t = 30$ s) will yield $B = 0.667$. Substituting $A = 0$ and $B = 0.667$ back into Eq. (*i*) will yield the function relating the speed of the car and time in phase 1:

$$v = 0.667\, t \tag{ii}$$

Note that since we already converted speed measurements into meter per second and time into seconds, the speed in Eq. (*ii*) is in meters per second and time is in seconds.

Now, Eqs. (11.7) and (11.8) can be utilized to determine the displacement and acceleration of the car in phase 1. If we measure displacements relative to the starting point, then the initial position of the car was $x_0 = 0$. Therefore:

$$x = x_0 + \int_0^t v\, dt = \int_0^t (0.667\, t)\, dt = 0.667 \left[\frac{t^2}{2} \right]_0^t = 0.333\, t^2 \tag{iii}$$

$$a = \frac{dv}{dt} = \frac{d}{dt}(0.67\, t) = 0.67 \tag{iv}$$

From Eq. (*iv*), the acceleration of the car in phase 1 was constant at 0.667 m/s^2. The total distance traveled by the car at the end of phase 1 can be determined by substituting $t = 30$ s into Eq. (*iii*). This will yield:

$$x_1 = 0.333\, t^2 = 0.333\,(30)^2 = 300 \text{ m}$$

Phase 2.   Phase 2 starts when time is $t_1 = 30$ s and ends when it is $t_2 = 102$ s. In phase 2, the speed of the car was constant at 20 m/s. Therefore, the function representing the speed in phase 2 is:

$$v = 20 \tag{v}$$

The total distance traveled by the car in phase 1 was computed as $x_1 = 300$ m. $x_1 = 300$ m also represents the initial position of the car in phase 2. Phase 1 ended when time was $t_1 = 30$ s. Therefore, phase 2 began when time was $t_1 = 30$ s. We can now write Eq. (11.7) relative to t_1 and x_1:

$$x = x_1 + \int_{t_1}^t v\, dt = 300 + \int_{30}^t 20\, dt = 300 + 20 \left[t \right]_{30}^t$$

$$x = -300 + 20\, t \tag{vi}$$

The acceleration of the car in phase 2 can be determined using Eq. (11.8):

$$a = \frac{dv}{dt} = 0 \tag{vii}$$

From Eq. (*vii*), the acceleration of the car in phase 2 is zero. The total distance traveled by the car at the end of phase 2 can be determined by substituting $t = 120$ s into Eq. (iv). This will yield:

$$x_2 = -300 + 20\,(120) = 2100 \text{ m}$$

Phase 3. Between $t_2 = 120$ s and $t_3 = 180$ s, the speed of the car decreased linearly with time and to zero in 60 s. The function representing the relationship between speed of the car and time in phase 3 can be determined using Eq. (i). The coefficients A and B in Eq. (i) can be calculated by taking into consideration two conditions related to phase 3. For example, $v = 20$ m/s when $t = 120$ s and $v = 0$ when $t = 180$ s. Substituting the second condition in Eq. (i) will yield $A + 180B = 0$ or $A = -180B$. Substituting the first condition and $A = -180B$ into Eq. (i) will yield $B = -0.333$. Since $A = -180B$ and $B = -0.333$, $A = 60$. Therefore, the function that relates the speed of the car and time in phase 3 is:

$$v = 60 - 0.333\,t \qquad\qquad (viii)$$

Phase 3 begins when time is $t_2 = 120$ s and the initial position of the car at phase 3 is $x_2 = 2100$ m. Using Eq. (11.7):

$$x = x_2 + \int_{t_2}^{t} v_3\,dt = 2100 + \int_{120}^{t} (60 - 0.333\,t)\,dt$$

$$= 2100 + \left[60\,t - 0.167\,t^2\right]_{120}^{t}$$

$$= 2100 + \left[60\,t - 0.167\,t^2\right] - \left[60\,(120) - 0.167\,(120)^2\right]$$

$$x = -2700 + 60\,t - 0.167\,t^2 \qquad\qquad (ix)$$

Using Eq. (11.8):

$$a = \frac{dv}{dt} = \frac{d}{dt}(60 - 0.333\,t) = -0.333 \qquad\qquad (x)$$

The total distance traveled by the car can be determined by substituting $t = 180$ s into Eq. (ix). This will yield:

$$x_3 = -2700 + 60\,(180) - 0.167\,(180)^2 = 2690 \text{ m}$$

In Figure 11.10, the functions derived for the displacement and acceleration of the car in different phases are used to plot displacement and acceleration versus time graphs (solid curves). In all phases, the car is moving in the positive x direction. Therefore, the displacement of the car is positive throughout. In phase 1, the acceleration of the car is positive, indicating increasing speed in the positive x direction. In phase 2, the acceleration of the car is zero and the speed is constant. In phase 3, the acceleration of the car is negative, indicating deceleration in the positive x direction.

Note that the displacement and acceleration versus time graphs (solid curves) in Figure 11.10 are not continuous. For example, the slope of the x versus t curve at the end of phase 1 is not necessarily equal to the slope of the x versus t curve at the

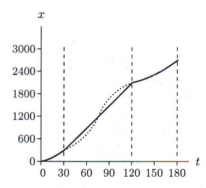

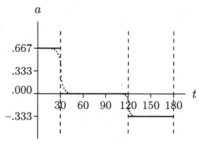

Figure 11.10 *Displacement, x (m), and acceleration, a (m/s^2), versus time, t (s), graphs for the car.*

beginning of phase 2. The discontinuity is more significant for the a versus t graph. This is because we approximated the actual speed versus time graph of the car with three straight lines in three regions. In reality, as illustrated by the dotted curves in Figure 11.10, the variations in the slopes of these curves would be less marked and more continuous.

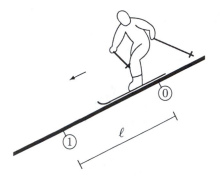

Figure 11.11 *A skier is moving down the slope with a constant acceleration of $a_0 = 2$ m/s^2.*

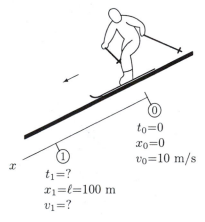

Figure 11.12 *Conditions at positions 0 and 1.*

Example 11.3 Consider the skier illustrated in Figure 11.11 descending a straight slope. Assume that the skier is moving down the slope at a constant acceleration of 2 m/s^2 and that the speed of the skier at position 0 is observed to be 10 m/s.

Calculate the speed v_1 of the skier when the skier is at position 1, which is at a distance $\ell = 100$ m from position 0 measured parallel to the slope. Also, calculate the time t_1 it took for the skier to cover the distance between positions 0 and 1.

Solution: Since the skier is moving with a constant acceleration, Eqs. (11.11) and (11.12) can be used to analyze this problem. In Figure 11.12, the direction parallel to the slope or the direction in which the skier is moving is identified by the x axis. For the sake of simplicity, the origin of the x axis is placed to coincide with position 0 so that $x_0 = 0$. Furthermore, we can make all time measurements relative to the instant when the skier was at position 0. That is, $t_0 = 0$. As indicated in Figure 11.12, the speed of the skier at position 0 is $v_0 = 10$ m/s. What we know about position 1 is the fact that it is located at a distance $\ell = 100$ m from position 0. Therefore, the position of the skier at 1 is $x_1 = \ell = 100$ m. The time t_1 it took for the skier to cover the distance between positions 0 and 1, and the speed v_1 of the skier at position 1 are unknowns to be determined.

Use Eq. (11.12) first. Writing this equation between positions 0 and 1:

$$x_1 = x_0 + v_0\, t_1 + \frac{1}{2}\, a_0\, t_1{}^2$$

Here, $x_0 = 0$, $v_0 = 10$ m/s, and $a_0 = 2$ m/s^2 is the constant acceleration of the skier. Substituting these parameters into the above equation and rearranging the order of terms will yield:

$$t_1{}^2 + 10\, t_1 - 100 = 0$$

Note that this is a quadratic equation. Solutions of quadratic equations are discussed in Appendix C.5. This equation has two solutions for t_1, one positive and one negative. Since negative time does not make any sense, the positive solution must be adopted, which is $t_1 = 6.18$ s. (For the validity of $t_1 = 6.18$ s, substitute it back into the quadratic equation and check whether the equilibrium is satisfied.) In other words, it took 6.18 s for the skier to cover the distance between positions 0 and 1.

Now, we can use Eq. (11.11) to calculate the speed of the skier at position 1:

$$v_1 = v_0 + a_0 \, t_1$$

Substituting $v_0 = 10$ m/s, $a_0 = 2$ m/s, and $t_1 = 6.18$ s into the above equation and carrying out the calculations will yield $v_1 = 22.36$ m/s.

Example 11.4 One of the most common examples of uniformly accelerated motion is that of an object allowed to fall vertically downward, which is called *free fall*. Free fall is a consequence of the effect of gravitational acceleration on the mass of the object. If the possible effects of air resistance are ignored (assuming that the motion occurs under vacuum), then the object released from a height would move downward with a constant acceleration equal to the magnitude of the gravitational acceleration, which is about 9.8 m/s^2.

As illustrated in Figure 11.13, consider a person holding a ball at a height $h = 1.5$ m above the ground level. If the ball is released to descend, how much time would it take for the ball to hit the ground and what would be its impact velocity?

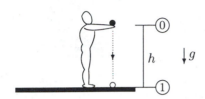

Figure 11.13 *Free fall.*

Solution: This is another example of uniaxial motion with constant acceleration. Once the ball is released, it moves downward with constant acceleration $a_0 = 9.8$ m/s^2, which is the magnitude of the gravitational acceleration. In Figure 11.14, the direction of motion of the ball is identified with the y axis such that positive y direction is downward. Since the direction of gravitational acceleration is also downward, the acceleration of the ball is positive.

In Figure 11.14, the origin of the y axis is chosen to coincide with the initial position, 0, of the ball. Therefore, the initial position of the ball is $y_0 = 0$. The initial time is $t_0 = 0$ because all time measurements are made relative to the instant when the ball was released. The initial speed of the ball is $v_0 = 0$ because the ball was initially at rest. The ground level is identified as position 1, which is at a vertical distance $y_1 = h = 1.5$ m away from position 0. The task is to determine the time t_1 it took for the ball to cover the vertical distance between positions 0 and 1, and the impact speed v_1 of the ball.

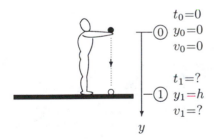

Figure 11.14 *Conditions at positions 0 and 1.*

Since the acceleration is constant at $a_0 = 9.8$ m/s^2, we can use Eqs. (11.11) and (11.12). However, we must replace x in these equations with y. Writing Eq. (11.12) between positions 0 and 1:

$$y_1 = y_0 + v_0 \, t_1 + \frac{1}{2} a_0 \, t_1{}^2$$

Substituting $y_1 = 1.5$ m, $y_0 = 0$, $v_0 = 0$, and $a_0 = 9.8$ m/s^2 into this equation and solving it for t_1 will yield $t_1 = 0.55$ s. That is, in the absence of air resistance, it would take only 0.55 s for the ball to hit the ground when it is released from a height of 1.5 m from the ground level.

The impact speed can be calculated using Eq. (11.11):

$$v_1 = v_0 + a_0\, t_1$$

Substituting $v_0 = 0$, $a_0 = 9.8$ m/s^2, and $t_1 = 0.55$ into the above equation and solving it for v_1 will yield $v_1 = 5.39$ m/s.

11.7 Biaxial Motion

Biaxial or *two-dimensional motion* is one in which the movement occurs on a plane surface. One-dimensional linear motion characteristics of an object are completely known if, for example, the position of the object in the direction of motion is known as a function of time. The concepts introduced earlier for uniaxial motion analysis can be expanded to analyze two- and three-dimensional linear movements. This can be achieved by considering the properties of position, velocity, and acceleration as vector quantities.

11.8 Position, Velocity, and Acceleration Vectors

For one-dimensional problems, the position of an object is defined by using a single coordinate axis. For plane problems, two coordinates must be specified to define the position of an object uniquely. As shown in Figure 11.15, let x and y represent the usual Cartesian (rectangular) coordinate directions with unit vectors $\underline{i}$ and $\underline{j}$ indicating positive x and y directions, respectively. The origin of the coordinate system is located at O. The *position vector $\underline{r}$* of a point P in this xy-plane is a vector drawn from O toward P. The position vector can be represented in terms of x and y coordinates of point P:

$$\underline{r} = x\,\underline{i} + y\,\underline{j} \tag{11.15}$$

The magnitude r of the position vector is equal to the length of the line connecting O and P, which can be calculated as:

$$r = \sqrt{x^2 + y^2} \tag{11.16}$$

If point P represents the position of a moving object at some time t, then $\underline{r}$ represents the *instantaneous position* of that object at that time. This implies that $\underline{r}$ can change with time or x and y coordinates of the moving object are functions of time.

By definition, velocity is the time rate of change of position. Therefore, the *velocity vector* is equal to the derivative of the

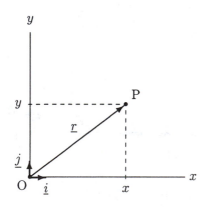

Figure 11.15 $\underline{r}$ *is the position vector of point P.*

position vector with respect to time:

$$\underline{v} = \frac{d}{dt}(\underline{r}) = \frac{d}{dt}(x\underline{i} + y\underline{j}) = \frac{dx}{dt}\underline{i} + \frac{dy}{dt}\underline{j} \qquad (11.17)$$

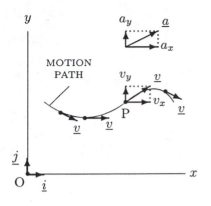

For two-dimensional problems, the velocity vector may have up to two components (Figure 11.16). If v_x and v_y refer to the scalar components of $\underline{v}$ in the x and y directions, respectively, then the velocity vector can also be expressed as:

$$\underline{v} = v_x\underline{i} + v_y\underline{j} \qquad (11.18)$$

By comparing Eqs. (11.17) and (11.18), we can conclude that:

Figure 11.16 *Velocity vector is always tangent to the path of motion.*

$$v_x = \frac{dx}{dt} = \dot{x} \qquad (11.19)$$

$$v_y = \frac{dy}{dt} = \dot{y} \qquad (11.20)$$

v_x and v_y are also known as the *rectangular components* of $\underline{v}$, and they indicate how fast the object is moving in the x and y directions, respectively. If the components v_x and v_y of the velocity vector are known, then the magnitude v of their resultant can also be determined:

$$v = \sqrt{v_x^2 + v_y^2} \qquad (11.21)$$

Note that v is a scalar quantity also known as the *speed*. As illustrated in Figure 11.16, it is very important to remember that the direction of the velocity vector is always tangent to the path of the motion and pointing in the direction of motion.

By definition, acceleration is the time rate of change of velocity. Therefore, if the velocity vector of an object is known as a function of time, then its *acceleration vector $\underline{a}$* can also be determined by considering the derivative of $\underline{v}$ with respect to time:

$$\underline{a} = \frac{d}{dt}(\underline{v}) = \frac{d}{dt}(v_x\underline{i} + v_y\underline{j}) = \frac{dv_x}{dt}\underline{i} + \frac{dv_y}{dt}\underline{j} \qquad (11.22)$$

The acceleration vector can also be expressed in terms of its components in the x and y directions (Figure 11.16):

$$\underline{a} = a_x\underline{i} + a_y\underline{j} \qquad (11.23)$$

By comparing Eqs. (11.22) and (11.23), the rectangular components of the acceleration vector can alternatively be written as:

$$a_x = \frac{dv_x}{dt} = \frac{d^2x}{dt^2} = \ddot{x} \qquad (11.24)$$

$$a_y = \frac{dv_y}{dt} = \frac{d^2y}{dt^2} = \ddot{y} \qquad (11.25)$$

If needed, the magnitude a of the acceleration vector can be calculated as:

$$a = \sqrt{a_x{}^2 + a_y{}^2} \tag{11.26}$$

The extension of these concepts to analyze three-dimensional movements is straightforward. For three-dimensional motion analyses, there is a need for a third dimension, namely z. For example, the position vector of a point in space can be expressed as:

$$\underline{r} = x\,\underline{i} + y\,\underline{j} + z\,\underline{k}$$

Here, $\underline{k}$ is the unit vector indicating the positive z direction. Similarly, the velocity and acceleration vectors in space can be expressed as:

$$\underline{v} = v_x\,\underline{i} + v_y\,\underline{j} + v_z\,\underline{k}$$
$$\underline{a} = a_x\,\underline{i} + a_y\,\underline{j} + a_z\,\underline{k}$$

For example, v_z is the speed of the object in the z direction and is equal to the time rate of change of position in the z direction, and a_z is the scalar component of the acceleration vector in the z direction and is equal to the time rate of change of the speed of the object in the z direction.

11.9 Biaxial Motion with Constant Acceleration

Two-dimensional linear motion of an object in the xy-plane can be analyzed in two stages by first considering its motion in the x and y directions separately and then combining the results using the vectorial properties of the parameters involved. The parameters defining the motion in the x direction are x, its first time derivative v_x, and its second time derivative a_x. Similarly, y, v_y, and a_y are the parameters that define the motion of the object in the y direction. If the acceleration of an object undergoing two-dimensional linear motion is constant, then a_x and a_y must be constants. The details of uniaxial motion with constant acceleration were analyzed in the previous sections. The results of these analyses can readily be adopted to analyze two-dimensional motions with constant acceleration.

In the x direction, Eqs. (11.11) and (11.12) can be rewritten in the following more specific forms:

$$v_x = v_{x0} + a_{x0}\,t \tag{11.27}$$

$$x = x_0 + v_{x0}\,t + \frac{1}{2}\,a_{x0}\,t^2 \tag{11.28}$$

Similarly, in the y direction:

$$v_y = v_{y0} + a_{y0}\,t \tag{11.29}$$

$$y = y_0 + v_{y0}\,t + \frac{1}{2}\,a_{y0}\,t^2 \tag{11.30}$$

Here, x_0 and y_0 are the initial coordinates of the object, v_{x0} and v_{y0} are the initial velocity components in the x and y directions, and a_{x0} and a_{y0} are the constant components of the acceleration vector in the x and y directions, respectively. For given x_0, y_0, v_{x0}, v_{y0}, a_{x0}, and a_{y0}, Eqs. (11.27) through (11.30) can be used to calculate the relative position of the moving object and its velocity components at any time t.

11.10 Projectile Motion

When an object is thrown into the air in any direction other than the vertical, it will move in a curved path under the influence of gravity and air resistance. The gravity of Earth will pull the object downward with a constant gravitational acceleration of about $9.8 \, \text{m/s}^2$, and the air resistance will retard its motion in a direction opposite to the direction of motion. This very common form of motion, called *projectile motion*, is relatively simple to analyze once the effect of air resistance is neglected.

Projectile motion is a particular form of two-dimensional linear motion with constant acceleration. To be able to define the basic parameters involved in all projectile motions, consider the motion of a cannonball fired into the air (Figure 11.17). Assume that the cannonball leaves the barrel and lands on the ground at the same elevation. As illustrated in Figure 11.18, the cannonball ascends, reaches a peak, starts descending, and finally lands on the ground. The curved flight path of the cannonball is called the *trajectory* of motion. 0 represents the initial position of the cannonball, 1 is the peak it reaches, and 2 is the location of landing. v_0 is the magnitude of the initial velocity of the cannonball, which is called the *speed of release* or *takeoff speed*. θ is the angle the initial velocity vector makes with the horizontal and is called the *angle of release*. The vertical distance h between 0 and 1 is the *maximum height* that the cannonball reaches, and the horizontal distance ℓ between 0 and 2 is called the *horizontal range of motion*. The total time the cannonball remains in the air is called the *time of flight*.

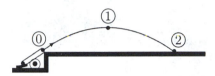

Figure 11.17 *A cannonball fired into the air will undergo a projectile motion.*

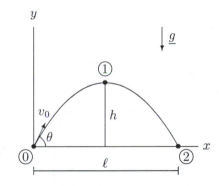

Figure 11.18 *Projectile motion.*

The equations necessary to analyze projectile motions can be derived from Eqs. (11.27) through (11.30). For example, if the speed and angle of release of the projectile are known, then components of the velocity vector along the horizontal (x) and vertical (y) directions can be calculated:

$$v_{x0} = v_0 \cos\theta$$

$$v_{y0} = v_0 \sin\theta$$

Assuming that the air resistance on the cannonball is negligible, the acceleration of the cannonball in the x direction is zero throughout the motion. That is, $a_{x0} = 0$. The gravitational acceleration g acts downward. Assuming that the y axis is positive upward, the gravitational acceleration acts in the negative y direction. To account for the negative direction of gravitational acceleration, the plus signs in front of the terms carrying a_{y0} in Eqs. (11.29) and (11.30) must be changed to minus. Under these considerations, Eqs. (11.27) through (11.30) take the following special forms for projectile motion:

$$x = x_0 + (v_0 \cos\theta)\, t \qquad (11.31)$$

$$y = y_0 + (v_0 \sin\theta)\, t - \frac{1}{2}\, g\, t^2 \qquad (11.32)$$

$$v_x = v_0 \cos\theta \qquad (11.33)$$

$$v_y = v_0 \sin\theta - g\, t \qquad (11.34)$$

Here, if the origin of the xy coordinate frame is chosen to coincide with the initial position of the ball, then $x_0 = 0$ and $y_0 = 0$. Note from Eq. (11.33) that the magnitude of the horizontal component of the velocity vector is not a function of time. Therefore, $v_x = v_{x0}$ remains constant throughout the projectile motion (Figure 11.19). The magnitude of the vertical component of the velocity vector is a linear function of time. It is positive (upward) initially, decreases in time as the object ascends, and drops to zero at the peak. After reaching the peak, the vertical component of the velocity vector changes its direction from upward to downward, while its magnitude increases until it lands on the ground. At any instant during the flight, the resultant velocity vector is tangent to the trajectory of the projectile motion.

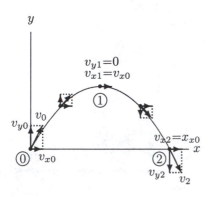

Figure 11.19 *Horizontal component of the velocity remains constant through the motion. Velocity vector is always tangent to the trajectory of motion.*

Another important aspect of the projectile motion is that if both the takeoff and landing occur at the same elevation, then the motion is symmetric with respect to a plane that passes through the peak and cuts the plane of motion at right angles. In other words, the time it takes for the object to ascend is equal to the time it takes to descend.

In some cases, the objective of the projectile motion may be to increase the horizontal range of motion to a maximum. This is particularly true for a ski jumper, for example. Other situations may require a control over the height, which is the case for a high-jumper. Therefore, it may be useful to derive some expressions for the horizontal range and maximum height of the projectile motion. Such derivations will be performed within the context of the following example problem.

Example 11.5 Consider Figure 11.20, which shows the trajectory of a projectile motion along with some of the parameters involved. $x_0 = 0$ and $y_0 = 0$ because the origin of the xy coordinate frame is chosen to coincide with the initial position of the object. Let t_1 be the time it takes to reach the peak, and t_2 be the total time of flight. At the peak, $y_1 = h$, $v_{y1} = 0$, and $x_1 = \ell/2$ from the symmetry of the motion. At the location where the object lands, $x_2 = \ell$ and $y_2 = 0$. The maximum height h reached by the object can be determined by noting that $v_{y1} = 0$ at the peak.

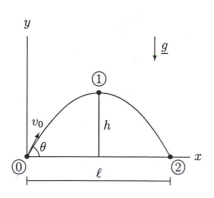

Figure 11.20 *h is the maximum elevation and ℓ is the horizontal range of motion.*

Writing Eq. (11.34) between 0 and 1:

$$0 = v_0 \sin\theta - g\,t_1$$

Solving this equation for t_1 will yield:

$$t_1 = \frac{v_0 \sin\theta}{g} \qquad\qquad (i)$$

Between 0 and 1, Eq. (11.32) will take the following form:

$$h = 0 + (v_0 \sin\theta)\,t_1 - \frac{1}{2}g\,t_1^2$$

Substituting Eq. (i) into this equation will yield:

$$h = \frac{v_0^2 \sin^2\theta}{2g} \qquad\qquad (ii)$$

Again from the symmetry of motion, the time lapses during the ascent must be equal to the time lapses during the descent. In other words, $t_2 = 2t_1$. This can be proven by writing Eq. (11.32) between 0 and 2, and solving it for t_2:

$$y_2 = y_0 + (v_0 \sin\theta)\,t_2 - \frac{1}{2}g\,t_2^2$$

$$0 = 0 + (v_0 \sin\theta)\,t_2 - \frac{1}{2}g\,t_2^2$$

$$0 = v_0 \sin\theta - \frac{1}{2}g\,t_2$$

Solving this equation for t_2 will yield:

$$t_2 = \frac{2\,v_0 \sin\theta}{g} \qquad\qquad (iii)$$

By comparing Eq. (iii) with Eq. (i), one can conclude that:

$$t_2 = 2t_1$$

Between 0 and 2, Eq. (11.31) can be written as:

$$\ell = 0 + (v_0 \cos\theta)\,t_2$$

Substituting Eq. (iii) into the above equation, and noting that $2\cos\theta \sin\theta = \sin(2\theta)$ (see Appendix C.4):

$$\ell = \frac{v_0^2 \sin(2\theta)}{g} \qquad\qquad (iv)$$

Eqs. (i) through (iv) are special forms of more general equations for projectile motions as given in Eqs. (11.31) through (11.34). From Eqs. (ii) and (iv), it is clear that for a given v_0, the maximum height and the horizontal range of motion of the projectile are

Table 11.3 *Variations of* $\sin^2\theta$ *and* $\sin(2\theta)$ *for* $0° \leq \theta \leq 90°$. *The maximum height h of the projectile is a function of* $\sin^2\theta$, *and range* ℓ *depends on* $\sin(2\theta)$.

θ	$\sin^2\theta$	$\sin(2\theta)$
0°	0.000	0.000
15°	0.067	0.500
30°	0.250	0.866
45°	0.500	1.000
60°	0.750	0.866
75°	0.933	0.500
90°	1.000	0.000

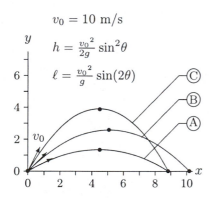

Figure 11.21 *Trajectories for* $v_0 = 10$ m/s *and* $\theta = 30, 45, 60°$ *represented as A, B, and C. (Both x and y are in meters.)*

Figure 11.22 *Long jumper.*

functions of the angle of release θ. To see the variations of h and ℓ with θ, $\sin^2\theta$ and $\sin(2\theta)$ are computed for θ between 0° and 90°. The values calculated are listed in Table 11.3. In Figure 11.21, a value of 10 m/s is assigned to v_0 and corresponding trajectories are calculated for $\theta = 30°$, 45°, and 60°. The significance of these results is that, for given v_0, the range of motion ℓ is maximum when $\theta = 45°$. Therefore, if the purpose is to maximize the horizontal range of motion of the projectile, then the angle of takeoff must be close to 45°.

Sometimes it is easier to measure the range of motion ℓ and the maximum height h of the projectile. In such cases, the unknowns are the takeoff speed v_0 and the angle of release θ. The relationship between these parameters can be derived as:

$$\theta = \arctan\left(\frac{4h}{\ell}\right) \qquad (v)$$

$$v_0 = \frac{\sqrt{2gh}}{\sin\theta} \qquad (vi)$$

Note that the results obtained in this example are valid for cases in which takeoff and landing occur at the same elevation. Also, these results are not meant to be "memorized." One of the objectives of this example was to demonstrate how Eqs. (11.31) through (11.34) that govern projectile motions could be manipulated.

11.11 Applications to Athletics

The concept of projectile motion may have many applications in athletics and sports mechanics. These applications include the motion analyses of athletes doing long jumping, high jumping, ski jumping, diving, and gymnastics, and the motion analyses of the discus, javelin, shot, baseball, basketball, football, and golf ball. The following examples are aimed to illustrate some of these applications. It should be reiterated here that we are ignoring the fundamental approximation in projectile motion analyses that possible effects of air resistance can influence the motion characteristics.

Example 11.6 Based on the assumption that the air resistance is negligible, it is suggested that the overall motion characteristic of a long jumper may be analyzed by assuming that the center of gravity of the athlete undergoes a projectile motion (Figure 11.22).

Consider an athlete who jumps a horizontal distance of 8 m. If the athlete was airborne for 1 s, calculate the takeoff speed, takeoff angle of the center of gravity of the athlete, and the maximum height the athlete's center of gravity reached.

Solution: Figure 11.23 illustrates the motion of the center of gravity of the long jumper. 0 represents takeoff, 1 represents the peak, and 2 is landing. Note that the origin of the xy coordinate system is placed at 0 so that all horizontal distance and elevation measurements are relative to the position of the center of gravity of the athlete at takeoff. $x_2 = \ell = 8$ m is the horizontal range of motion and $t_2 = 1$ s is the total time the athlete was airborne. The task is to calculate the takeoff speed v_0, takeoff angle θ, and maximum height h. Note that since t_2 is known and the motion is symmetric with respect to the peak, the time it took for the athlete to reach to the peak is $t_1 = t_2/2 = 0.5$ s.

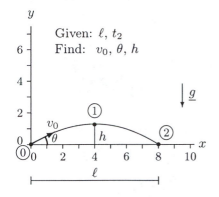

Figure 11.23 *Trajectory of the center of gravity of the athlete. (Both x and y are in meters.)*

We know ℓ and t_2. The equation that governs the displacement in the horizontal (x) direction is Eq. (11.31). Writing this equation for position 2 relative to position 0 and substituting the known parameters:

$$x_2 = x_0 + v_{x0} t_2$$
$$\ell = 0 + v_{x0} t_2$$
$$8 = v_{x0} (1)$$

Solving this equation for the unknown parameter will yield the horizontal component of the takeoff velocity, $v_{x0} = 8.0$ m/s.

We also know that at landing the elevation of the center of gravity is zero. That is, $y_2 = 0$. The equation that governs the displacement in the vertical (y) direction is Eq. (11.32). Writing this equation for position 2 relative to position 0 and substituting the known parameters:

$$y_2 = y_0 + v_{y0} t_2 - \frac{1}{2} g\, t_2^2$$
$$0 = 0 + v_{y0}(1) - \frac{1}{2}(9.8)(1)^2$$
$$0 = v_{y0} - 4.9$$

Solving this equation for the vertical component of the takeoff velocity will yield $v_{y0} = 4.9$ m/s.

Now that we know the magnitudes of the horizontal and vertical components of the takeoff velocity, we can calculate the takeoff speed:

$$v_0 = \sqrt{v_{x0}^2 + v_{y0}^2} = \sqrt{(8.0)^2 + (4.9)^2} = 9.4 \text{ m/s}$$

Note that $v_{x0} = v_0 \cos\theta$. Since we know v_{x0} and v_0, we can calculate the takeoff angle:

$$\theta = \cos^{-1}\left(\frac{v_{x0}}{v_0}\right) = \cos^{-1}\left(\frac{8.0}{9.4}\right) = \cos^{-1}(0.85) = 31.7°$$

To calculate the maximum height reached, we can again utilize Eq. (11.32). Writing this equation for position 1 relative to 0 and solving it for h will yield:

$$y_1 = y_0 + v_{y0} t_1 - \frac{1}{2} g t_1^2$$

$$h = 0 + (4.9)(0.5) - \frac{1}{2}(9.8)(0.5)^2$$

$$h = 1.2 \text{ m}$$

Therefore, at the peak, the center of gravity of the athlete was 1.2 meters above the level it was at the takeoff.

Example 11.7 During a practice, a shot-putter puts the shot at a distance $\ell = 6$ m. At the instant the athlete releases the shot, the elevation of the shot is $h_0 = 1.8$ m as measured from the ground level, and the angle of release is $\theta = 30°$ (Figure 11.24).

Determine the speed at which the athlete released the shot, the landing speed of the shot, and the total time the shot was in the air.

Solution: Eqs. (11.31) and (11.32) can be utilized to solve this problem. In Figure 11.25, the origin of the xy coordinate frame is located at the ground level directly under the point of release, which is designated as 0. The shot ascends, reaches a peak at 1, and lands on the field at 2. With respect to the coordinate frame adopted, the initial and landing coordinates of the shot are: $x_0 = 0$, $y_0 = h_0$, $x_2 = \ell$, and $y_2 = 0$. If t_2 refers to the total time the shot was in the air, then Eq. (11.31) can be written between 0 and 2 as:

$$x_2 = x_0 + (v_0 \cos \theta) t_2$$

$$\ell = 0 + (v_0 \cos \theta) t_2$$

Solving this equation for t_2 and substituting the known parameters:

$$t_2 = \frac{\ell}{v_0 \cos \theta} = \frac{6}{v_0 \cos 30°} = \frac{6.93}{v_0} \tag{i}$$

Similarly, writing Eq. (11.32) between 0 and 2 and substituting the known parameters:

$$y_2 = y_0 + (v_0 \sin \theta) t_2 - \frac{1}{2} g t_2^2$$

$$0 = h_0 + (v_0 \sin \theta) t_2 - \frac{1}{2} g t_2^2$$

$$0 = 1.8 + (v_0 \sin 30°) t_2 - \frac{1}{2}(9.8) t_2^2$$

$$0 = 1.8 + 0.5 v_0 t_2 - 4.9 t_2^2 \tag{ii}$$

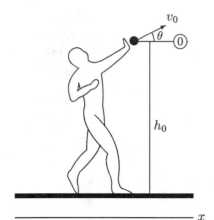

Figure 11.24 *Shot-putter.*

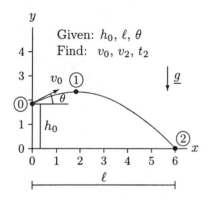

Given: h_0, ℓ, θ
Find: v_0, v_2, t_2

Figure 11.25 *Trajectory of the shot. (x and y are in meters.)*

Substituting Eq. (*i*) into Eq. (*ii*):

$$0 = 1.8 + 0.5v_0 \left(\frac{6.93}{v_0}\right) - 4.9\left(\frac{6.93}{v_0}\right)^2$$

Simplifying the second term on the right-hand side of this equation by eliminating v_0, carrying out the calculations, and solving this equation for v_0 will yield:

$$v_0 = 6.69 \text{ m/s}$$

Knowing v_0, time t_2 can be calculated using Eq. (*i*). This will yield $t_2 = 1.04$ s.

We can utilize Eqs. (11.33) and (11.34) to calculate the landing speed v_2 of the shot. Since the horizontal component of the velocity vector is constant throughout the projectile motion:

$$v_{x2} = v_{x0} = v_0 \cos\theta = (6.69)(\cos 30°) = 5.79 \text{ m/s}$$

Writing Eq. (11.34) between 0 and 2:

$$v_{y2} = v_{y0} - g\,t_2 = v_0 \sin\theta - g\,t_2$$

Substituting the known parameters and carrying out the calculations will yield:

$$v_{y2} = (6.69)(\sin 30°) - (9.8)(1.04) = -6.85$$

Note that we obtained a negative value for a scalar quantity, which is not permitted. Here, the negative sign implies direction. We adopted the upward direction to be positive for the y axis. The negative value calculated above indicates that the direction of the vertical component of the landing velocity is downward (opposite to that of positive y axis). Now, we can rewrite v_{y2} as:

$$v_{y2} = 6.85 \text{ m/s} \quad (\downarrow)$$

Knowing the magnitudes of the horizontal and vertical components of the landing velocity enables us to calculate the landing speed:

$$v_2 = \sqrt{v_{x2}^2 + v_{y2}^2} = \sqrt{(5.79)^2 + (6.85)^2} = 8.97 \text{ m/s}$$

Example 11.8 The diver illustrated in Figure 11.26 undergoes both translational and rotational, or general motion. The overall translational motion of the diver can be analyzed by observing the trajectory of the diver's center of gravity which can be assumed to undergo a projectile motion.

Consider a case in which a diver takes off from a diving board located at a height $h_0 = 10$ m above the water level and enters

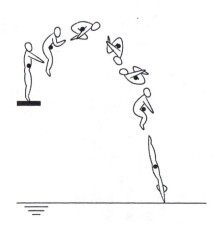

Figure 11.26 *A diver.*

the water at a horizontal distance $\ell = 5$ m from the end of the board. If the total time the diver remains in the air is $t_2 = 2.5$ s, calculate the speed and angle of takeoff of the diver's center of gravity.

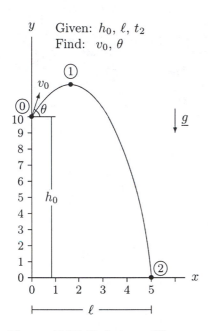

Given: h_0, ℓ, t_2
Find: v_0, θ

Figure 11.27 *Trajectory of the center of gravity of the diver. (Both x and y are in meters.)*

Solution: The trajectory of the center of gravity of the diver is shown in Figure 11.27. In this case, speed and angle of takeoff (v_0 and θ) are not known, but are to be determined. In Figure 11.27, 0, 1, and 2 represent the takeoff, peak, and entry into the water stages of motion, respectively. The origin of the xy coordinate frame is located at the water level directly under 0. Therefore, the coordinates of position 0 are: $x_0 = 0$ and $y_0 = h_0 = 10$ m. We know the coordinates of position 2 as well: $x_2 = \ell = 2.5$ m and $y_0 = 0$. We know another parameter associated with position 2, which is $t_2 = 2.5$ s.

We can utilize Eqs. (11.31) and (11.32). Writing Eq. (11.31) between 0 and 2, and substituting the known parameters:

$$x_2 = x_0 + (v_0 \cos \theta)\, t_2$$
$$5 = 0 + (v_0 \cos \theta)(2.5)$$

Solving this equation for $v_0 \cos \theta$:

$$v_0 \cos \theta = \frac{5}{2.5} = 2 \qquad (i)$$

Similarly, writing Eq. (11.32) between 0 and 2, and substituting the known parameters:

$$y_2 = y_0 + (v_0 \sin \theta)\, t_2 - \frac{1}{2} g\, t_2^2$$
$$0 = 10 + (v_0 \sin \theta)(2.5) - \frac{1}{2}(9.8)(2.5)^2$$

Solving this equation for $v_0 \sin \theta$:

$$v_0 \sin \theta = 8.25 \qquad (ii)$$

Noting that $v_0 \sin \theta$ over $v_0 \cos \theta$ is equal to $\tan \theta$, divide Eq. (*ii*) by Eq. (*i*):

$$\tan \theta = \frac{8.25}{2} = 4.125$$

Considering the inverse tangent of the value calculated above will yield:

$$\theta = 76.4°$$

The speed of takeoff can now be determined from Eq. (*i*):

$$v_0 = \frac{2}{\cos \theta} = \frac{2}{\cos 76.4°} = 8.5 \text{ m/s}$$

11.12 Exercise Problems

Problem 11.1 Consider a person throwing a ball upward into the air with an initial speed of $v_0 = 10\,\text{m/s}$ (Figure 11.28). Assume that at the instant when the ball is released, the person's hand is at a height $h_0 = 1.5$ m above the ground level.

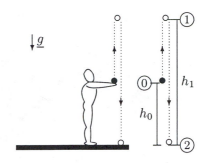

Neglecting the possible effects of air resistance, determine the maximum height h_1 that the ball reached, the total time t_2 it took for the ball to ascend and descend, and the speed v_2 of the ball just before it hit the ground.

Note that this problem must be handled in two phases: ascend and descend. Also note that the speed of the ball at the peak was zero.

Answers: $h_1 = 6.6$ m, $t_2 = 1.16$ s, $v_2 = 11.4$ m

Figure 11.28 *Problem 11.1.*

Problem 11.2 Consider the car shown in Figure 11.29. At position 0, the car is stationary on a hill that makes an angle θ with the horizontal. Assume that the gear of the car is at "neutral" and that at time $t_0 = 0$ the brakes of the car are released. Under the effect of the gravitational acceleration g, the car will start moving down the hill. After some time, the car will be at position 1, which is at a distance d from position 0 measured parallel to the hill.

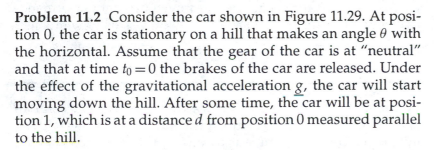

Figure 11.29 *Problem 11.2.*

Show that time t_1 to cover the distance between positions 0 and 1, and speed v_1 of the car at position 1 can be expressed as:

$$t_1 = \sqrt{\frac{2\,d}{g\,\sin\theta}}$$

$$v_1 = \sqrt{2\,g\,d\,\sin\theta}$$

Problem 11.3 Based on the assumption that the air resistance is negligible, it is suggested that the overall motion characteristic of a long jumper may be analyzed by assuming that the center of gravity of the athlete undergoes a projectile motion (Figure 11.30).

Figure 11.30 *Problem 11.3.*

Consider an athlete who jumps a horizontal distance of 9 m after reaching a maximum height of 1.5 m. What was the takeoff speed v_0 of the athelete? Discuss how the athlete can improve his/her performance.

Answer: $v_0 = 7.28$ m/s

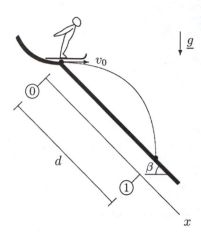

Figure 11.31 *Problem 11.4.*

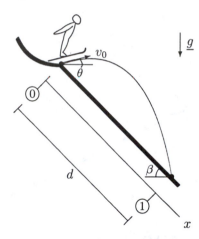

Figure 11.32 *Problem 11.5.*

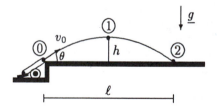

Figure 11.33 *Problem 11.6.*

Problem 11.4 The ski jumper shown in Figure 11.31, leaves the ramp with a horizontal speed of v_0 and lands on a slope that makes an angle $\beta = 45°$ with the horizontal.

Neglecting air resistance (the effect of which may be quite significant), determine the takeoff speed v_0, landing speed v_1, and the total time t_1, that the ski jumper was airborne if the skier touched down at a distance $d = 50$ m from the ramp measured parallel to the slope.

Answers: $v_0 = 13.2$ m/s, $v_1 = 29.6$ m/s, $t_1 = 2.7$ s

Problem 11.5 Assume that in another trial, the ski jumper in the previous problem manages to maintain the takeoff speed at $v_0 = 13.2$ m/s, but leaves the ramp at an angle $\theta = 10°$ with the horizontal (Figure 11.32).

How far from the ramp would the ski jumper land on the slope? Discuss whether the ski jumper improved his/her performance as compared to the trial in Problem 11.4. If yes, by how much?

Answers: $d = 57$ m, 14% improvement

Problem 11.6 Figure 11.33 illustrates the trajectory of a cannonball. Assume that the cannonball was fired into the air with an initial speed of $v_0 = 100$ m/s at position 0. The cannonball landed at position 2 that is at a horizontal distance $\ell = 1000$ m measured from position 0.

Calculate the angle of takeoff θ, time of flight t_2, and maximum height h, that the cannonball reached.

Note that $2 \sin \theta \cos \theta = \sin(2\theta)$ and take $g = 10$ m/s².

Answers: $\theta = 45°$, $t_2 = 14$ s, $h = 250$ m

Chapter 12

Linear Kinetics

12.1 Overview

As studied in the previous chapter, kinematic analyses are concerned with the description of time-dependent aspects of motion in terms of displacement, velocity, and acceleration without dealing with the factors causing the motion. The field of *kinetics*, on the other hand, is based on kinematics and incorporates into the analysis the effects of forces that cause the motion.

Based on the type of motion involved, the field of kinetics can be divided into linear (translational) and angular (rotational) kinetics. Translation is caused by the net force applied on an object, whereas rotation is the consequence of the net torque. An object will translate and rotate simultaneously (undergo a general motion) if there is both a net force and a net moment acting on it. In addition to classifying a motion as translational, rotational, or general, the field of kinetics can be further distinguished as the *kinetics of particles* and the *kinetics of rigid bodies*. Particle kinetics is easier to implement than rigid body kinetics that introduces the size and shape of the bodies into the analyses. The distinction between a particle and a rigid body is particularly important if the object is undergoing a rotational motion. If the object is sufficiently small or it is undergoing translational motion only, then the geometric characteristics of the object may be ignored and the object can be treated as a particle located at its center of gravity with a mass equal to the total mass of the object. For example, what is significant for a person pushing a block on a flat surface is the total mass of the block, not its size or shape.

Kinetic analyses utilize Newton's second law of motion, which can be formulated in various ways. One way of representing Newton's second law of motion is in terms of the equations of motion, which are particularly suitable for solving problems requiring the analysis of acceleration. The use of the equations of motion for linear kinetics will be discussed next. Another way of formulating Newton's second law is through work and energy methods that will also be discussed in this chapter within the context of linear kinetics. There are also methods based on impulse and momentum that will be presented in Chapter 15.

12.2 Equations of Motion

A body accelerates if there is a nonzero net force acting on it. Newton's second law of motion states that the magnitude of the acceleration of a body is directly proportional to the magnitude of the resultant force and inversely proportional to its mass. The direction of the acceleration is the same as the direction of the resultant force.

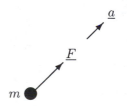

Consider a particle of mass m that is acted upon by a force F and let a be the resulting acceleration of the particle (Figure 12.1).

Figure 12.1 *An object will accelerate in the direction of the applied force.*

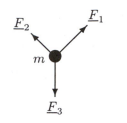

Figure 12.2 *The net force is the vector sum of all forces acting on the object.*

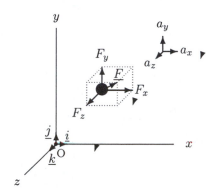

Figure 12.3 *Rectangular components of the force and acceleration vectors.*

If Newton's second law of motion can be expressed as:

$$\underline{F} = m\,\underline{a} \qquad (12.1)$$

there is more than one force acting on the particle (Figure 12.2), then $\underline{F}$ in Eq. (12.1) must be replaced by the net or the resultant of all forces acting on it. The resultant of a system of forces can be determined by considering the vector sum of all forces. Therefore:

$$\sum \underline{F} = m\,\underline{a} \qquad (12.2)$$

This is known as the *equation of motion*. Since force and acceleration are vector quantities, they can be expressed in terms of their components in reference to a chosen coordinate frame. For translational motion analyses, it is best to use the Cartesian (rectangular) coordinate system that consists of the x, y, and z axes with $\underline{i}$, $\underline{j}$, and $\underline{k}$ unit vectors indicating the positive x, y, and z directions, respectively (Figure 12.3). If there is a single force acting on an object, then the force vector and the resulting acceleration vector can be expressed in terms of their components along the rectangular coordinate directions:

$$\underline{F} = F_x\,\underline{i} + F_y\,\underline{j} + F_z\,\underline{k} \qquad (12.3)$$
$$\underline{a} = a_x\,\underline{i} + a_y\,\underline{j} + a_z\,\underline{k} \qquad (12.4)$$

Substituting Eqs. (12.3) and (12.4) into Eq. (12.1):

$$F_x\,\underline{i} + F_y\,\underline{j} + F_z\,\underline{k} = m\,a_x\,\underline{i} + m\,a_y\,\underline{j} + m\,a_z\,\underline{k} \qquad (12.5)$$

For the equilibrium of this vector equation, the following conditions must be satisfied:

$$
\begin{aligned}
F_x &= m\,a_x \\
F_y &= m\,a_y \\
F_z &= m\,a_z
\end{aligned}
\qquad (12.6)
$$

If there is more than one force acting on the object, then F_x, F_y, and F_z must be replaced by the sum of all forces acting in the x, y, and z directions, respectively:

$$
\begin{aligned}
\sum F_x &= m\,a_x \\
\sum F_y &= m\,a_y \\
\sum F_z &= m\,a_z
\end{aligned}
\qquad (12.7)
$$

Equations (12.7) state that the sum of all forces acting in one direction is equal to the mass times the acceleration of the body in that direction. Note that for one-dimensional motion analysis, only one of these equations need to be considered. For a two-dimensional case, two of the above equations are sufficient to analyze the problem.

12.3 Special Cases of Translational Motion

A force can be applied in various ways. For example, an applied force may be constant or it may vary over time. Applied forces can be measured in various ways as well. The magnitude of a force vector can be measured as a function time, as a function of the relative position of the object upon which it is applied, or as a function of velocity. Some of these cases will be discussed next. To illustrate the methods of handling these cases in a concise manner, it will be assumed that the motion is along a straight line in the x direction and under the effect of only one applied force. Using the vectorial properties of the parameters involved, these methods can be easily expanded to analyze two- and three-dimensional translational motions under the action of more than one force.

Note that the derivations provided in this section are aimed to demonstrate that different cases can be handled through proper mathematical manipulations. The mathematics involved for the cases in which the applied force is a function of displacement may be beyond the scope of this text, and can be omitted without losing the continuity of the topics to be covered in the following sections.

12.3.1 Force is constant

If a force applied on an object has a constant magnitude and direction, the object will move with a constant acceleration in the direction of the applied force. Assume that a force with magnitude F_x is applied on an object with mass m. The magnitude a_x of the constant acceleration of the object in the x direction can be calculated using the equation of motion in the x direction:

$$a_{x0} = \frac{F_x}{m} = \text{constant}$$

Once the acceleration of the object is determined, the kinematic equations can be utilized to calculate the velocity and displacement of the object as well:

$$v_x = v_{x0} + \int_0^t a_x \, dt = v_{x0} + \frac{F_x}{m} t \qquad (12.8)$$

$$x = x_0 + \int_0^t v_x \, dt = x_0 + v_{x0} \, t + \frac{1}{2} \frac{F_x}{m} t^2 \qquad (12.9)$$

Here, v_{x0} and x_0 are the initial speed and displacement of the object in the x direction at time $t = 0$. Note that these results can be used for a situation in which there is a second force with constant magnitude F_y acting on the object in the y direction simply by replacing x with y throughout the equations.

12.3.2 Force is a function of time

If the magnitude of a force applied on an object is a function of time, then $F_x = F_x(t)$. The resulting acceleration of the object is also a function of time:

$$a_x(t) = \frac{F_x(t)}{m}$$

The velocity and displacement of the object can now be determined using the kinematic relationships:

$$v_x = v_{x0} + \int_0^t a_x(t)\, dt \tag{12.10}$$

$$x = x_0 + \int_0^t v_x(t)\, dt \tag{12.11}$$

The function $F_x(t)$ must be provided so that the integral in Eq. (12.10) can be evaluated.

12.3.3 Force is a function of displacement

Sometimes it is more convenient to express force as a function of displacement, in which case $F_x = F_x(x)$. By definition, acceleration is equal to the time rate of change of velocity. Therefore, the equation of motion in the x direction can be expressed as:

$$\frac{dv_x}{dt} = \frac{F_x(x)}{m}$$

Employing the chain rule of differentiation (see Appendix C.2.6), the time derivative of velocity can be expressed as:

$$\frac{dv_x}{dt} = \frac{dv_x}{dx}\frac{dx}{dt} = \frac{dv_x}{dx} v_x$$

Therefore, the equation of motion in the x direction is:

$$v_x \frac{dv_x}{dx} = \frac{F_x(x)}{m}$$

Multiplying both sides by dx:

$$v_x\, dv_x = \frac{F_x(x)}{m}\, dx$$

The left-hand side of this equation is a function of v_x only and can be integrated with respect to v_x, and the right-hand side is a function of x only and can be integrated with respect to x:

$$\int_{v_{x0}}^{v_x} v_x\, dv_x = \int_{x_0}^{x} \frac{F_x(x)}{m}\, dx$$

Evaluating the integral on the left-hand side:

$$\frac{1}{2}\left(v_x^2 - v_{x0}^2\right) = \frac{1}{m}\int_{x_0}^{x} F_x(x)\,dx$$

Rearranging the order of terms:

$$v_x^2 = v_{x0}^2 + \frac{2}{m}\int_{x_0}^{x} F_x(x)\,dx \qquad (12.12)$$

F_x must be provided as a function of x, so that the integral in Eq. (12.10) can be evaluated. For given $F_x(x)$, Eq. (12.12) will yield v_x as a function of x. Once v_x is known, the acceleration of the object can be determined using:

$$a_x = v_x \frac{dv_x}{dx} \qquad (12.13)$$

12.4 Procedure for Problem Solving in Kinetics

The procedure for analyzing kinetic characteristics of objects undergoing translational motion using the equations of motion can be outlined as follows:

• Draw a simple, neat diagram of the system to be analyzed.

• Isolate the bodies of interest from their surroundings and draw their free-body diagrams by showing all external forces acting on them. Indicate the correct directions for the known forces. If the direction of a force vector is not known, assume a positive direction for it. If that force appears to have a negative value in the solution, it would mean that the assumed direction for the force vector was incorrect.

• Designate the direction of motion of each object on the sidelines (not as parts of the free-body diagrams). It is particularly important to be consistent with the assumed direction of the motion throughout the analyses.

• Choose a convenient coordinate system. For two-dimensional cases, rectangular coordinates x and y are usually the most convenient.

• Apply the equations of motion. For two-dimensional motion analysis there are two governing equations, and therefore, the number of unknowns to be determined cannot be more than two. In linear kinetics, the unknowns are either forces or accelerations.

• Include the correct directions of forces and accelerations in the solution, along with their units.

• The kinematic relations between position, velocity, and acceleration can also be utilized if the information about the velocity and/or position of the object analyzed is given or required.

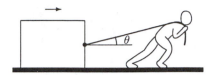

Figure 12.4 *A block is being pulled on a horizonal surface.*

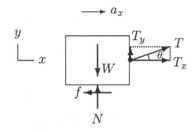

Figure 12.5 *The free-body diagram of the block.*

Example 12.1 As illustrated in Figure 12.4, consider a block of mass $m = 50$ kg, which is being pulled on a rough, horizontal surface by a person using a rope. Assume that the person is applying a constant force of $T = 150$ N on the block, the rope makes an angle $\theta = 30°$ with the horizontal, and the coefficient of kinetic friction between the block and the horizontal surface is $\mu = 0.2$.

Determine the acceleration of the block if the bottom surface of the block remains in full contact with the floor throughout the motion.

Solution: The free-body diagram of the block is shown in Figure 12.5. The positive x direction is chosen in the direction of motion of the block. W is the weight of the block, f is the magnitude of the frictional force acting in the direction opposite to the direction of motion, and N is the magnitude of the reaction force applied by the floor on the block. T is the force exerted by the person, which is transmitted to the block through the rope. The rope makes an angle $\theta = 30°$ with the horizontal. Therefore, T has components in the x and y directions:

$$T_x = T \, \cos\theta \qquad (\rightarrow)$$
$$T_y = T \, \sin\theta \qquad (\uparrow)$$

The weight of the block is due to the gravitational effect of Earth on the mass of the block, and can be expressed as:

$$W = m \, g \qquad (\downarrow)$$

The magnitude of the frictional force is proportional to the magnitude of the normal force, and they are related through the coefficient of friction between the surfaces in contact:

$$f = \mu \, N \qquad (\leftarrow) \qquad \qquad (i)$$

Equations of motion in the x and y directions can now be applied to determine an expression for the acceleration of the block. The block has no motion in the y direction, and therefore, the acceleration of the block in the y direction is zero ($a_y = 0$). The equation of motion in the y direction is:

$$\sum F_y = 0 : \quad N + T_y - W = 0$$

Solving this equilibrium equation for force N will yield:

$$N = W - T_y = m \, g - T \, \sin\theta \qquad \qquad (ii)$$

Substituting Eq. (ii) into Eq. (i) will yield:

$$f = \mu \, (m \, g - T \, \sin\theta) \qquad \qquad (iii)$$

Now, the equation of motion in the x direction can be considered:

$$\sum F_x = m\,a_x: \quad T_x - f = m\,a_x$$

Solving this equation for a_x will yield

$$a_x = \frac{1}{m}(T_x - f) \qquad (iv)$$

Substituting Eq. (iii) and $T_x = T\cos\theta$ into Eq. (iv):

$$a_x = \frac{1}{m}\left[T\,\cos\theta - \mu(m\,g - T\,\sin\theta)\right]$$

Substituting the numerical values $m = 50$ kg, $T = 150$ N, $\theta = 30°$, $\mu = 0.2$, and $g = 9.8$ m/s, and carrying out the calculations will yield $a_x = 0.94$ m/s^2 ($\rightarrow$).

12.5 Work and Energy Methods

The fundamental method of analyzing the kinetic characteristics of bodies is based on the equations of motion, which are mathematical representations of Newton's second law of motion. Using the equations of motion, one can determine accelerations. In some cases, particularly when the forces involved are not constant, the solution of equations of motion may be difficult. To handle such situations, alternative methods are developed that are based on the concepts of work and energy. These methods are also derived from Newton's laws, and can be applied to analyze the forces, velocities, and displacements involved in relatively complex systems without resorting to the equations of motion.

12.6 Mechanical Work

By definition, *mechanical work* is the product of force and corresponding displacement. Work is a scalar quantity. There is no direction associated with work.

12.6.1 Work done by a constant force

To explore the definition of work, consider the block in Figure 12.6. Assume that a constant, horizontal force $\underline{F}$ is applied on the block so as to move it from position 1 to position 2, which are s distance apart. The work done, W, by force $\underline{F}$ on the block to move the block from position 1 to 2 is equal to the magnitude of the force vector times the displacement:

$$W = F\,s \qquad (12.14)$$

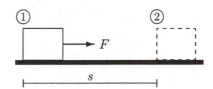

Figure 12.6 *A constant force applied on the block displaces it from position 1 to position 2.*

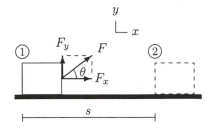

Figure 12.7 *A constant force that makes an angle θ with the horizontal is applied on the block.*

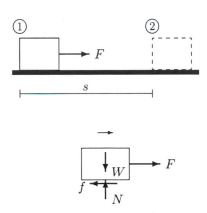

Figure 12.8 *Frictional forces do negative work.*

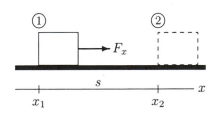

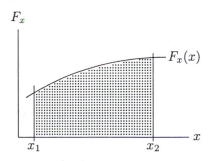

Figure 12.9 *Work is equal to the area under the force versus displacement curve.*

Consider the same block, which is pulled from position 1 to 2 by another constant force $\underline{F}$ that makes an angle θ with the horizontal (Figure 12.7). The work done by $\underline{F}$ on the block is equal to the magnitude of the force component in the direction of displacement times the displacement itself. Since the component of $\underline{F}$ along the horizontal is $F_x = F \cos \theta$, the work done by $\underline{F}$ to move the block from position 1 to 2 is:

$$W = F_x \, s = F \, s \, \cos \theta \qquad (12.15)$$

Note that Eqs. (12.14) and (12.15) are consistent with each other since $\cos \theta = 1$ when $\theta = 0°$.

For a force to do work, the body on which the force is applied must undergo a displacement and the force vector must have a nonzero component in the direction of displacement. For example, the vertical component, $F_y = F \sin \theta$, of the force vector in Figure 12.7 does no work because the block is not displaced in the vertical direction.

Work done can be positive or negative. The work done by a force is positive if the force is applied in the same direction as the displacement. If the applied force and displacement have opposite directions, then the work done by that force is negative. A typical example of negative work is the one done by a frictional force. As illustrated in Figure 12.8, assume that a block is pulled by a force $\underline{F}$ toward the right to displace the block by a distance s. The work done W_f by the frictional force f on the block while the block was displaced by a distance s is:

$$W_f = -f \, s \qquad (12.16)$$

If there is more than one external force acting on a body in motion, then there is one work done for each force. The net work done is the algebraic sum of work done by individual forces. For example, the net work done for the case illustrated in Figure 12.8 is:

$$W = F \, s - f \, s$$

12.6.2 Work done by a varying force

Equation (12.14) can only be used to calculate the work done by a constant force. If an applied force is a function of displacement, then the work done can be calculated by considering the integral of the force over the distance it is applied.

As illustrated in Figure 12.9, consider a block pulled along the x direction by a force F_x that varies with the displacement of the block in the x direction. That is, $F_x = F_x(x)$. Assume that the block that was originally located at position 1 moves to position 2, which are s distance apart. Let x_1 and x_2 represent the initial and final positions of the block, respectively. If the variation of F_x with respect to x is known, then the work done by F_x to move the block from position 1 to 2 can be determined using:

$$W = \int_{x_1}^{x_2} F_x \, dx \qquad (12.17)$$

Note that the evaluation of the definite integral in Eq. (12.17) will yield the total area under the force versus position curve, the x axis, and vertical lines passing through $x = x_1$ and $x = x_2$ (shaded area in Figure 12.9). Also note that if the force F_x is constant, then the integration in Eq. (12.17) will yield $W = F_x(x_2 - x_1) = F_x s$, which is consistent with Eq. (12.14).

12.6.3 Work as a scalar product

For some applications, it may be convenient to utilize the definition of work as the dot (scalar) product of the force and displacement vectors. As discussed in Appendix B.14, the dot product of any two vectors is a scalar quantity equal to the product of magnitudes of the two vectors multiplied by the cosine of the smaller angle between the two. In the case of work done by a constant force $\underline{F}$ on a body whose displacement vector is given by $\underline{s}$:

$$W = \underline{F} \cdot \underline{s} \qquad (12.18)$$

If θ is the smaller angle between vectors $\underline{F}$ and $\underline{s}$, then:

$$W = \underline{F} \cdot \underline{s} = F \, s \, \cos\theta \qquad (12.19)$$

Force and displacement vectors can be expressed in terms of their rectangular components:

$$\underline{F} = F_x \underline{i} + F_y \underline{j} + F_z \underline{k} \qquad (12.20)$$
$$\underline{s} = x \underline{i} + y \underline{j} + z \underline{k} \qquad (12.21)$$

The dot product of unit vectors are such that $\underline{i} \cdot \underline{i} = \underline{j} \cdot \underline{j} = \underline{k} \cdot \underline{k} = 1$ and $\underline{i} \cdot \underline{j} = \underline{j} \cdot \underline{k} = \underline{k} \cdot \underline{i} = 0$. Substituting Eqs. (12.20) and (12.21) into Eq. (12.18) and carrying out the dot products of unit vectors will yield:

$$W = F_x \, x + F_y \, y + F_z \, z \qquad (12.22)$$

Equation (12.22) is significant in that it represents the total work done by the components of the force vector in the x, y, and z directions. For example, the work done in the x direction is equal to the magnitude of the force component in the x direction times the displacement in the same direction. Note that for a biaxial motion in the xy-plane, Eq. (12.22) reduces to $W = F_x x + F_y y$ and $W = F_x x$ for a uniaxial motion in the x direction.

12.7 Mechanical Energy

The term *energy* is used to describe the capacity of a system to do work on another system. Energy can take various forms such as mechanical, thermal, chemical, and nuclear. The field of mechanics is primarily concerned with the mechanical form of energy. Mechanical energy can be categorized as potential energy and kinetic energy. Energy is also a scalar quantity.

12.7.1 Potential energy

The *potential energy* of a system is associated with its position or elevation. It is the energy stored in the system that can be converted into kinetic energy. The concept of potential energy comes from the perception that an object located at a height can do useful work if it is allowed to descend. The potential of an object to do work due to the relative height of its center of gravity is defined as *gravitational potential energy*. Consider the object with weight $W = mg$ shown in Figure 12.10. The object is at position 1, which is located at a height h measured relative to position 2. The gravitational potential energy, $\mathcal{E}_P$, of the object at position 1 relative to position 2 is:

$$\mathcal{E}_P = W h = m g h \qquad (12.23)$$

Note that Wh is essentially the work that the force of gravity would do on the object to move it from position 1 to position 2, which are h distance apart.

12.7.2 Kinetic energy

Kinetic energy is associated with motion. Every moving object has a kinetic energy. The kinetic energy, $\mathcal{E}_K$, of an object with mass m moving with a speed v is equal to the product of one half of the mass and the square of the speed of the object:

$$\mathcal{E}_K = \frac{1}{2} m v^2 \qquad (12.24)$$

12.8 Work–Energy Theorem

There is a relationship between the kinetic energy and the work done. The net work done, W_{12}, on an object to displace the object from position 1 to position 2 is equal to the change in kinetic energy, $\Delta \mathcal{E}_K$, of the object between positions 1 and 2. This is known as the *work–energy theorem* and can be expressed as:

$$W_{12} = \Delta \mathcal{E}_K = \mathcal{E}_{K2} - \mathcal{E}_{K1} \qquad (12.25)$$

12.9 Conservation of Energy Principle

Forces may be conservative and nonconservative. A force is conservative if the work done by that force to move an object

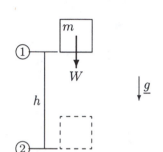

Figure 12.10 *Gravitational potential energy.*

between two positions is independent of the path taken. A typical example of conservative forces is the gravitational force. The frictional force, on the other hand, is a nonconservative force. Nonconservative forces dissipate energy as heat.

The net work done on a system by conservative forces is converted into kinetic and potential energies in such a manner that the total energy of the system (sum of kinetic and potential energies) remains constant throughout the motion. This is known as the *principle of conservation of mechanical energy*, and between any two positions 1 and 2 it can be stated as:

$$\mathcal{E}_{K1} + \mathcal{E}_{P1} = \mathcal{E}_{K2} + \mathcal{E}_{P2} \tag{12.26}$$

12.10 Dimension and Units of Work and Energy

Mechanical work and energy have the same dimension and units. By definition, work done is force times displacement. Therefore, work has the dimension of force times the dimension of length.

$$[\text{WORK}] = [\text{FORCE}]\,[\text{DISPLACEMENT}] = M\frac{L^2}{T^2}$$

The units of work and energy in different systems of units are provided in Table 12.1.

Table 12.1 *Units of work and energy.*

SYSTEM	UNITS OF WORK AND ENERGY	SPECIAL NAME
SI	Newton-meter (N-m)	Joule (J)
c-g-s	dyne-centimeter (dyn-cm)	erg
British	pound-foot (lb-ft)	

12.11 Power

Power, $\mathcal{P}$, is defined as the time rate of work done:

$$\mathcal{P} = \frac{d\mathcal{W}}{dt} \tag{12.27}$$

The work done by a constant force on an object can be determined by considering the dot product of the force and displacement vectors ($\mathcal{W} = \underline{F} \cdot \underline{s}$):

$$\mathcal{P} = \frac{d}{dt}(\underline{F} \cdot \underline{s})$$

If the force vector $\underline{F}$ is constant, then:

$$\mathcal{P} = \underline{F} \cdot \frac{d\underline{s}}{dt} = \underline{F} \cdot \underline{v} \tag{12.28}$$

In Eq. (12.28), $\underline{v}$ is the velocity vector of the object. If the applied force is collinear with the velocity, then $\mathcal{P} = F v$. Power is a scalar quantity, and has the dimension of force times velocity. The units of power are given in Table 12.2.

Table 12.2 *Units of power.* $(1 \text{ hp} = 550 \text{ lb-ft/s} = 746 \text{ W})$

System	Units of Power	Special Name
SI	N-m/s = J/s	Watt (W)
c-g-s	dyn-cm/s = erg/s	
British	lb-ft/s	horsepower (hp)

12.12 Applications of Energy Methods

The work–energy theorem stated in Eq. (12.23) and the principle of conservation of energy stated by Eq. (12.24) provide alternative methods of problem solving in dynamics. The work–energy theorem can be used to analyze problems involving nonconservative forces. On the other hand, the principle of conservation of energy is useful only when the forces involved are conservative. As compared to the applications of the equations of motion, these methods are easier to apply and are particularly useful when the information provided or to be determined is in terms of velocities rather than accelerations. Definitions of important concepts introduced in this chapter and various methods of analyses in kinetics are summarized in Table 12.3. The following examples will demonstrate some of the applications of these methods.

Table 12.3 *Summary of equations and formulas.*

WORK DONE BY A VARYING FORCE	$W = \int_{x_1}^{x_2} F_x dx$
WORK DONE BY A CONSTANT FORCE	$W = F_x(x_2 - x_1) = F_x s$
POTENTIAL ENERGY	$\mathcal{E}_P = m g h$
KINETIC ENERGY	$\mathcal{E}_K = \frac{1}{2} m v^2$
CONSERVATION OF ENERGY PRINCIPLE	$\mathcal{E}_{K1} + \mathcal{E}_{P1} = \mathcal{E}_{K2} + \mathcal{E}_{P2}$
WORK-ENERGY THEOREM	$W_{12} = \mathcal{E}_{K2} - \mathcal{E}_{K1}$
EQUATION OF MOTION	$\sum F_x = m a_x$

Example 12.2 A 20 kg block is pushed up a rough, inclined surface by a constant force of $P = 150 \text{ N}$ that is applied parallel

to the incline (Figure 12.11). The incline makes an angle $\theta = 30°$ with the horizontal and the coefficient of friction between the incline and the block is $\mu = 0.2$.

If the block is displaced by $\ell = 10$ m, determine the work done on the block by force $\underline{P}$, by the force of friction, and by the force of gravity. What is the net work done on the block?

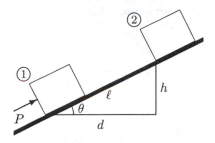

Solution: The free-body diagram of the block is shown in Figure 12.12. W is the weight of the block, f is the frictional force at the bottom surface of the block, and N is the reaction force applied by the incline on the block. The x and y directions are chosen in such a manner that the motion occurs in the positive x direction, and there is no displacement of the block in the y direction. Force $\underline{P}$ is applied in the same direction as the displacement of the block. Therefore, the work done by $\underline{F}$ to displace the block by a distance of ℓ along the incline is:

Figure 12.11 *A block is pushed from position 1 to 2.*

$$W_F = P\,\ell \qquad (i)$$

The weight $\underline{W}$ of the block has components along the x and y directions, such that $W_x = W \sin\theta$ and $W_y = W \cos\theta$. Since there is no motion in the y direction, the block is in equilibrium in the y direction. The equilibrium in the y direction requires that $N = W_y$, or since $W = mg$, $N = mg \cos\theta$. The relationship between the frictional force and the normal force at the surfaces of contact is such that $f = \mu N = \mu mg \cos\theta$. The frictional force acts in a direction parallel to the incline but opposite to that of the displacement of the block. Therefore, the work done by f on the block as the block is displaced by a distance ℓ is:

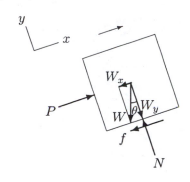

Figure 12.12 *Free-body diagram of the block.*

$$W_f = -f\,\ell = -\mu\,m\,g\,\ell\,\cos\theta \qquad (ii)$$

The force of gravity (weight) always acts downward. In this case, it has a component parallel to the incline in the negative x direction with magnitude $W_x = mg \sin\theta$. The work done by W_x on the block as it moves from position 1 to position 2 is:

$$W_g = -m\,g\,\ell\,\sin\theta \qquad (iii)$$

Knowing the work done by individual forces acting on the block, we can determine the net work done on the block:

$$W = W_F + W_f + W_g = F\,\ell - m\,g\,\ell\,(\sin\theta + \mu\,\cos\theta) \qquad (iv)$$

Substituting the numerical values of the parameters involved into Eqs. (i) through (iv) and carrying out the calculations will yield:

$$W_F = (150)(10) = 1500\ \text{J}$$
$$W_f = -(0.2)(20)(9.8)(10)(\cos 30°) = -340\ \text{J}$$
$$W_g = -(20)(9.8)(10)(\sin 30°) = -980\ \text{J}$$
$$W = 1500 - 340 - 980 = 180\ \text{J}$$

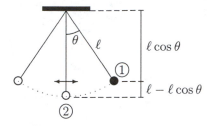

Figure 12.13 *The pendulum.*

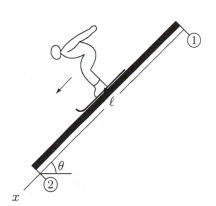

Figure 12.14 *A ski jumper.*

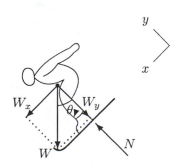

Figure 12.15 *The free-body diagram of the ski jumper.*

Example 12.3 Figure 12.13 illustrates a pendulum with mass m and length ℓ. The mass is pulled to position 1 that makes an angle θ with the vertical and is released to swing.

Assuming that frictional effects and air resistance are negligible, determine the speed v_2 of the mass when it is at position 2.

Solution: The mass has zero speed and no kinetic energy at the instant of release (position 1). If we choose position 2 to be the datum from which heights are measured, then the mass is located at a height $h_1 = \ell(1 - \cos\theta)$ at position 1, and the height of the mass at position 2 is zero. Applying the conservation of energy principle between positions 1 and 2:

$$\mathcal{E}_{K1} + \mathcal{E}_{P1} = \mathcal{E}_{K2} + \mathcal{E}_{P2}$$

$$\frac{1}{2}\,m\,v_1{}^2 + m\,g\,h_1 = \frac{1}{2}\,m\,v_2{}^2 + m\,g\,h_2$$

Substituting $v_1 = 0$, $h_1 = \ell(1 - \cos\theta)$, and $h_2 = 0$ into this equation and solving it for the speed of the mass at position 2 will yield:

$$v_2 = \sqrt{2\,g\,\ell\,(1 - \cos\theta)}$$

Example 12.4 As illustrated in Figure 12.14, consider a ski jumper moving down a track to acquire sufficient speed to accomplish the ski jumping task. The length of the track is $\ell = 25$ m and the track makes an angle $\theta = 45°$ with the horizontal.

If the skier starts at the top of the track with zero initial speed, determine the takeoff speed of the skier at the bottom of the track using (a) the work–energy theorem, (b) the conservation of energy principle, and (c) the equation of motion along with the kinematic relationships. Assume that the effects of friction and air resistance are negligible.

Solution (a): Work–energy method

The free-body diagram of the ski jumper is shown in Figure 12.15. The forces acting on the ski jumper are the gravitational force $\underline{W}$ and the reaction force applied by the track on the skis in a direction perpendicular to the track. The x direction is chosen to coincide with the direction of motion and y is perpendicular to the track. Therefore, the weight of the ski jumper has components along the x and y directions, such that $W_x = W \sin\theta = mg \sin\theta$ and $W_y = W \cos\theta = mg \cos\theta$. On the other hand, $\underline{N}$ acts in the y direction. Note that W_x is the driving force for the skier.

Since there is only one force component in the x direction, the work done by that force component is also the net work done on the ski jumper. Labeling the top and the bottom of the track as positions 1 and 2, the work done by W_x to move the skier from position 1 to 2 that are ℓ distance apart is:

$$W_{12} = W_x \, \ell = m \, g \, \ell \, \sin\theta \qquad (i)$$

According to the work–energy theorem, W_{12} must be equal to the change in kinetic energy of the skier between positions 1 and 2:

$$W_{12} = \mathcal{E}_{K2} - \mathcal{E}_{K1} = \frac{1}{2} \, m \, v_2{}^2 - \frac{1}{2} \, m \, v_1{}^2 \qquad (ii)$$

The second term on the right-hand side of Eq. (ii) is zero because the initial speed of the skier is $v_1 = 0$. Substituting Eq. (i) into Eq. (ii), eliminating the repeated parameter m (the mass of the ski jumper), and solving Eq. (ii) for the takeoff speed v_2 of the ski jumper will yield:

$$v_2 = \sqrt{2 \, g \, \ell \, \sin\theta} \qquad (iii)$$

Solution (b): *Conservation of energy method*

Since the effects of nonconservative forces due to friction and air resistance are assumed to be negligible, this problem can also be analyzed by utilizing the principle of conservation of energy. Between positions 1 and 2 of the ski jumper:

$$\mathcal{E}_{K1} + \mathcal{E}_{P1} = \mathcal{E}_{K2} + \mathcal{E}_{P2}$$

$$\frac{1}{2} \, m \, v_1{}^2 + m \, g \, h_1 = \frac{1}{2} \, m \, v_2{}^2 + m \, g \, h_2 \qquad (iv)$$

In Eq. (iv), the first term on the left-hand side is zero since $v_1 = 0$. If we measure heights relative to the bottom of the track (or selecting 2 to be the datum as shown in Figure 12.16), then height $h_2 = 0$ and the height of the top of the track is $h_1 = \ell \sin\theta$. Therefore, the second term on the right-hand side of Eq. (iv) is zero as well. Substituting $h_1 = \ell \sin\theta$ into Eq. (iv), eliminating the repeated parameter m, and solving Eq. (iv) for v_2 will again yield Eq. (iii).

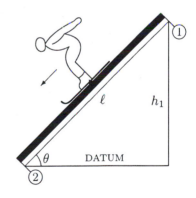

Figure 12.16 $h_1 = \ell \sin\theta$ and $h_2 = 0$.

Solution (c): *Using the equation of motion*

The equation of motion in the direction of motion (x) is:

$$\sum F_x = m \, a_x : \qquad W_x = m \, a_x \qquad (v)$$

Substituting $W_x = mg \sin\theta$ into Eq. (v), eliminating m, and solving Eq. (v) for the acceleration of the ski jumper in the x direction will yield:

$$a_x = g \, \sin\theta \qquad (vi)$$

Since the acceleration of the ski jumper is due to gravity only, what we have is a one-dimensional motion with constant acceleration. By definition, acceleration is the time rate of change of velocity, or velocity is the integral of acceleration with respect to time. Since a_x is constant and the initial velocity of the ski jumper at position 1 is zero, we can write:

$$v_x = a_x\, t \qquad\qquad (vii)$$

The kinematic relationship between the velocity and displacement is such that displacement is equal to the integral of velocity. If we measure the displacement relative to the initial position of the ski jumper, then the initial displacement is zero. Therefore, the equation relating displacement, acceleration, and time is:

$$x = \frac{1}{2}\, a_x\, t^2 \qquad\qquad (viii)$$

Eq. (vii) can be solved for time $t = v_x/a_x$, which can then be substituted into Eq. $(viii)$ so as to eliminate t. This will yield:

$$x = \frac{1}{2}\frac{v_x{}^2}{a_x}$$

Solving this equation for v_x will give:

$$v_x = \sqrt{2\, x\, a_x} \qquad\qquad (ix)$$

This is a general solution relating the acceleration, speed, and displacement of the ski jumper when the ski jumper is anywhere along the track. $x = \ell$ and $v_x = v_2$ when the ski jumper reaches the bottom of the track, and the acceleration of the ski jumper is always $a_x = g\sin\theta$. Substituting these parameters into Eq. (ix) will again yield Eq. (iii).

Finally, substituting the numerical values of $g = 9.8$ m/s^2, $\ell = 25$ m, and $\theta = 45°$ into Eq. (iii) and carrying out the calculations will yield $v_2 = 18.6$ m/s.

Remark:

• It is clear that for problems involving displacement, speed, and force, applications of the methods based on the work–energy theorem and the conservation of energy principle are more straightforward as compared to the application of equations of motion. In general, one should try work–energy or conservation of energy methods first before resorting to the equations of motion.

• Since the effects of nonconservative forces due to friction and air resistance are neglected, the solution of the problem is independent of the shape of the track or how the skier covers the distance between the top and bottom of the track. The most important parameter in this problem affecting the takeoff speed of the skier is the total vertical distance between 1 and 2. This implies that the problem could be simplified by noting that the skier undergoes a "free fall" between 1 and 2, which are $h_1 = \ell \sin\theta$ distance apart. This is illustrated in Figure 12.17. Applying the principle of conservation of energy between 1 and 2 will again yield Eq. (iii).

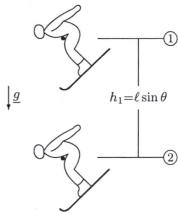
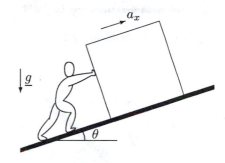

Figure 12.17 *The solution of the problem is independent of the path of motion.*

12.13 Exercise Problems

Problem 12.1 Figure 12.18 shows a person pushing a block of mass m on a surface that makes an angle θ with the horizontal. The coefficient of kinetic friction between the block and the inclined surface is μ.

If the person is applying a force with constant magnitude P and in a direction parallel to the incline, show that the acceleration of the block in the direction of motion can be expressed as:

$$a_x = \frac{P}{m} - g\left(\mu\,\cos\theta + \sin\theta\right)$$

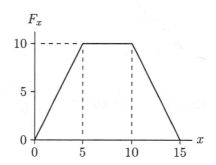

Figure 12.18 *Problem 12.1.*

Problem 12.2 Consider that a force with magnitude F_x that varies with displacement along the x direction is applied on an object. Assume that the variation of the force is as shown in Figure 12.19 where force is measured in Newtons and displacement is measured in meters.

Determine the work done by F_x on the object as the object moved from $x = 0$ to $x = 15$ m.

Answer: 100 J

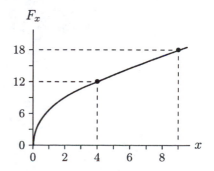

Figure 12.19 *Problem 12.2.*

Problem 12.3 A force with varying magnitude F_x is applied on an object and the displacement of the object is recorded in terms of x. The applied force is then plotted as a function of displacement and the curve shown in Figure 12.20 is obtained. It is observed that between $x = 0$ and $x = 9$ m, the force is proportional to the square root of displacement:

$$F_x = c\sqrt{x}$$

Here, F_x is measured in Newtons and x in meters, and the constant of proportionality between F_x and x is estimated to be $c = 6$.

Figure 12.20 *Problem 12.3.*

Determine the work done by F_x on the object to move the object from:

(a) $x=0$ to $x=4$ m and
(b) $x=0$ to $x=9$ m.

Answers: (a) 32 J and (b) 108 J.

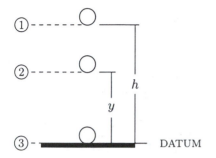

Figure 12.21 *Problem 12.4.*

Problem 12.4 As illustrated in Figure 12.21, a ball is dropped from a height h measured from the ground level. If the air resistance is neglected, show that the speed of the ball as a function of height y measured from the ground level can be expressed as:

$$v = \sqrt{2g(h-y)}$$

Here, g is the magnitude of the gravitational acceleration.

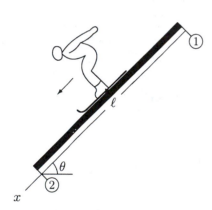

Figure 12.22 *Problem 12.5.*

Problem 12.5 As illustrated in Figure 12.22, consider a ski jumper moving down a track to acquire sufficient speed to accomplish the ski jumping task. The length of the track is ℓ, the track makes an angle θ with the horizontal, and the coefficient of friction between the track and the skis is μ.

If the ski jumper starts at the top of the track with zero initial speed, determine expressions for:

(a) the takeoff speed v_2 of the ski jumper at the bottom of the track using the work–energy theorem, and
(b) the acceleration a_x of the ski jumper using the equation of motion.

Assume that effects of air resistance are negligible.

Answers:

(a) $v_2 = \sqrt{2\ell g(\sin\theta - \mu\cos\theta)}$
(b) $a_x = g(\sin\theta - \mu\cos\theta)$

Chapter 13

Angular Kinematics

13.1 Polar Coordinates

Two-dimensional angular motions of bodies are commonly described in terms of a pair of parameters, r and θ (theta), which are called the *polar coordinates*. Polar coordinates are particularly well suited for analyzing motions restricted to circular paths. As illustrated in Figure 13.1, let O and P be two points on a two-dimensional surface. The location of P with respect to O can be specified in many different ways. For example, in terms of rectangular coordinates, P is a point with coordinates x and y. Point P is also located at a distance r from point O making an angle θ with the horizontal. Both x and y, and r and θ specify the position of P with respect to O uniquely, and O forms the origin of both the rectangular and polar coordinate systems. Note that these pairs of coordinates are not mutually independent. If one pair is known, then the other pair can be calculated because they are associated with a right-triangle: r is the hypotenuse, θ is one of the two acute angles, and x and y are the lengths of the adjacent and opposite sides of the right-triangle with respect to angle θ. Therefore:

$$x = r \cos\theta$$
$$y = r \sin\theta \qquad (13.1)$$

Expressing r and θ in terms of x and y:

$$r = \sqrt{x^2 + y^2}$$
$$\theta = \arctan\left(\frac{y}{x}\right) \qquad (13.2)$$

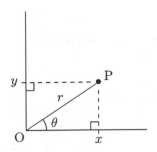

Figure 13.1 *Rectangular and polar coordinates of point P.*

13.2 Angular Position and Displacement

Consider an object undergoing a rotational motion in the xy-plane about a fixed axis. Let O be a point in the xy-plane along the axis of rotation of the object, and P be a fixed point on the object located at a distance r from O (Figure 13.2). Point P will move in a circular path of radius r and center located at O. Assume that at some time t_1, the point is located at P_1 which makes an angle θ_1 with the horizontal. At a later time t_2, the point is at P_2 which makes an angle θ_2 with the horizontal. Angles θ_1 and θ_2 define the *angular positions* of the point at times t_1 and t_2, respectively. If θ denotes the change in angular position of the point in the time interval between t_1 and t_2, then $\theta = \theta_2 - \theta_1$ is called the *angular displacement* of the point in the same time interval between. In the same time interval, the point travels a distance s measured along the circular path. The equation relating the radius r of the circle, angle θ, and arc length s is:

$$s = r\theta \qquad \text{or} \qquad \theta = \frac{s}{r} \qquad (13.3)$$

In Eq. (13.3), angle θ must be measured in radians, rather than in degrees. As reviewed in Appendix C, radians and degrees

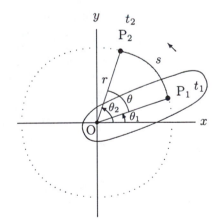

Figure 13.2 $\theta = \theta_2 - \theta_1$ is the angular displacement in the time interval between t_1 and t_2.

are related in that there are 360 degrees in a complete circle that must correspond to an arc length equal to the circumference, $s = 2\pi r$, of the circle, with $\pi = 3.14$ approximately. Therefore, $\theta = s/r = 2\pi r/r = 2\pi$ for a complete circle, or $360° = 2\pi$. One radian is then equal to $360°/2\pi = 57.3°$. The following formula can be used to convert angles given in degrees to corresponding angles in radians:

$$\theta \ (\text{radians}) = \frac{\pi}{180} \theta \ (\text{degrees})$$

Selected angles and their equivalents in radians are listed in Table 13.1.

Table 13.1 *Selected angles in degrees and radians.*

Degrees (°)	Radians (rad)
30	$\pi/6 = 0.524$
45	$\pi/4 = 0.785$
60	$\pi/3 = 1.047$
90	$\pi/2 = 1.571$
180	$\pi = 3.142$
270	$3\pi/2 = 4.712$
360	$2\pi = 6.283$

13.3 Angular Velocity

The time rate of change of angular position is called *angular velocity*, and it is commonly denoted by the symbol ω (omega). If the angular position of an object is known as a function of time, its angular velocity can be determined by taking its derivative with respect to time:

$$\omega = \frac{d\theta}{dt} = \dot{\theta} \tag{13.4}$$

The *average angular velocity* ($\bar{\omega}$) of an object in the time interval between t_1 and t_2 is defined by the ratio of change in angular position of the object divided by the time interval:

$$\bar{\omega} = \frac{\Delta\theta}{\Delta t} = \frac{\theta_2 - \theta_1}{t_2 - t_1} \tag{13.5}$$

In Eq. (13.5), θ_1 and θ_2 are the angular positions of the object at times t_1 and t_2, respectively.

13.4 Angular Acceleration

The angular velocity of an object may vary during motion. The time rate of change of angular velocity is called *angular acceleration*, usually denoted by the symbol α (alpha). If the angular velocity of a body is given as a function of time, then its angular acceleration can be determined by considering the derivative of the angular velocity with respect to time:

$$\alpha = \frac{d\omega}{dt} \tag{13.6}$$

The *average angular acceleration*, $\bar{\alpha}$, is equal to the change in angular velocity over the time interval in which the change occurs. If ω_1 and ω_2 are the instantaneous angular velocities of a body measured at times t_1 and t_2, respectively, then the average angular acceleration of the body in the time interval between t_1 and t_2 is:

$$\bar{\alpha} = \frac{\Delta\omega}{\Delta t} = \frac{\omega_2 - \omega_1}{t_2 - t_1} \tag{13.7}$$

Note that using the definition of angular velocity in Eq. (13.4), angular acceleration can alternatively be expressed in the following forms:

$$\alpha = \frac{d\omega}{dt} = \frac{d}{dt}\left(\frac{d\theta}{dt}\right) = \frac{d^2\theta}{dt^2} = \ddot{\theta} \qquad (13.8)$$

Also note that Eqs. (13.4) and (13.6) are the kinematic equations relating angular quantities θ, ω, and α.

Angular displacement, velocity, and acceleration are vector quantities. Therefore, their directions must be stated as well as their magnitudes. For two-dimensional problems, the motion is either in the clockwise or counterclockwise direction. Angular displacement and velocity are positive in the direction of motion. Angular acceleration is positive when angular velocity is increasing over time, and it is negative when angular velocity is decreasing over time.

13.5 Dimensions and Units

From Eq. (13.3), the angular displacement θ of an object undergoing circular motion is equal to the ratio of the arc length s and radius r of the circular path. Both arc length and radius have the dimension of length. Therefore, the dimension of angular displacement is 1, or it is a *dimensionless* quantity:

$$[\text{ANGULAR DISPLACEMENT}] = \frac{L}{L} = 1$$

By definition, angular velocity is the time rate of change of angular position, and angular acceleration is the time rate of change of angular velocity. Therefore, angular velocity has the dimension of 1 over time, and angular acceleration has the dimension of angular velocity divided by time, or 1 over time squared.

Note that angular quantities θ, ω, and α differ dimensionally from their linear counterparts x, v, and a by a length factor.

The units of angular quantities in different unit systems are the same. Angular displacement is measured in radians (rad), angular velocity is measured in radians per second (rad/s) or s^{-1}, and angular acceleration is measured in radians per second squared (rad/s^2) or s^{-2}.

13.6 Definitions of Basic Concepts

To be able to define concepts common in angular motions, consider the simple pendulum illustrated in Figure 13.3. The pendulum consists of a mass attached to a string. The string is fixed to the ceiling at one end and the mass is free to swing. Assume that ℓ is the length of the string and it is attached to the ceiling at O. If the mass is simply released, it would stretch the string and

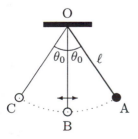

Figure 13.3 *Pendulum.*

come to a rest at B that represents the *neutral* or *equilibrium position* of the mass. If the mass is pulled to the side, to position A, so that the string makes an angle θ_0 with the vertical and is then released, the mass will oscillate or swing back and forth about its neutral position in a circular arc path of radius ℓ. Owing to internal friction and air resistance, the oscillations will die out over time and eventually the pendulum will come to a stop at its neutral position. An analysis of the motion characteristics of this relatively simple system may give us considerable insight into the nature of other more complex dynamic systems.

For the sake of simplicity, we ignore the air resistance and frictional effects, and assume that once the pendulum is excited, it will oscillate forever. Also, assume that there is a roll of paper behind the pendulum that moves in a prescribed manner. (For example, 10 mm of paper rolls up in each second.) Furthermore, the mass has a dye on it that marks the position of the mass on the paper. In other words, as the mass swings back and forth, it draws its motion path on the paper that would look like the one illustrated in Figure 13.4. In Figure 13.4, θ represents the angle that the pendulum makes with the vertical and t is time. Angle θ is a measure of the instantaneous angular position of the pendulum.

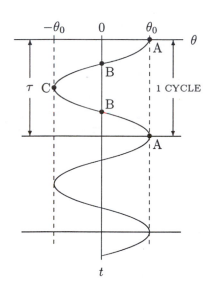

Figure 13.4 *Simple harmonic motion.*

The motion described in Figure 13.4 is known as the *simple harmonic motion*. At time $t = 0$ that corresponds to the instant when the mass is first released, the mass is located at A, which makes an angle θ_0 with the vertical. The mass swings, passes through B where $\theta = 0$, and reaches C where $\theta = -\theta_0$. Here, it is assumed that θ is positive between A and B, zero at B, and negative between B and C. At C, the mass momentarily stops and then reverses its direction of motion from clockwise to counterclockwise. It passes through B again and returns to A, thus completing one full *cycle* in a time interval of τ (tau) seconds, which is called the *period* of harmonic motion. The total angle covered by the pendulum between A and C is called the *range of motion* and, in this case, it is equal to $2\theta_0$. Also, the entire motion of the pendulum is confined between $+\theta$ and $-\theta$ that set the limits of the range of motion. Half of the range of motion is called the *amplitude* of the oscillations measured in radians and here is equal to θ_0. Note that in this case, both the amplitude and period of the harmonic motion are constants. Also note that since the effects of friction and air resistance are neglected, the series of events between A, B, C, B, and A are repeated forever in τ time intervals.

From Figure 13.4, it is clear that angular position θ is a function of time t. Furthermore, θ is a harmonic, cyclic function of t that must remind us of trigonometric functions. As discussed in Appendix C, the θ versus t graph in Figure 13.4 can be compared to the graphs of known functions to establish the functions that relate θ and t. It can be shown that:

$$\theta = \theta_0 \cos(\phi \, t)$$

In this equation, the parameter θ_0 multiplied with the cosine function is the amplitude of the harmonic motion and ϕ (phi) is called the *angular frequency* measured in radians per second (rad/s). The period and angular frequency are related:

$$\phi = \frac{2\pi}{\tau} \qquad (\pi = 3.1416)$$

For oscillatory motions, the reciprocal of the period is called the *frequency*, f, measured in Hertz (Hz) that represents the total number of cycles occurring per second:

$$f = \frac{1}{\tau} = \frac{\phi}{2\pi}$$

Note that for the simple harmonic motion discussed herein the parameters involved (range of motion, amplitude, period, and frequency) are constants. Also note that the validity of the function relating angular position and time can be checked by assigning values to t and calculating corresponding θ values. For example, at A: $t = 0$, $\phi t = 0$, $\cos(0) = 1$, and $\theta = \theta_0$. At B: $t = \tau/4$, $\phi t = \pi/2 = 90°$, $\cos(90) = 0$, and $\theta = 0$. At C: $t = \tau/2$, $\phi t = \pi = 180°$, $\cos(180) = -1$, and $\theta = -\theta_0$. All of these are consistent with the observations in Figure 13.4.

Now that we have defined most of the important parameters involved, we can also determine the angular velocity and angular acceleration of the pendulum. Utilizing Eqs. (13.4) and (13.6):

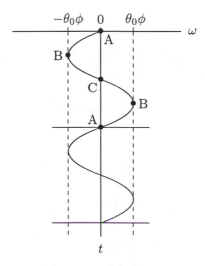

Figure 13.5 *Angular velocity ω versus time t.*

$$\omega = \frac{d\theta}{dt} = -\theta_0 \, \phi \, \sin(\phi \, t)$$

$$\alpha = \frac{d\omega}{dt} = -\theta_0 \, \phi^2 \, \cos(\phi \, t)$$

These functions are plotted in Figures 13.5 and 13.6. Note that the amplitude of the angular velocity of the pendulum is $\theta_0\phi$, and the amplitude of its angular acceleration is $\theta_0\phi^2$. They are the terms multiplied by the sine and cosine functions. At A, the angular velocity is zero. Between A and B, the mass accelerates and the magnitude of its angular velocity increases in the clockwise direction. The angular velocity reaches a peak value of $\theta_0\phi$ at B. The angular velocity is negative and the angular acceleration is positive between B and C. Therefore, the mass decelerates (its angular velocity decreases in the clockwise direction) between B and C. The angular velocity reduces to zero at C. In the meantime, the magnitude of the angular acceleration reaches its peak value of $\theta_0\phi^2$. Between C and B, the mass

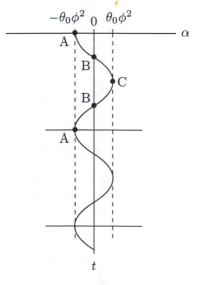

Figure 13.6 *Angular acceleration α versus time t.*

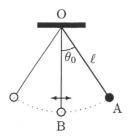

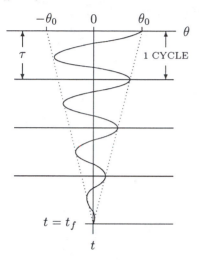

Figure 13.7 *Pendulum under the effect of air resistance.*

Figure 13.8 *Damped oscillations.*

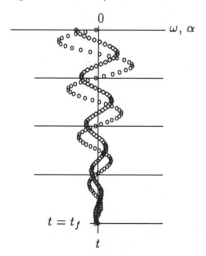

Figure 13.9 *Angular velocity ω ($\circ$) and angular acceleration α ($\diamond$) versus time t.*

accelerates in the counterclockwise direction, the magnitude of its angular velocity returns to a peak at B, slows down between B and C, and momentarily comes to rest at A. This series of events is repeated over time.

Next, consider that the mass is again pulled to A so that the pendulum makes an angle θ_0 with the vertical and is released (Figure 13.7). The mass will oscillate about its neutral position in a circular arc path of radius ℓ. Owing to internal friction and air resistance, the oscillations will die out over time and eventually the pendulum will come to a rest at its neutral position, B. This type of motion is called *damped oscillations*. To help understand some aspects of damped oscillations, consider the angular position θ versus time t graph shown in Figure 13.8. The pendulum completes four full cycles in t_f seconds before coming to a stop. The period of each cycle is equal, but the amplitude of the harmonic oscillations decreases linearly with time and to zero at time t_f. That is, we have a harmonic motion with a constant period but varying amplitude. For measured θ_0, τ, and t_f, the θ versus time graph shown in Figure 13.8 can be represented as:

$$\theta = \theta_0 \left(1 - \frac{t}{t_f}\right)\cos(\phi\, t)$$

Here, ϕ is again the angular frequency of harmonic oscillations and is equal to $2\pi/\tau$. What is different in this case is that the harmonic oscillations of the pendulum are confined between two converging straight lines that can be represented by the functions $\theta = \theta_0(1 - t/t_f)$ and $\theta = -\theta_0(1 - t/t_f)$, and that the oscillations of the pendulum are "damped-out" by friction and air resistance. Knowing the angular position of the pendulum as a function of time enables us to determine the angular velocity and acceleration of the pendulum. Using Eqs. (13.4) and (13.6), and applying the product and chain rules of differentiation (see Appendix C):

$$\omega = \frac{d\theta}{dt} = -\frac{\theta_0}{t_f}\cos(\phi\, t) - \theta_0\,\phi\left(1 - \frac{t}{t_f}\right)\sin(\phi\, t)$$

$$\alpha = \frac{d\omega}{dt} = \frac{2\,\theta_0\,\phi}{t_f}\sin(\phi\, t) - \theta_0\,\phi^2\left(1 - \frac{t}{t_f}\right)\cos(\phi\, t)$$

These functions are relatively complex. Their graphs are shown in Figure 13.9, which are obtained simply by assuming a value for τ, assigning values to t, calculating corresponding ω and α, and plotting them.

Example 13.1   *Shoulder abduction*

Figure 13.10 shows a person doing shoulder abduction in the frontal plane. O represents the axis of rotation of the shoulder joint in the frontal plane, line OA represents the position of the arm when it is stretched out parallel to the ground (horizontal), line OB represents the position of the arm when the hand is at its highest elevation, and line OC represents the position of the arm when the hand is closest to the body. In other words, for this activity, OB and OC are the arm's limits of range of motion. Assume that the angle between OA and OB is equal to the angle between OA and OC, which are represented by angle θ_0. The motion of the arm is symmetric with respect to line OA. Also assume that the time it takes for the arm to cover the angles between OA and OB, OB and OA, OA and OC, and OC and OA are approximately equal.

Derive expressions for the angular displacement, velocity, and acceleration of the arm. Take the period of angular motion of the arm to be 3 s and the angle θ_0 to be 80°.

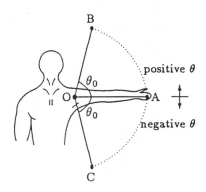

Figure 13.10 *Shoulder abduction.*

Solution: Note the similarities between the motion of the arm in this example and the simple harmonic motion of the pendulum discussed previously. In this case, angle θ_0 represents the amplitude of the angular displacement of the arm while undergoing a harmonic motion about line OA. The range of motion (ROM) of the arm is equal to twice that of angle θ_0. The period of the angular motion is given as $\tau = 3$ s, and the angular frequency of harmonic oscillations of the arm about line OA (the horizontal) can be calculated as $\phi = 2\pi/\tau = 2.09$ rad/s. If we let θ represent the angular displacement of the arm measured relative to the position defined by line OA, then θ can be written as a sine function of time:

$$\theta = \theta_0 \, \sin(\phi \, t) \qquad (i)$$

The angular displacement of the arm as given in Eq. (*i*) is plotted as a function of time in Figure 13.11. Note that θ is zero when the arm is at position A. θ assumes positive values between A and B, and it is negative while the arm is between A and C. θ reaches its peak at B and C, and θ_0 is the amplitude of angular displacement of the arm. Since all of these are consistent with the information provided in the statement of the problem, Eq. (*i*) does represent the angular displacement of the arm.

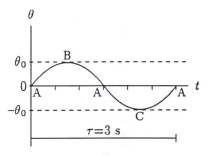

Figure 13.11 *Graph of function* $\theta = \theta_0 \sin(\phi t)$ *with* $\phi = 2\pi/\tau$.

To derive expressions for the angular velocity and acceleration of the arm, we have to consider time derivatives of the function given in Eq. (*i*). The time rate of change of angular displacement is defined as angular velocity:

$$\omega = \frac{d\theta}{dt} = \theta_0 \, \phi \, \cos(\phi \, t) \qquad (ii)$$

The time rate of change of angular velocity is angular acceleration:

$$\alpha = \frac{d\omega}{dt} = -\theta_0 \, \phi^2 \, \sin(\phi \, t) \qquad (iii)$$

Equations (*ii*) and (*iii*) can alternatively be written as:

$$\omega = \omega_0 \, \cos(\phi \, t) \qquad (iv)$$

$$\alpha = -\alpha_0 \, \sin(\phi \, t) \qquad (v)$$

Here, ω_0 is the amplitude of the angular velocity and α_0 is the amplitude of the angular acceleration of the arm, such that:

$$\omega_0 = \theta_0 \, \phi = \theta_0 \frac{2\pi}{\tau}$$

$$\alpha_0 = \theta_0 \, \phi^2 = \theta_0 \frac{4\pi^2}{\tau^2}$$

Note that the amplitude of the angular velocity is a linear function of the angular frequency, and the amplitude of angular acceleration is a quadratic function of angular frequency. Angular frequency, on the other hand, is inversely proportional with the period of harmonic oscillations. Therefore, low period indicates high frequency, which indicates high angular velocity and acceleration amplitudes.

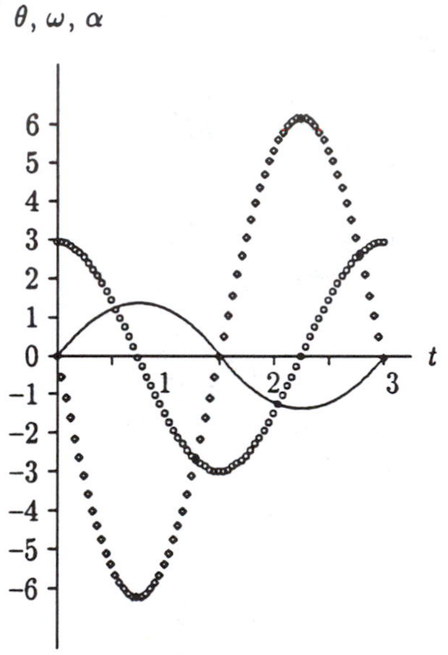

We can use the numerical values of $\theta_0 = 80° = 1.40$ rad and $\phi = 2.09$ rad/s to calculate ω_0 and α_0 as 2.93 rad/s and 6.12 rad/s^2, respectively. Eqs. (*i*), (*iv*), and (*v*) can now be expressed as:

$$\theta = 1.40 \, \sin(2.09 \, t) \qquad (vi)$$

$$\omega = 2.93 \, \cos(2.09 \, t) \qquad (vii)$$

$$\alpha = -6.12 \, \sin(2.09 \, t) \qquad (viii)$$

Equations (*vi*) through (*viii*) can be used to calculate the instantaneous angular position, velocity, and acceleration of the arm at any time t. These equations can also be used to plot θ, ω, and α versus t graphs for the arm by assigning values to time and calculating corresponding θ, ω, and α values. A set of sample graphs are shown in Figure 13.12 for a single cycle.

Figure 13.12 *Angular position, velocity (○) and acceleration (◇) versus time. (θ in rad, ω in rad/s, α in rad/s², and t is seconds.)*

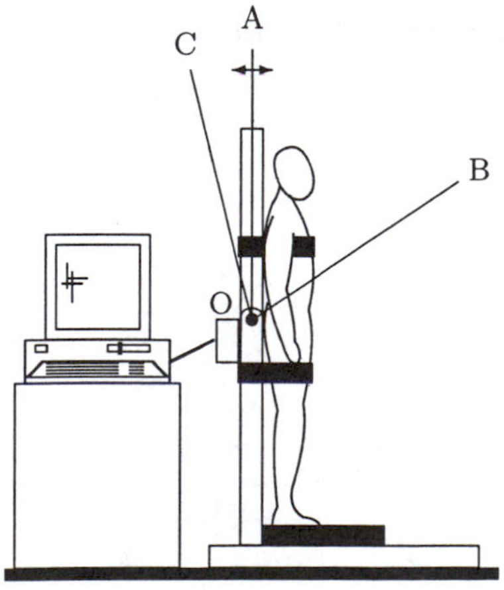

Figure 13.13 *Dynamometer.*

Example 13.2   *Flexion–extension test*

Figure 13.13 illustrates a computer-controlled dynamometer that can be used to measure angular displacement, angular velocity, and torque output of the trunk. During a repetitive flexion–extension test in the sagittal plane (plane that passes through the chest and divides the body into right-hand and left-hand parts), a subject is placed in the dynamometer, positioned in the machine so that the subject's fifth lumbar vertebra (L5/S1) is aligned with the flexion–extension axis (indicated as O) of the machine, tied to the equipment firmly, and asked to perform trunk flexion and extension as long as possible, exerting as much effort as possible. The angular position of the subject's trunk relative to the upright position is measured and recorded. The data collected are then plotted to obtain an angular displacement θ versus time t graph. The curves obtained for this particular subject in different cycles are observed to be qualitatively and quantitatively similar except for the first and the last few cycles. A couple of sample cycles are provided in Figure 13.14, in which the angular displacement of the trunk measured in degrees is plotted as a function of time measured in seconds.

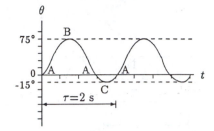

Figure 13.14 *Angular position versus time graph. (θ is in degrees and t is in seconds.)*

The angular position measurements are made relative to the upright position in which the angular displacement of the trunk is zero. The subject flexes between A and B, and reaches a peak flexion at B. The extension phase is identified with the motion of the trunk from B toward A. The angular displacement of the trunk is positive between A and B. Between A and C, the trunk undergoes hyperextension and reaches a peak extension at C. In this range, the angular displacement of the trunk assumes negative values.

The purpose of this example is to illustrate the means of analyzing experimentally collected data. The specific task is to find a function that can express the angular displacement of the subject's trunk as a function of time, from which we can derive expressions for the angular velocity and acceleration of the trunk.

Solution: The problem may be easier to visualize if we form an analogy between the upper body and a mechanical system called the *inverted pendulum*, shown in Figure 13.15. An inverted pendulum consists of a concentrated mass m attached to a very light rod of length ℓ that is hinged to the ground through an axis about which it is allowed to rotate. In this case, the concentrated mass represents the total mass of the upper body. The hinge corresponds to the disk between the fifth lumbar vertebra and the sacrum, about which the upper body rotation occurs in the sagittal plane. Length ℓ is the distance between the fifth lumbar vertebra and the center of gravity of the upper body.

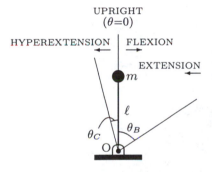

Figure 13.15 *Inverted pendulum.*

It is clear from Figure 13.14 that θ is a harmonic (sine or cosine) function of time. From Figure 13.14, it is possible to read the peak angles the trunk makes with the upright position during flexion and extension phases, and the period of harmonic motions. However, it is not easy to determine exactly how θ varies with time. To obtain a function relating θ and t, we must work through several steps.

Let τ be the period of harmonic oscillations, and θ_B and θ_C the peak angular displacements of the trunk in the flexion and extension phases, respectively. From Figure 13.14 or using the experimentally obtained raw data, $\tau = 2$ s, $\theta_B = 75°$, and $\theta_C = -15°$. Knowing the period, the angular frequency of the harmonic oscillations can be determined:

$$\phi = \frac{2\pi}{\tau} = \frac{2\pi}{2} = \pi \ \text{rad/s}$$

Using θ_B and θ_C, we can also calculate the range of motion of the

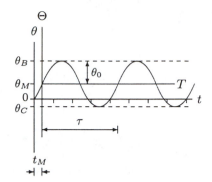

Figure 13.16 *Translating the θ versus t coordinate frame to Θ versus T coordinate frame.*

trunk. By definition, range of motion is the total angle covered by the rotating object. Therefore:

$$\text{ROM} = \theta_B - \theta_C = 75° + 15° = 90°$$

Figure 13.14 is redrawn in Figure 13.16 in which two sets of coordinates are used. In addition to θ and t, we have a second set of coordinates Θ (capital theta) and T that is obtained by translating the origin of the θ versus t coordinate system to a point with coordinates $t = t_M$ and $\theta = \theta_M$. Here, θ_M designates the mean angular displacement that can be calculated as:

$$\theta_M = \frac{\theta_B + \theta_C}{2} = \frac{75° - 15°}{2} = 30°$$

Time t_M corresponds to the time when $\theta = \theta_M$. t_M can be determined from the experimentally collected data. In this case, $t_M = 0.232$ s.

We define a second set of coordinates so that, with respect to Θ and T, the function representing the angular displacement versus time curve is simply a sine function:

$$\Theta = \theta_0 \, \sin(\phi \, T) \qquad (i)$$

In Eq. (*i*), θ_0 is the amplitude of the harmonic oscillations and is equal to one half of the range of motion:

$$\theta_0 = \frac{\text{ROM}}{2} = \frac{90°}{2} = 45° \quad \left(\frac{\pi}{4} \text{ rad}\right)$$

We now have a function representing the experimentally obtained curve in terms of Θ and T. If we can relate Θ to θ and T to t, then we can derive a function in terms of θ and t. This can be achieved by employing *coordinate transformation*. Note that $\Theta = 0$ when $\theta = \theta_M$. Therefore:

$$\Theta = \theta - \theta_M \qquad (ii)$$

Also note that $T = 0$ when $t = t_M$. Hence:

$$T = t - t_M \qquad (iii)$$

Substituting Eqs. (*ii*) and (*iii*) into Eq. (*i*) will yield:

$$\theta = \theta_M + \theta_0 \, \sin[\phi(t - t_M)] \qquad (iv)$$

In Eq. (*iv*), the angular displacement of the trunk is defined as a function of time, representing the experimentally obtained curve shown in Figure 13.14. We can also obtain expressions for the angular velocity and acceleration of the trunk by considering the time derivatives of θ in Eq. (*iv*):

$$\omega = \frac{d\theta}{dt} = \theta_0 \phi \, \cos[\phi(t - t_M)] \tag{v}$$

$$\alpha = \frac{d\omega}{dt} = -\theta_0 \phi^2 \, \sin[\phi(t - t_M)] \tag{vi}$$

The numerical values of θ_M, θ_0, t_M, and ϕ can be substituted into the above equations to obtain:

$$\theta = \frac{\pi}{6} + \frac{\pi}{4} \, \sin[\pi(t - 0.232)] \tag{vii}$$

$$\omega = \frac{\pi^2}{4} \, \cos[\pi(t - 0.232)] \tag{viii}$$

$$\alpha = -\frac{\pi^3}{4} \, \sin[\pi(t - 0.232)] \tag{ix}$$

These functions are plotted in Figure 13.17 to obtain angular displacement, velocity, and acceleration versus time graphs for the trunk.

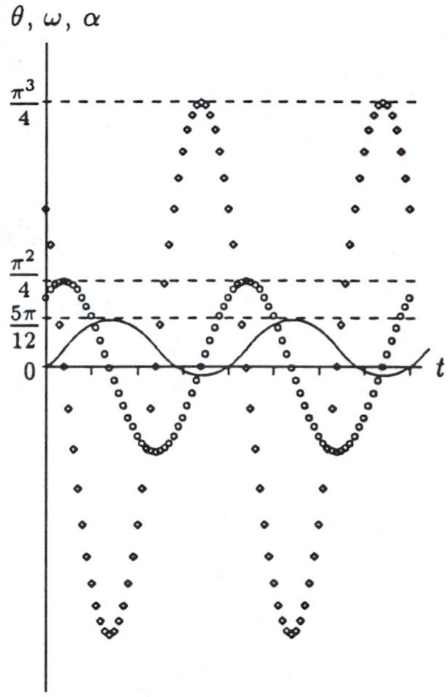

Note that the validity of Eq. (*vii*) can be checked by assigning values to t and calculating corresponding θ values using Eq. (*vii*). For example, $\theta = 0$ when $t = 0$ and $t = \tau = 2$ s, and $\theta = \pi/6 = 0.52$ rad or 30° when $t = t_M = 0.232$ s. These are consistent with the initial data presented in Figure 13.14.

Also note that angular velocity is a cosine function of time. The amplitude of the ω versus t curve shown in Figure 13.17 is equal to the coefficient $\pi^2/4 = 2.47$ rad/s in front of the cosine function in Eq. (*viii*). Similarly, the amplitude of the angular acceleration is $\pi^3/4 = 7.75$ rad/s^2.

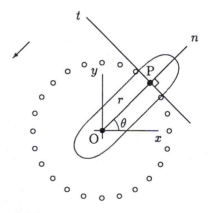

θ, ω, α

Figure 13.17 *Angular position, velocity (○), and acceleration (◇) versus time.*

13.7 Rotational Motion About a Fixed Axis

Consider the arbitrarily shaped object in Figure 13.18. Assume that the object is undergoing a rotational motion in the xy-plane about a fixed axis that is perpendicular to the xy-plane. Let O and P be two points in the xy-plane, such that O is along the axis of rotation of the object and P is a fixed point on the rotating object located at a distance r from O. Owing to the rotation of the object, point P will experience a circular motion with r being the radius of its circular path.

To describe circular motions, it is usually convenient to define velocity and acceleration vectors with respect to two mutually perpendicular directions normal (radial) and tangential to the circular path of motion. These directions are indicated as n and t in Figure 13.18, and are also known as *local coordinates*. By

Figure 13.18 *n and t are the normal (radial) and tangential directions at point P.*

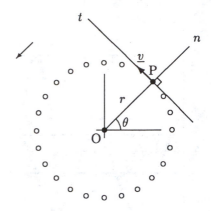

Figure 13.19 *Velocity vector $\underline{v}$ is always tangent to the path of motion.*

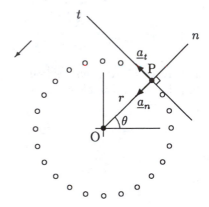

Figure 13.20 *$\underline{a}_t$ and $\underline{a}_n$ are the tangential and normal components of the acceleration vector.*

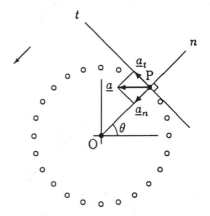

Figure 13.21 *$\underline{a}$ is the resultant linear acceleration vector.*

definition, the velocity vector $\underline{v}$ is always tangent to the path of motion. Therefore, for a circular motion, the velocity vector can have only one component tangent to the circular path of motion (Figure 13.19). $\underline{v}$ is called the *tangential* or *linear velocity*. The magnitude v of the velocity vector can be determined by considering the time rate of change of relative position of point P along the circular path:

$$v = \frac{ds}{dt} \tag{13.9}$$

For a circular motion, the acceleration vector can have both tangential and normal components (Figure 13.20). The *tangential acceleration a_t* is related to the change in magnitude of the velocity vector and has a magnitude:

$$a_t = \frac{dv}{dt} \tag{13.10}$$

The *normal acceleration $\underline{a}_n$* is related to the change in direction of the velocity vector and has a magnitude:

$$a_n = \frac{v^2}{r} \tag{13.11}$$

For an object undergoing a rotational motion, a_t is zero if the object is rotating with constant v. On the other hand, a_n is always present because it is associated with the direction of $\underline{v}$ that changes continuously throughout the motion.

The direction of $\underline{a}_t$ is the same as the direction of $\underline{v}$ if v is increasing, or opposite to that of $\underline{v}$ if v is decreasing over time. The normal component of the acceleration vector is also known as *radial* or *centripetal* (center-seeking), and it is always directed toward the center of rotation of the body.

If the tangential and normal acceleration components are known, then the net or resultant acceleration of a point on a body rotating about a fixed axis can also be determined (Figure 13.21). If $\underline{t}$ and $\underline{n}$ are unit vectors indicating positive tangential and normal directions, respectively, then the resultant acceleration vector can be expressed as:

$$\underline{a} = \underline{a}_t + \underline{a}_n = a_t\,\underline{t} + a_n\,\underline{n} \tag{13.12}$$

The magnitude of the resultant acceleration vector can be determined as:

$$a = \sqrt{a_t^2 + a_n^2} \tag{13.13}$$

On the other hand, the velocity vector can be expressed as:

$$\underline{v} = v\,\underline{t} \tag{13.14}$$

Note that v and a are linear quantities. v has the dimension of length divided by time, and both a_t and a_n have the dimension of length divided by time squared. Also note that it is customary to take the positive normal direction (the direction of $\underline{n}$) to be outward (from the center of rotation towards the rim), and the positive tangential direction (the direction of $\underline{t}$) to be counterclockwise.

13.8 Relationships Between Linear and Angular Quantities

Recall from Eq. (13.3) that $s = r\theta$. For a circular motion, radius r is constant and Eq. (13.9) can be evaluated as follows:

$$v = \frac{d}{dt}(r\,\theta) = r\,\frac{d\theta}{dt}$$

By definition, time rate of change of angular displacement is angular velocity. Therefore:

$$v = r\,\omega \tag{13.15}$$

Equation (13.15) states that the magnitude of the linear velocity of a point in a body that is undergoing a rotational motion about a fixed axis is equal to the distance of that point from the center of rotation multiplied by the angular velocity of the body. Note that at a given instant, every point on the body has the same angular velocity but may have different linear velocities. The magnitude of the linear velocity increases with increasing radial distance, or as one moves outward from the center of rotation toward the rim.

Using the relationship given in Eq. (13.15), Eq. (13.10) can be evaluated for a motion in a circular path as follows:

$$a_t = \frac{d}{dt}(r\,w) = r\,\frac{dw}{dt}$$

By definition, time rate of change of angular velocity is angular acceleration. Therefore:

$$a_t = r\,\alpha \tag{13.16}$$

Similarly, substituting Eq. (13.15) into Eq. (13.11) will yield:

$$a_n = r\,w^2 \tag{13.17}$$

Equations (13.15), (13.16), and (13.17) relate linear quantities v, a_t, and a_n to angular quantities r, ω, and α. Eq. (13.16) states that the tangential component of linear acceleration of a point on a body rotating about a fixed axis is equal to the distance of that point from the axis of rotation times the angular acceleration of the body.

13.9 Uniform Circular Motion

Uniform circular motion occurs when the angular velocity of an object undergoing a rotational motion about a fixed axis is constant. When angular velocity is constant, angular acceleration is zero. Therefore, for a point located at a radial distance r from the center of rotation of an object undergoing uniform circular motion:

$$v = r\,\omega \qquad \text{(constant)}$$
$$a_t = 0$$
$$a_n = r\,\omega^2 \qquad \text{(constant)}$$

13.10 Rotational Motion with Constant Acceleration

In Chapter 11, a set of kinematic equations [Eqs. (11.11) through (11.14)] were derived to analyze the motion characteristics of bodies undergoing translational motion with constant acceleration. Similar equations can also be derived for rotational motion about a fixed axis with constant angular acceleration:

$$\omega = \omega_0 + \alpha_0\,t \qquad (13.18)$$

$$\theta = \theta_0 + \omega_0\,t + \frac{1}{2}\,\alpha_0\,t^2 \qquad (13.19)$$

$$\theta = \theta_0 + \frac{1}{2}\,(\omega + \omega_0)\,t \qquad (13.20)$$

$$\omega^2 = \omega_0^2 + 2\,\alpha_0\,(\theta - \theta_0) \qquad (13.21)$$

In Eqs. (13.18) through (13.21), α_0 is the constant angular acceleration, and θ_0 and ω_0 are the initial angular position and velocity of the object at time $t_0 = 0$, respectively.

13.11 Relative Motion

A motion observed in different frames of reference may be different. For example, the motion of a train observed by a stationary person would be different than the motion of the same train observed by a passenger in a moving car. The motion of a ball thrown up into the air by a person riding in a moving vehicle would have a vertical path as observed by the person riding in the same vehicle (Figure 13.22a), but a curved path for a second, stationary person watching the ball (Figure 13.22b). The general approach in analyzing such physical situations requires defining the motion of the moving body with respect to a convenient moving coordinate frame, defining the motion of this frame with respect to a fixed coordinate frame, and combining the two.

Assume that the motion of a point P in a moving body is to be analyzed. Let XYZ and xyz refer to two coordinate frames with origins at A and B, respectively (Figure 13.23). Assume that the XYZ frame is fixed (stationary) and the xyz frame is moving, such that the respective coordinate directions (for example, x

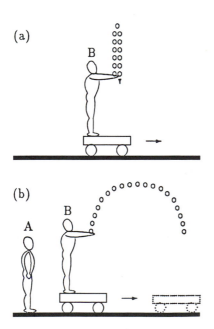

Figure 13.22 *A motion observed by different observers in different reference frames may be different.*

and X) remain parallel throughout the motion. This implies that the xyz coordinate frame is undergoing a translational motion only, and that the same set of unit vectors $\underline{i}$, $\underline{j}$, and $\underline{k}$ can be used in both reference frames. The motion of the moving xyz frame can be identified by specifying the motion of its origin B. If $\underline{r}_B$ denotes the position vector of B with respect to the fixed coordinate frame, then the velocity and acceleration vectors of B with respect to the XYZ coordinate frame are:

$$\underline{v}_B = \frac{d}{dt}(\underline{r}_B) = \underline{\dot{r}}_B \qquad \underline{a}_B = \frac{d}{dt}(\underline{v}_B) = \underline{\ddot{r}}_B$$

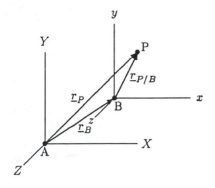

Figure 13.23 *XYZ is a fixed and xyz is a moving coordinate frame.*

Similarly, the motion of point P with respect to the moving coordinate frame xyz can be defined by the position vector $\underline{r}_{P/B}$ of point P relative to the origin B of the xyz frame. The first and second time derivatives of $\underline{r}_{P/B}$ will yield the velocity and acceleration vectors of point P relative to the xyz frame:

$$\underline{v}_{P/B} = \frac{d}{dt}(\underline{r}_{P/B}) = \underline{\dot{r}}_{P/B} \qquad \underline{a}_{P/B} = \frac{d}{dt}(\underline{v}_{P/B}) = \underline{\ddot{r}}_{P/B}$$

Finally, the position vector $\underline{r}_P$, velocity vector $\underline{v}_P$, and acceleration vector $\underline{a}_P$ of point P with respect to the fixed coordinate frame XYZ can be obtained by superposition:

$$\underline{r}_P = \underline{r}_B + \underline{r}_{P/B} \qquad (13.22)$$

$$\underline{v}_P = \underline{v}_B + \underline{v}_{P/B} \qquad (13.23)$$

$$\underline{a}_P = \underline{a}_B + \underline{a}_{P/B} \qquad (13.24)$$

The motion of point B (which happens to be the origin of the moving coordinate frame xyz) with respect to the fixed XYZ coordinate frame is called the *absolute motion* of B and is denoted by the subscript B. Similarly, the motion of point P observed relative to the XYZ frame is the absolute motion of P. The motion of point P with respect to the moving coordinate frame is called the *relative motion* of P and is denoted by the subscript P/B. Note here that the position vector $\underline{r}_{P/B}$ refers to a vector drawn from point P to point B. Also note that the position vector of point P relative to the XYZ coordinate frame could also be expressed as $\underline{r}_{P/A}$. However, by convention, $\underline{r}_P$ implies that the position vector is defined relative to the fixed coordinate frame.

Example 13.3 Consider the motion described in Figure 13.22. A person (B) riding on a vehicle that is moving toward the right by a constant speed of 2 m/s throws a ball straight up into the air with an initial speed of 10 m/s.

Describe the motion of the ball as observed by a stationary person (A) in the time interval between when the ball is first released and when it reaches its maximum elevation.

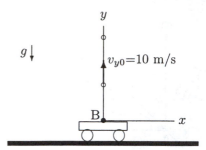

Figure 13.24 *Relative to the xy frame, the ball is undergoing a translational motion in the y direction.*

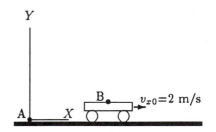

Figure 13.25 *Relative to the XY frame, the vehicle is undergoing a translational motion in the X direction with constant velocity.*

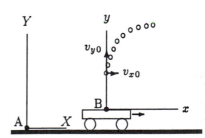

Figure 13.26 *Relative to the XY frame, the ball moves both in the X and Y directions.*

Solution: This is a two-dimensional problem and can be analyzed in three steps. First, let x and y represent a coordinate frame moving with the vehicle. With respect to the xy frame, the ball thrown up into the air will undergo one-dimensional linear motion (translation) in the y direction (Figure 13.24). Because of the constant downward gravitational acceleration, the ball will decelerate in the positive y direction, reach its maximum elevation, change its direction of motion, and begin to descend. With respect to the xy coordinate frame moving with the vehicle, or as observed by person B moving with the vehicle, the speed of the ball in the y direction between the instant of release and when the ball reaches its peak elevation can be determined from (see Chapter 11):

$$v_y = v_{y0} - g\,t$$

Here, $v_{y0} = 10\,\text{m/s}$ is the initial speed of the ball and $g \approx 10\,\text{m/s}^2$ is the magnitude of gravitational acceleration. This equation is valid in the time interval between $t = 0$ (the instant of release) and $t = v_{y0}/g = 10/10 = 1\,\text{s}$ (the time it takes for the ball to reach its maximum elevation where $v_y = 0$). As observed by person B, the ball has no motion in the x direction. Therefore, the velocity $\underline{v}_{P/B}$ of the ball relative to person B can be expressed as:

$$\underline{v}_{P/B} = v_y\,\underline{j}$$

Next, let X and Y represent a coordinate frame fixed to the ground. With respect to the XY frame, or with respect to the stationary person A, the vehicle is moving in the positive X direction with a constant speed of $v_{x0} = 2\,\text{m/s}$ (Figure 13.25). Therefore:

$$\underline{v}_B = v_{x0}\,\underline{i}$$

Finally, to determine the velocity of the ball relative to person A, we have to add velocity vectors $\underline{v}_B$ and $\underline{v}_{P/B}$ together:

$$\underline{v}_P = \underline{v}_B + \underline{v}_{P/B} = v_{x0}\,\underline{i} + v_y\,\underline{j}$$

Or, by substituting the known parameters:

$$\underline{v}_P = 2\,\underline{i} + (10 - 10\,t)\,\underline{j}$$

For example, half a second after the ball is released, the ball has a velocity:

$$\underline{v}_P = 2\,\underline{i} + 5\,\underline{j}$$

That is, according to person A or relative to the XY coordinate frame, the ball is moving to the right with a speed of 2 m/s and upward with a speed of 5 m/s (Figure 13.26). At this instant, the magnitude of the net velocity of the ball is $v_P = \sqrt{(2)^2 + (5)^2} = 5.4\,\text{m/s}$.

13.12 Linkage Systems

A *linkage system* is composed of several parts connected to each other and/or to the ground by means of hinges or joints, such that each part constituting the system can undergo motion relative to the other segments. An example of such a system is the *double pendulum* shown in Figure 13.27. A double pendulum consists of two bars hinged together and to the ground. Linkage systems are also known as *multilink systems*.

If the angular velocity and acceleration of individual parts are known, then the principles of relative motion can be applied to analyze the motion characteristics of each part constituting the multilink system. The following example will illustrate the procedure of analyzing the motion of a double pendulum. However, the procedure to be introduced can be generalized to analyze any multilink system.

An important concept associated with linkage systems is the number of independent coordinates necessary to describe the motion characteristics of the parts constituting the system. The number of independent parameters required defines the *degrees of freedom* of the system. For example, the two-dimensional motion characteristics of the simple pendulum shown in Figure 13.28 can be fully described by θ that defines the location of the pendulum uniquely. Therefore, a simple pendulum has one degree of freedom. On the other hand, parameters θ_1 and θ_2 are necessary to analyze the coplanar motion of bar BC of the double pendulum shown in Figure 13.27, and therefore, a double pendulum has two degrees of freedom.

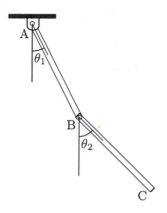

Figure 13.27 *Double pendulum.*

Figure 13.28 *Pendulum.*

Example 13.4 *Double pendulum.*

Assume that arms AB and BC of the double pendulum shown in Figure 13.29 are undergoing coplanar motion. Let $\ell_1 = 0.3$ m and $\ell_2 = 0.3$ m be the lengths of arms AB and BC, and θ_1 and θ_2 be the angles arms AB and BC make with the vertical. The angular velocity and acceleration of arm AB are measured as $\omega_1 = 2$ rad/s (counterclockwise) and $\alpha_1 = 0$ relative to point A. The angular velocity and acceleration of arm BC is measured as $\omega_2 = 4$ rad/s (counterclockwise) and $\alpha_2 = 0$ relative to point B.

Determine the linear velocity and acceleration of point B on arm AB and point C on arm BC at an instant when $\theta_1 = 30°$ and $\theta_2 = 45°$.

Solution: Let X and Y refer to a set of rectangular coordinates with origin located at A, and x and y be a second set of rectangular coordinates with origin at B. The XY coordinate frame is stationary, while the xy frame can move as point B moves.

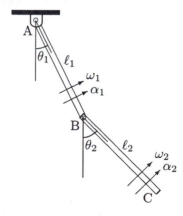

Figure 13.29 *Double pendulum.*

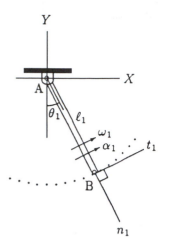

Figure 13.30 *Circular motion of B as observed from point A.*

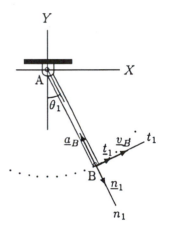

Figure 13.31 *Tangential velocity and normal acceleration of B.*

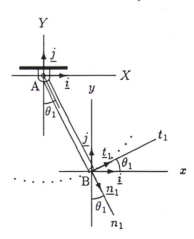

Figure 13.32 *Expressing unit vectors $\underline{n}_1$ and $\underline{t}_1$ in terms of Cartesian unit vectors $\underline{i}$ and $\underline{j}$.*

Since the angular velocity and acceleration of arm AB are given relative to point A, the motion characteristics of any point on arm AB can be determined with respect to the XY coordinate frame. Similarly, the motion of any point on arm BC can easily be analyzed relative to the xy coordinate frame.

Motion of point B as observed from point A:

Every point on arm AB undergoes a rotational motion about a fixed axis passing through point A with constant angular velocity of $\omega_1 = 2$ rad/s. Every point on arm AB experiences a uniform circular motion in the counterclockwise direction. As illustrated in Figure 13.30, point B moves in a circular path of radius ℓ_1. Magnitudes of linear velocity in the tangential direction and linear acceleration in the normal direction of point B can be determined using:

$$v_B = \ell_1\, \omega_1$$
$$a_B = \ell_1\, \omega_1{}^2$$

The magnitude of the tangential component of the acceleration vector is zero since ω_1 is constant or since $\alpha_1 = 0$. Therefore, v_B and a_B are essentially the magnitudes of the resultant linear velocity and acceleration vectors. To express these quantities in vector forms, let n_1 and t_1 represent the normal and tangential directions to the circular path of point B when arm AB makes an angle θ_1 with the horizontal (Figure 13.31). Also let $\underline{n}_1$ and $\underline{t}_1$ be unit vectors in the positive n_1 and t_1 directions, such that the positive $\underline{n}_1$ direction is outward (i.e. from A toward B) and positive $\underline{t}_1$ direction is pointing in the direction of motion (i.e. counterclockwise). The normal (centripetal) acceleration is always directed toward the center of motion, and is acting in the negative $\underline{n}_1$ direction:

$$\underline{v}_B = v_B\, \underline{t}_1 = \ell_1\, \omega_1\, \underline{t}_1$$
$$\underline{a}_B = -a_B\, \underline{n}_1 = -\ell_1\, \omega_1{}^2\, \underline{n}_1$$

Note that directions defined by unit vectors $\underline{n}_1$ and $\underline{t}_1$ change continuously as point B moves along its circular path. That is, $\underline{n}_1$ and $\underline{t}_1$ define a set of *local coordinate* directions that vary in time. By employing proper coordinate transformations, we can express these unit vectors in terms of Cartesian unit vectors $\underline{i}$ and $\underline{j}$. Cartesian coordinate directions, which are *global* as opposed to local, are not influenced by the motion of point B. The coordinate transformation can be done by expressing unit vectors $\underline{n}_1$ and $\underline{t}_1$ in terms of Cartesian unit vectors $\underline{i}$ and $\underline{j}$. It can be observed from the geometry of the problem that (Figure 13.32):

$$\underline{n}_1 = \sin\theta_1\, \underline{i} - \cos\theta_1\, \underline{j}$$
$$\underline{t}_1 = \cos\theta_1\, \underline{i} + \sin\theta_1\, \underline{j}$$

Therefore, the velocity and acceleration vectors of point B with respect to the XY coordinate frame and in terms of Cartesian unit vectors are:

$$\underline{v}_B = \ell_1\, \omega_1\, (\cos\theta_1\, \underline{i} + \sin\theta_1\, \underline{j})$$

$$\underline{a}_B = -\ell_1\, \omega_1{}^2\, (\sin\theta_1\, \underline{i} - \cos\theta_1\, \underline{j})$$

If we substitute the numerical values of $\ell_1 = 0.3$ m, $\theta_1 = 30°$,

$$\underline{v}_B = 0.52\,\underline{i} + 0.30\,\underline{j} \qquad (i)$$

$$\underline{a}_B = -0.60\,\underline{i} + 1.04\,\underline{j} \qquad (ii)$$

Motion of point C as observed from point B:

The motion of point C as observed from point B is similar to the motion of point B as observed from point A. Point C rotates with a constant angular velocity of ω_2 in a circular path of radius ℓ_2 about point B (Figure 13.33). Therefore, the derivation of velocity and acceleration vectors for point C relative to the xy coordinate frame follows the same procedure outlined for the derivation of velocity and acceleration vectors for point B relative to the XY coordinate frame. The magnitudes of the tangential velocity and normal acceleration vectors of point C relative to B are:

$$v_{C/B} = \ell_2\, \omega_2$$

$$a_{C/B} = \ell_2\, \omega_2{}^2$$

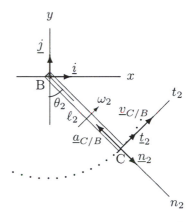

Figure 13.33 *Circular motion of C as observed from point B.*

If $\underline{n}_2$ and $\underline{t}_2$ are unit vectors in the normal and tangential directions to the circular path of C when arm BC makes an angle θ_2 with the vertical, then:

$$\underline{v}_{C/B} = v_{C/B}\, \underline{t}_2 = \ell_2\, \omega_2\, \underline{t}_2$$

$$\underline{a}_{C/B} = -a_{C/B}\, \underline{n}_2 = -\ell_2\, \omega_2{}^2\, \underline{n}_2$$

From Figure 13.34, unit vectors $\underline{n}_2$ and $\underline{t}_2$ can be expressed in terms of Cartesian unit vectors $\underline{i}$ and $\underline{j}$ as:

$$\underline{n}_2 = \sin\theta_2\, \underline{i} - \cos\theta_2\, \underline{j}$$

$$\underline{t}_2 = \cos\theta_2\, \underline{i} + \sin\theta_2\, \underline{j}$$

Therefore, the velocity and acceleration vectors of point C relative to the xy coordinate frame can be written as:

$$\underline{v}_{C/B} = \ell_2\, \omega_2\, (\cos\theta_2\, \underline{i} + \sin\theta_2\, \underline{j})$$

$$\underline{a}_{C/B} = -\ell_2\, \omega_2{}^2\, (\sin\theta_2\, \underline{i} - \cos\theta_2\, \underline{j})$$

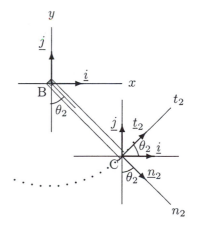

Figure 13.34 *Expressing unit vectors $\underline{n}_2$ and $\underline{t}_2$ in terms of Cartesian unit vectors $\underline{i}$ and $\underline{j}$.*

Substituting the numerical values of $\ell_2 = 0.3$ m, $\theta_2 = 45°$, and $\omega_2 = 4$ rad/s, and carrying out the necessary calculations we obtain:

$$\underline{v}_{C/B} = 0.85\,\underline{i} + 0.85\,\underline{j} \qquad (iii)$$

$$\underline{a}_{C/B} = -3.39\,\underline{i} + 3.39\,\underline{j} \qquad (iv)$$

Motion of point C as observed from point A:

We determined the velocity and acceleration of point C relative to B, and velocity and acceleration of point B with respect to A. Now, we can apply the principles of relative motion to determine the velocity and acceleration of point C as observed from point A or with respect to the XY coordinate frame:

$$\underline{v}_C = \underline{v}_B + \underline{v}_{C/B} \qquad\qquad (v)$$

$$\underline{a}_C = \underline{a}_B + \underline{a}_{C/B} \qquad\qquad (vi)$$

Since we have already expressed $\underline{v}_B$, $\underline{v}_{C/B}$, $\underline{a}_B$, and $\underline{a}_{C/B}$ in terms of Cartesian unit vectors, we can simply substitute Eqs. (*i*) and (*iii*) into Eq. (*v*), and Eqs. (*ii*) and (*iv*) into Eq. (*vi*):

$$\underline{v}_C = (0.52\,\underline{i} + 0.30\,\underline{j}) + (0.85\,\underline{i} + 0.85\,\underline{j})$$

$$\underline{a}_C = (-0.60\,\underline{i} + 1.04\,\underline{j}) + (-3.39\,\underline{i} + 3.39\,\underline{j})$$

Collecting the horizontal and vertical components together:

$$\underline{v}_C = 1.37\,\underline{i} + 1.15\,\underline{j}$$

$$\underline{a}_C = -3.99\,\underline{i} + 4.43\,\underline{j}$$

The magnitudes of the velocity and acceleration vectors are:

$$v_C = \sqrt{(1.37)^2 + (1.15)^2} = 1.79 \text{ m/s}$$

$$a_C = \sqrt{(3.99)^2 + (4.43)^2} = 5.96 \text{ m/s}^2$$

Chapter 14

Angular Kinetics

14.1   Kinetics of Angular Motion

The kinetic characteristics of objects undergoing translational motion were discussed in Chapter 12. Kinetic analyses utilize Newton's second law of motion that can be formulated in terms of the equations of motion and work and energy methods. Similar methods can be employed to analyze the kinetic characteristics of objects undergoing rotational motion.

Consider an object undergoing a rotational motion in the xy-plane about a fixed point O (Figure 14.1). Let P be a point on the object located at a distance r from O. As the object rotates, point P moves in a circular path of radius r and center located at O. To be able to analyze the kinetic characteristics of point P using the equations of motion, forces acting on the object and the acceleration of point P can be expressed in terms of their components normal and tangential to the circular path of motion. If n and t designate the normal (radial) and tangential directions at point P, then the equations of motion can be expressed as:

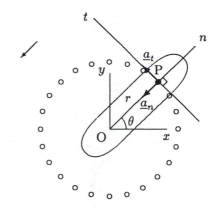

Figure 14.1 *Rotational motion.*

$$\sum F_n = m\, a_n \tag{14.1}$$

$$\sum F_t = m\, a_t \tag{14.2}$$

Here, $\sum F_n$ is the net force acting in the normal direction, $\sum F_t$ is the net force acting in the tangential direction, a_n is the magnitude of the centripetal acceleration (always directed toward the center of rotation), and a_t is the magnitude of tangential acceleration. While applying Eq. (14.1), the forces acting toward the center of rotation (centripetal forces) must be taken to be positive, and the forces directed outward (centrifugal forces) must be negative.

For rotational motion about a fixed axis, the motion characteristics are completely known if the linear velocity (with magnitude v and direction tangent to the circular path) and the radius r of the circular path are known. Since $a_n = v^2/r$ and $a_t = dv/dt$, Eqs. (14.1) and (14.2) can alternatively be written as:

$$\sum F_n = m\, \frac{v^2}{r} \tag{14.3}$$

$$\sum F_t = m\, \frac{dv}{dt} \tag{14.4}$$

Note that $v = r\omega$ and $dv/dt = r\alpha$. Therefore, if the angular velocity and acceleration are known, then the kinetic characteristics of the problem can be analyzed using:

$$\sum F_n = m\, r\, w^2 \tag{14.5}$$

$$\sum F_t = m\, r\, \alpha \tag{14.6}$$

Note that for rotational motion there is always a normal component of the acceleration vector, and therefore, a force acting in the normal direction, but the tangential components of the force and acceleration vectors may or may not exist.

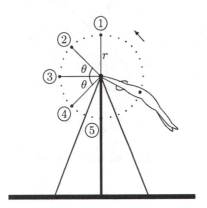

Figure 14.2 *A gymnast on the high bar.*

Example 14.1 Figure 14.2 illustrates a 60 kg gymnast swinging on a high bar. The rotational motion of the gymnast may be simplified by modeling the gymnast as a particle attached to a string such that the mass of the particle is equal to the mass of the gymnast and the length of the string is equal to the distance between the high bar and the center of gravity of the gymnast. As the gymnast moves, the center of gravity undergoes a circular motion.

Assume that the center of gravity of the gymnast is located at a distance $r = 1$ m from the high bar, the speed of the center of gravity at position 1 is almost zero, and that the effects of air resistance are negligible. Position 1 is directly above the high bar and it represents the highest elevation reached by the center of gravity of the gymnast

(a) By using the conservation of energy principle, calculate the speeds of the gymnast's center of gravity at positions 2, 3, 4, and 5. As shown in Figure 14.2, positions 2 and 4 make an angle $\theta = 45°$ with the horizontal, position 3 is along the same horizontal line as the high bar, and position 5 is directly under the high bar.

(b) Calculate the angular velocities of the gymnast at positions 1, 2, 3, 4, and 5.

(c) Calculate the normal component of the linear accelerations of the gymnast's center of gravity at positions 1, 2, 3, 4, and 5.

(d) Calculate the forces applied on gymnast's arms at positions 1, 2, 3, 4, and 5.

(e) Calculate the tangential component of the linear accelerations of the gymnast's center of gravity at positions 1, 2, 3, 4, and 5.

(f) Calculate the angular accelerations of the gymnast at positions 1, 2, 3, 4, and 5.

Solution:

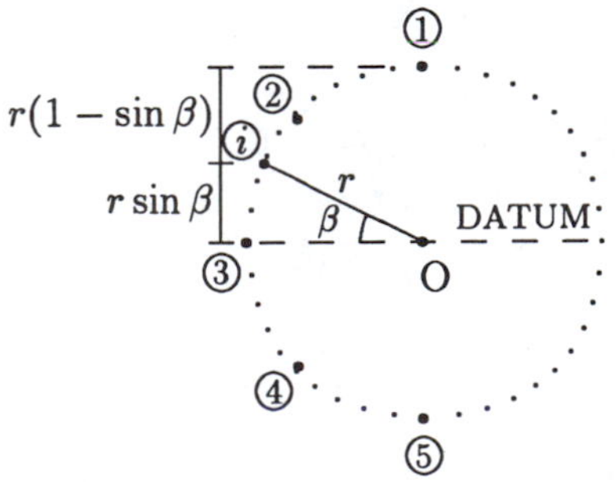

(a) Different positions of the gymnast's center of gravity are illustrated in Figure 14.3. Point O in Figure 14.3 represents the high bar. The conservation of energy principle states that if an object is moving under the effect of conservative forces, then the total energy (sum of potential and kinetic energies) will remain constant. Between any two positions 1 and 2:

Figure 14.3 *For different β values, i represents positions 1, 2, 3, 4, and 5.*

$$\mathcal{E}_{P1} + \mathcal{E}_{K1} = \mathcal{E}_{P2} + \mathcal{E}_{K2}$$

$$m\,g\,h_1 + \frac{1}{2}\,m\,v_1{}^2 = m\,g\,h_2 + \frac{1}{2}\,m\,v_2{}^2$$

Here, $m = 60$ kg is the total mass of the gymnast, $g = 9.8$ m/s^2 is the magnitude of gravitational acceleration, $v_1 = 0$ is the speed of the gymnast at position 1, v_2 is the unknown speed of the gymnast at position 2, and h_1 and h_2 are the heights of positions 1 and 2 relative to a datum. Once a datum is chosen, h_1 and h_2 can be calculated and the above equation can be solved for the unknown parameter v_2. To calculate v_3, v_4, and v_5, the above procedure must be repeated.

Instead of carrying out the same procedure four times, consider the position of the gymnast labeled as "i" in Figure 14.3. Assume that the line connecting O and position i makes an angle β with the horizontal that passes through O such that $\beta = 90°$ at position 1, $\beta = 45°$ at position 2, $\beta = 0°$ at position 3, $\beta = -45°$ at position 4, and $\beta = -90°$ at position 5. We can apply the conservation of energy principle between positions 1 and i:

$$m\,g\,h_1 + \frac{1}{2}\,m\,v_1{}^2 = m\,g\,h_i + \frac{1}{2}\,m\,v_i{}^2$$

Since $v_1 = 0$, the second term on the left-hand side of this equation is zero. Also, m is a common parameter in all of the terms and can be eliminated. To calculate heights h_1 and h_i, we need to choose a datum. If we choose the datum to coincide with the level of high bar, then from the geometry of the problem $h_1 = r$ and $h_i = r\sin\beta$. We can substitute h_1 and h_i into the above equation and solve it for v_i:

$$v_i = \sqrt{2\,g\,r\,(1 - \sin\beta)}$$

This is a general solution valid for any position on the circular path of motion of the gymnast's center of gravity. For example, when angle $\beta = 45°$, $i = 2$ and $v_2 = \sqrt{2(9.8)(1)(1 - \sin 45)} = 2.39$ m/s. When $\beta = 90°$, $i = 3$ and $v_3 = \sqrt{2(9.8)(1)(1 - \sin 90)} = 4.43$ m/s. Similarly, $v_4 = 5.77$ m/s and $v_5 = 6.26$ m/s. These results are used to plot a speed versus angular position (measured in terms of angle β) graph in Figure 14.4.

(b) The relationship between the angular velocity and linear velocity of a point on an object undergoing a rotational motion about a fixed axis is:

$$\omega = \frac{v}{r}$$

For example, for position 2, $\omega_2 = v_2/r = 2.39/1 = 2.39$ rad/s. Similarly, $\omega_1 = 0$, $\omega_3 = 4.43$ rad/s, $\omega_4 = 5.77$ rad/s, and $\omega_5 = 6.26$ rad/s.

(c) The normal (radial) component of the acceleration of the gymnast's center of gravity can be calculated using:

$$a_n = \frac{v^2}{r} = r\,\omega^2$$

For example, $a_{n2} = r\omega_2{}^2 = (1)(2.39)^2 = 5.74$ m/s^2 for position 2. On the other hand, $a_{n1} = 0$, $a_{n3} = 19.62$ m/s^2, $a_{n4} = 33.41$ m/s^2, and $a_{n5} = 39.19$ m/s^2.

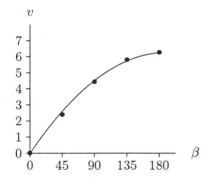

Figure 14.4 *Speed measured in m/s versus angle β in degrees.*

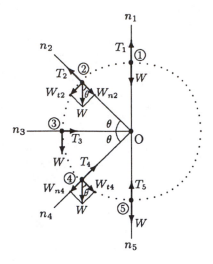

Figure 14.5 *Forces acting on the gymnast's center of gravity.*

(d) To calculate the forces applied on the gymnast's arms, consider Figure 14.5 that shows the free-body diagrams of the gymnast's center of gravity. The forces acting on the gymnast are those applied by gravity and on the gymnast's arms by the high bar. The force applied by gravity is always directed vertically downward. The force applied by the high bar is in the radial direction. The equation of motion in the normal direction is:

$$\sum F_n = m\, a_n$$

At position 1, $a_{n1} = 0$ and the gymnast is in equilibrium. Therefore, since the weight of the gymnast is acting downward, the force, T_1, applied by the high bar on the gymnast's arms must be upward. Applying the equation of equilibrium:

$$\sum F_{n1} = 0: \quad W - T_1 = 0$$
$$T_1 = W = m\, g$$
$$T_1 = (60)(9.8) = 588\ \text{N}$$

At position 2, forces acting in the normal direction are the radial component, W_{n2}, of the gymnast's weight and T_2 applied by the high bar on the gymnast's arms. From the geometry of the problem, $W_{n2} = W\cos(45)$ and the direction of $\underline{W}_{n2}$ is toward the center of rotation. At this point, we do not know the direction of $\underline{T}_2$. We can assume that it is centrifugal. Now, we can apply the equation of motion in the normal direction:

$$\sum F_{n2} = m\, a_{n2}: \quad W_{n2} - T_2 = m\, a_{n2}$$
$$T_2 = W_{n2} - m\, a_{n2}$$
$$T_2 = m\, g\, \sin(45) - m\, a_{n2}$$
$$T_2 = (60)(9.8)(\sin 45) - (60)(5.74) = 71\ \text{N}$$

Since we calculated a positive value for T_2, the direction we assumed for $\underline{T}_2$ was correct. In other words, the effect of the high bar is such that it is "pushing" the arms of the gymnast at position 2.

At position 3, the gymnast's weight has no component in the radial direction. The force acting in the normal direction is T_3 applied by the high bar. In this case, assume that $\underline{T}_3$ is centripetal (toward O). Writing the equation of motion for position 3:

$$\sum F_{n3} = m\, a_{n3}: \quad T_3 = m\, a_{n3}$$
$$T_3 = (60)(19.62) = 1177\ \text{N}$$

Again, since we calculated a positive value for T_3, the direction we assumed for $\underline{T}_3$ was correct. At position 3, the tendency of the gymnast is to move away from the center of rotation and what is holding the gymnast in the circular path of motion is the "pulling" effect of the high bar.

At position 4, forces acting in the radial direction are T_4 applied by the high bar and $W_{n4} = W\sin(45)$ component of the gymnast's weight. From the geometry of the problem, $\underline{W}_{n4}$ is centrifugal. Assuming that $\underline{T}_4$ is centripetal and applying the equation of motion:

$$\sum F_{n4} = m\,a_{n4}: \quad T_4 - W_{n4} = m\,a_{n4}$$
$$T_4 = W_{n4} + m\,a_{n4}$$
$$T_4 = m\,g\,\sin(45) + m\,a_{n4}$$
$$T_4 = (60)(9.8)(\sin 45) + (60)(33.41)$$
$$T_4 = 2420 \text{ N}$$

Similarly at position 5:

$$\sum F_{n5} = m\,a_{n5}: \quad T_5 - W = m\,a_{n5}$$
$$T_5 = m\,g + m\,a_{n5}$$
$$T_5 = (60)(9.8) + (60)(39.19) = 2939 \text{ N}$$

In Figure 14.6, these results are used to plot a force applied by the high bar on the arms of the gymnast versus angular position (measured in terms of angle β in Figure 14.3) graph. Note that between positions 1 and 2, the high bar has a pushing effect on the arms. In other words, the force applied by the high bar on the arms is compressive. Just after position 2, the force applied by the high bar is zero, and thereafter it has a pulling or tensile effect on the arms.

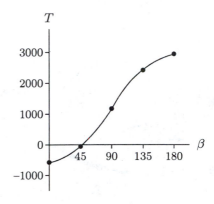

Figure 14.6 *Force applied by the high bar on the gymnast's arms (measured in Newtons) versus angle β in degrees.*

(e) The equation of motion in the tangential direction is:

$$\sum F_t = m\,a_t \quad \text{or} \quad a_t = \frac{\sum F_t}{m}$$

Forces acting in the tangential direction for different positions of the gymnast's center of gravity are shown in Figure 14.5. At position 1, there is no force in the tangential direction, and therefore, $a_{t1} = 0$. At position 2, $W_{t2} = W\cos(45)$ is the only tangential force. Therefore:

$$a_{t2} = \frac{W_{t2}}{m} = \frac{m\,g\,\cos(45)}{m} = g\,\cos(45) = 6.93 \text{ m/s}^2$$

At position 3, $W = mg$ is the only tangential force:

$$a_{t3} = \frac{W_{t3}}{m} = \frac{m\,g}{m} = g = 9.80 \text{ m/s}^2$$

At position 4, $W_{t4} = W\cos(45)$ is the only tangential force:

$$a_{t4} = \frac{W_{t4}}{m} = \frac{m\,g\,\cos(45)}{m} = g\,\cos(45) = 6.93 \text{ m/s}^2$$

There is no tangential force at position 5, and therefore, $a_{t5} = 0$.

(f) Now that we calculated the tangential components of the acceleration vector, we can also calculate the angular acceleration of the gymnast using:

$$\alpha = \frac{a_t}{r}$$

Note that $r = 1$ m. Therefore, $\alpha_1 = 0$, $\alpha_2 = 6.93 \, \text{rad}/\text{s}^2$, $\alpha_3 = 9.80 \, \text{rad}/\text{s}^2$, $\alpha_4 = 6.93 \, \text{rad}/\text{s}^2$, and $\alpha_5 = 0$.

Example 14.2 Figure 14.7 illustrates the circular end region of a ski jump track. The radius of curvature of the track at this region is $r = 50$ m. At the end of the track, the direction normal to the track coincides with the vertical and the direction tangential to the track coincides with the horizontal.

Consider a 70 kg ski jumper who is decelerating at a rate of 1.5 m/s^2 due to air resistance. If the friction on the track is negligible and the ski jumper reaches the end of the track with a horizontal velocity of $v = 20$ m/s, determine the forces applied on the skier by the air resistance and the track.

Solution: The free-body diagram of the ski jumper at the very end of the track (just before takeoff) is shown in Figure 14.8. The forces acting on the ski jumper are $\underline{W}$ due to gravity, $\underline{R}_1$ due to air resistance, and the reaction force $\underline{R}_2$ applied by the track on the skis. Note that $\underline{R}_2$ is applied in the vertical direction or direction normal to the track. It is assumed that $\underline{R}_1$ due to air resistance is applied in the horizontal direction or direction tangent to the track.

At the circular end region of the track, the ski jumper undergoes a motion in a circular path with radius $r = 50$ m. The speed of the ski jumper at the very end of the track is $v = 20$ m/s. Therefore, the magnitude of the ski jumper's acceleration in the normal direction is:

$$a_n = \frac{v^2}{r} = \frac{(20)^2}{50} = 8.0 \, \text{m}/\text{s}^2$$

Owing to air resistance, the ski jumper is decelerating at a rate of 1.5 m/s^2 in the direction of motion (toward the left). Or, the ski jumper is accelerating at a rate of 1.5 m/s^2 in the direction opposite to the direction of motion (toward the right). Therefore, the tangential acceleration of the ski jumper toward the right is:

$$a_t = 1.5 \, \text{m}/\text{s}^2$$

Now we can utilize the equations of motion. In the tangential direction:

$$\sum F_t = m \, a_t : \qquad R_1 = m \, a_t$$
$$R_1 = (70)(1.5) = 105 \, \text{N}$$

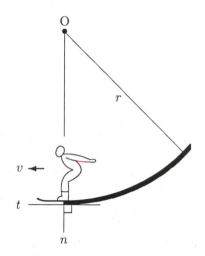

Figure 14.7 *Circular end region of a ski jump track.*

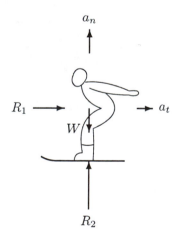

Figure 14.8 *Free-body diagram of the ski jumper before takeoff.*

Equation of motion in the normal direction:

$$\sum F_n = m\, a_n: \qquad R_2 - W = m\, a_n$$
$$R_2 = W + m\, a_n = m\, g + m\, a_n$$
$$R_2 = (70)(9.8) + (70)(8.0) = 1246\ \text{N}$$

Therefore, at the very end of the ski jump track, air resistance is applying a horizontal force of $R_1 = 105\,\text{N}$ to retard the motion of the skier and the track is applying a vertical force of $R_2 = 1246\,\text{N}$ on the skis. Note that R_2 includes the effects of the weight W and rotational inertia ma_n of the ski jumper.

14.2 Torque and Angular Acceleration

Torque is the quantitative measure of the ability of a force to rotate an object. The mathematical definition of torque is the same as that of moment, studied in detail in Chapter 3. Consider the bolt and wrench arrangement illustrated in Figure 14.9. Force $\underline{F}$ applied on the wrench rotates the wrench, which advances the bolt into the wall by rotating it in the clockwise direction. The magnitude of torque $\underline{M}$ due to force $\underline{F}$ about O is:

$$M = r\, F_t = r\, F \sin\phi \qquad (14.7)$$

The line of action of $\underline{M}$ is perpendicular to the plane of rotation and its direction can be determined by using the right-hand rule (in this case, clockwise).

An object would rotate about an axis if the rotational motion of the object is not constrained and if there is a net torque acting on the object about that axis. The angular acceleration of an object undergoing a rotational motion is directly proportional to the resultant torque acting on it. To derive the relationship between torque and angular acceleration, consider a particle of mass m undergoing a rotational motion about a fixed axis. Let O be a point on this axis, r be the radius of the circular path of motion, and $\underline{F}_t$ be the tangential force causing the rotational motion (Figure 14.10). The equation of motion in the tangential direction can be written as:

$$F_t = m\, a_t \qquad (14.8)$$

In Eq. (14.8), a_t is the magnitude of the tangential acceleration of the particle. If the angular acceleration, α, of the particle is known, then $a_t = r\alpha$. Replacing a_t by $r\alpha$ and multiplying both sides of Eq. (14.8) by r will yield:

$$r\, F_t = (m\, r^2)\, \alpha \qquad (14.9)$$

Note that the left-hand side of Eq. (14.9) is the magnitude M_o of the torque generated by force $\underline{F}_t$ about O. The term mr^2 on

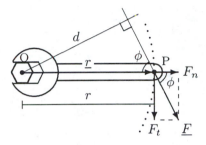

Figure 14.9 *Force $\underline{F}$ applied on the wrench produces a clockwise torque about the centerline of the bolt.*

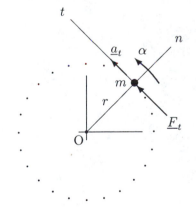

Figure 14.10 $F_t = ma_t$ and $M_o = I_o\alpha$.

the right-hand side is known as the *mass moment of inertia* of the particle about O. Denoting the mass moment of inertia with I_o, Eq. (14.9) can also be written as:

$$M_o = I_o\,\alpha \qquad (14.10)$$

If there is more than one torque-generating force applied to the particle, then M_o in Eq. (14.10) represents the net torque acting on the particle about O. The general form of Eq. (14.10) can be obtained by representing the torque and angular acceleration as vector quantities:

$$\underline{M} = I\,\underline{\alpha} \qquad (14.11)$$

This is the rotational analog of Newton's second law of motion, which states that angular acceleration is directly proportional to the net torque and inversely proportional to the mass moment of inertia.

14.3 Mass Moment of Inertia

In general, the term inertia implies resistance to change. When a rotation-causing force is applied to a pivoted body, its tendency to resist angular acceleration depends on its mass moment of inertia. The larger the mass moment of inertia of a body, the more difficult it is to accelerate it in rotation. For a particle of mass m, the mass moment of inertia about an axis is defined as the mass times the square of the shortest distance, r, between the particle and the axis about which the mass moment of inertia is to be determined:

$$I = m\,r^2 \qquad (14.12)$$

A rigid body that is not a particle can be assumed to consist of many particles, the sum of masses of which is equal to the total mass of the body itself. The mass moment of inertia of the entire body can be determined by considering the sum of the mass of each particle of the body multiplied by the square of its distance from the axis of rotation.

Note that the mass moment of inertia of a rigid body is proportional to its mass, which is a function of its density and volume. Therefore, the mass moment of inertia of a body depends upon its material and geometric properties as well as the location and orientation of the axis about which it is to be determined. The ability of a body to resist changes in its angular velocity is dependent not only upon the mass of the body, but also upon the distribution of the mass. The greater the concentration of mass at the periphery, the greater the mass moment of inertia and the more difficult it is to change the angular velocity. For a rigid body with a simple, symmetrical geometry and homogeneous composition, the mass moment of inertia about an axis coinciding with an axis of symmetry, called a *centroidal axis*, can be calculated relatively easily. In Table 14.1, moments of inertia for some geometric shapes are provided about their centroidal axes.

Table 14.1 *Moments of inertia of homogeneous rigid bodies with different geometries about their centroidal axes. Their volumes are also provided.*

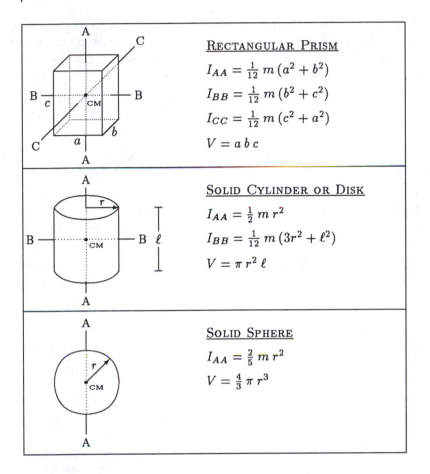

RECTANGULAR PRISM

$I_{AA} = \frac{1}{12} m (a^2 + b^2)$

$I_{BB} = \frac{1}{12} m (b^2 + c^2)$

$I_{CC} = \frac{1}{12} m (c^2 + a^2)$

$V = a\,b\,c$

SOLID CYLINDER OR DISK

$I_{AA} = \frac{1}{2} m r^2$

$I_{BB} = \frac{1}{12} m (3r^2 + \ell^2)$

$V = \pi r^2 \ell$

SOLID SPHERE

$I_{AA} = \frac{2}{5} m r^2$

$V = \frac{4}{3} \pi r^3$

The mass moment of inertia is a scalar quantity. It has a dimension $[M][L^2]$, and is measured in terms of kg-m² in SI.

14.4 Parallel-Axis Theorem

If the moment of inertia of a body about a centroidal axis is known, then the moment of inertia of the same body about any other axis parallel to that centroidal axis can be determined using the *parallel-axis theorem*. This theorem can be stated as:

$$I = I_c + m\, r_c^2 \qquad (14.13)$$

In Eq. (14.13), m is the total mass of the body, I_c is the mass moment of inertia of the body about one of its centroidal axes, I is the required mass moment of inertia about an axis parallel to the centroidal axis, and r_c is the shortest distance between

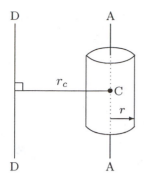

Figure 14.11 *According to the parallel-axis theorem,* $I_{DD} = I_{AA} + mr_c{}^2.$

the two axes. For example, consider the solid cylinder shown in Figure 14.11. From Table 14.1, the mass moment of inertia of the cylinder about AA is $I_{AA} = \frac{1}{2}mr^2$. The mass moment of inertia of the same cylinder about DD, which is parallel to AA and located at a distance r_c from AA, is:

$$I_{DD} = I_{AA} + mr_c{}^2 = \frac{1}{2}mr^2 + mr_c{}^2$$

Note that in the case of human body segments, each segment or limb rotates about the joints at either end of the moving segment rather than about its mass center or centroidal axes. Furthermore, mass moment of inertia measurements can only be made about a joint center. If needed, the parallel-axis theorem can be utilized to determine the mass moment of inertia of a segment about its mass center.

14.5 Radius of Gyration

Consider a rigid body of mass m. Let I be the mass moment of inertia of the rigid body about a given axis AA. Also consider a point mass m located at a distance ρ (rho) from the same axis such that its mass moment of inertia $m\rho^2$ about AA is equal to the mass moment of inertia I of the rigid body about AA. That is:

$$\rho = \sqrt{\frac{I}{m}} \tag{14.14}$$

ρ is called the *radius of gyration*, and for rotational motion analysis, the rigid body can be treated as a point mass equal to the total mass of the body located at a distance ρ from the axis of rotation.

14.6 Segmental Motion Analysis

The information provided in the previous sections can be utilized to develop mathematical models for analyzing the motion characteristics of human body segments. Here, the general procedure for developing a dynamic model of a body segment will be outlined, and then applied to analyze the rotational motion of the lower leg about the knee joint.

The first step of a dynamic model analysis involves defining the forces acting on the body segment. These may include the gravitational (weight), external, inertial, muscle, and joint reaction forces. The weight of the body segment can be assumed to act at its center of gravity, and therefore, the center of gravity of the segment must be known. The magnitude, point of application, and direction of any external force present must be specified. Inertial forces are those present due to the dynamics of the problem under consideration. One way of incorporating inertial effects into the model is through the use of the radius of gyration. Muscle and joint reaction forces are the unknowns to be determined as a result of these analyses. It is important to draw the free-body diagram of the segment to be analyzed, and to identify all the known and unknown forces acting on it.

The next step of dynamic analysis is the identification of measurable quantities. In general, both the angular displacement of the moving body segment and the net torque generated about its axis of rotation can be measured as functions of time over the range of segmental motion. The angular displacement measurement techniques include goniometric, dynamometric, and photogrammetric methods. Through the use of kinematic relationships, the angular displacement data can be used to calculate the angular velocity and angular acceleration of the moving segment. If the angular displacement θ is known as a function of time t, then the angular velocity ω and angular acceleration α can be determined by considering the first and second derivatives of θ with respect to t:

$$\omega = \frac{d\theta}{dt} \qquad \alpha = \frac{d\omega}{dt} = \frac{d^2\theta}{dt^2}$$

If it is not possible to find a function representing the relationship between the angular displacement and time, then numerical differentiation techniques can be employed. Once the angular acceleration is determined, the relationship between torque, mass moment of inertia, and angular acceleration can be used to calculate the net torque produced about the joint center, provided that the mass moment of inertia (or the radius of gyration) of the segment about the joint center is known.

For a two-dimensional (planar) motion analysis of a segment about its joint center, if the net torque M produced about the joint axis is measured as a function of time and the mass moment of inertia I of the moving segment about the same axis is known, then the angular acceleration of the segment can be determined from:

$$\alpha = \frac{M}{I}$$

If needed, the angular velocity and displacement of the moving segment can also be determined by considering the integral of the function representing angular acceleration with respect to time.

It is clear from this discussion that anthropometric information about the moving segment must be available. For this purpose, anthropometric data tables listing average segmental weights, lengths, and radius of gyration can be utilized. Another important consideration is that the instantaneous center of rotation of the moving segment must also be known. Note however that the instantaneous center of rotation about a given joint may vary.

The final step of dynamic model analyses involves the computation of muscle and joint reaction forces. Note that the net torque measured or calculated includes the effects of all forces acting on the moving segment. The torque generated by the muscles crossing the joint can be determined by subtracting the effects of external and gravitational forces from the net torque measured or calculated about the joint center. If the forces generated by individual muscles are required, then additional factors must be considered. For example, the locations of muscle attachments and the lines of action (lines of pull) of the muscle forces must be known. The distribution of forces among different muscles must also be specified. If the segmental motion is achieved primarily by a single muscle group, then the rotational component of the muscle force can be determined by applying Newton's second law of motion. Since the line of action of the muscle force is assumed to be known, the magnitude of the muscle force can also be determined.

Example 14.3   *Knee extension.*
The angular motion of the lower leg about the knee joint, and the forces and torques produced by the muscles crossing the knee joint during knee flexions and extensions have been investigated by a number of researchers utilizing different experimental techniques. One of these techniques is discussed here.

Consider the person illustrated in Figure 14.12. The test subject is sitting on a table, with the back placed against a back rest and the lower legs free to rotate about the knee joint. The subject's torso is strapped to the back rest and the right thigh is strapped firmly to the table. A well-padded sawhorse is placed in front of the subject to prevent hyperextension at the knee joint (not shown in Figure 14.12). An electrogoniometer is attached to the subject's right leg. The arms of the goniometer are aligned with the estimated long axes of the thigh and shank, and the axis of rotation of the goniometer is aligned with the estimated axis of rotation of the knee joint. The subject is then asked to extend the lower leg as rapidly as possible. The signals received from the electrogoniometer's potentiometer are stored in a computer, and are used to calculate the angular displacement θ of the lower leg as measured from its initial vertical position. Using a finite difference (numerical differentiation) technique, the angular velocity ω and angular acceleration α of the lower leg are also computed.

Some of the forces acting on the lower leg are shown in Figure 14.13, along with the geometric parameters of the model under consideration. This model is based on the assumption that the quadriceps muscle is the primary muscle group responsible for

Figure 14.12 *Knee extension.*

knee extension. Point O represents the instantaneous center of rotation of the knee joint. The patellar tendon is attached to the tibia at A. For the position of the lower leg relative to the upper leg shown in Figure 14.13, it is estimated that the line of pull of the patellar tendon force $\underline{F}_m$ makes an angle β with the long axis of the tibia. The lever arm of $\underline{F}_m$ relative to O can be represented by a distance a that changes as the lower leg moves up through the range of motion. The total weight of the lower leg is W and its center of gravity is located at B, which is at a distance b from O measured along the long axis of the tibia. The intended direction of motion is counterclockwise (extension).

At an instant when $\theta = 60°$, $\omega = 5$ rad/s, and $\alpha = 200$ rad/s^2, and assuming that $W = 50$ N, $a = 4$ cm, $b = 22$ cm, $\beta = 24°$, and the mass moment of inertia of the lower leg about the knee joint is $I_o = 0.25$ kg-m^2, determine:

(a) the net torque produced about the knee joint,
(b) the tension in the patellar tendon, and
(c) the reaction force at the knee joint.

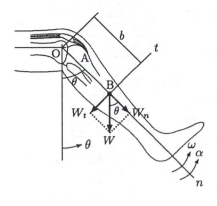

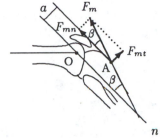

Figure 14.13 *Some of the forces acting on the lower leg.*

Solution:

(a) From Newton's second law of motion, the net torque M_o generated about the knee joint is:

$$M_o = I_o\,\alpha = (0.25)(200) = 50 \text{ N-m} \qquad \text{(ccw)}$$

(b) Note that M_o is the magnitude of the net torque about the knee joint and it includes the rotational effects of all of the external forces acting on the lower leg. Relative to the knee joint, there are two external forces with rotational effects: patellar tendon force $\underline{F}_m$ and weight of the lower leg $\underline{W}$. For the position of the lower leg that makes an angle $\theta = 60°$ with the vertical, the lever arm of the patellar tendon force relative to O is estimated to be $a = 0.04$ m. Therefore, the torque generated by $\underline{F}_m$ relative to O is $M_m = a\,F_m$ (counter-clockwise). On the other hand, $\underline{W}$ acts downward. The torque generated by $\underline{W}$ about O is $M_w = W_t\,b = W\,\sin\theta\,b$ (clockwise). Therefore, the magnitude of the net torque, M_o, about the knee joint due to $\underline{F}_m$ and $\underline{W}$ is:

$$M_o = M_m - M_w = a\,F_m - b\,W\,\sin\theta$$

This equation can be solved for the unknown force, F_m:

$$F_m = \frac{M_o + b\,W\,\sin\theta}{a}$$

Substituting the known parameters and carrying out the calculations will yield:

$$F_m = \frac{50 + (0.22)(50)(\sin 60°)}{0.04} = 1488 \text{ N}$$

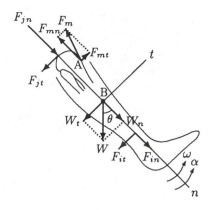

Figure 14.14 *Free-body diagram of the lower leg.*

(c) The free-body diagram of the lower leg is shown in Figure 14.14. W is the weight of the lower leg, F_m is the magnitude of the patellar tendon force applied on the tibia at A, F_{jn} is the component of the tibiofemoral joint reaction force along the long axis of the tibia, and F_{jt} is the component of the tibiofemoral joint reaction force in a direction perpendicular to the long axis of the tibia. The components of $\underline{W}$ and $\underline{F}_m$ along the long axis of the tibia ($\underline{F}_{in}$) and in the perpendicular direction ($\underline{F}_{it}$) are also shown in Figure 14.14.

One way of taking into account the inertial effects of a moving body, in this case a rotating body segment, is by means of d'Alembert's principle. If the mass m, distance r between the mass center and the axis of rotation, angular velocity ω, and angular acceleration α of the rotating body are known, then the magnitudes of inertial forces F_{in} and F_{it} that are normal and tangent to the path of motion can be calculated using Eqs. (14.5) and (14.6) provided in Section 14.1:

$$F_{in} = m\, a_n = m\, r\, \omega^2$$
$$F_{it} = m\, a_t = m\, r\, \alpha$$

In this case, we have $m = W/g = 50/9.8 = 5.1$ kg, $r = b = 0.22$ cm, $\omega = 12$ rad/s, and $\alpha = 200$ rad/s^2. a_n and a_t are the components of the acceleration vector of the lower leg in the directions normal and tangential to its path of motion. $\underline{a}_n$ is always toward the center of rotation, and since the motion is counterclockwise, $\underline{a}_t$ is counterclockwise. Therefore, the inertial forces $\underline{F}_{in}$ and $\underline{F}_{it}$ are such that $\underline{F}_{in}$ is centripetal (towards the center of rotation) and $\underline{F}_{it}$ is trying to rotate the leg in the counterclockwise direction. As illustrated in Figure 14.14, d'Alembert's principle can be applied by assuming that $\underline{F}_{in}$ is a centrifugal force (rather than centripetal) trying to pull the leg outward, $\underline{F}_{it}$ is trying to rotate the lower leg in the clockwise direction (rather than in the counterclockwise direction), and the system is in static equilibrium. The conditions of the equilibrium of the system can be represented by the following equations valid along the directions normal and tangential to the path of motion:

$$\sum F_n = 0: \qquad F_{jn} - F_{mn} + F_{in} + W_n = 0$$
$$\sum F_t = 0: \qquad F_{jt} - F_{mt} + F_{it} + W_t = 0$$

Solving these equations for the components of the joint reaction force will yield:

$$F_{jn} = F_{mn} - F_{in} - W_n$$
$$F_{jt} = F_{mt} - F_{it} - W_t$$

Substituting the known parameters by their mathematical expressions will yield:

$$F_{jn} = F_m \cos\beta - m\, b\, \omega^2 - W\cos\theta$$
$$F_{jt} = F_m \sin\beta - m\, b\, \alpha - W\sin\theta$$

Now, substituting the numerical values and carrying out the calculations will yield:

$$F_{jn} = (1488)(\cos 24) - (5.1)(0.22)(5)^2 - (50)(\cos 60) = 1306N$$

$$F_{jt} = (1488)(\sin 24) - (5.1)(0.22)(200) - (50)(\sin 60) = 338N$$

Therefore, the magnitude of the resultant force applied by the femur on the tibia is $F_j = \sqrt{(F_{jn})^2 + (F_{jt})^2} = 1349 \text{ N}$.

14.7 Rotational Kinetic Energy

Assume that the rigid body shown in Figure 14.15 is composed of many small particles and that the body rotates about a fixed axis with an angular velocity ω. If m_i and v_i are the mass and the speed of the ith particle in the body, respectively, then the kinetic energy of the particle is:

$$\mathcal{E}_{Ki} = \frac{1}{2} m_i v_i^2$$

At any instant, every particle in the body has the same angular velocity ω, but the linear velocity of each particle depends on its distance measured from the axis of rotation. If r_i is the perpendicular distance between the ith particle and the axis of rotation (i.e. radius of the circular motion path of the ith particle), then $v_i = r_i \omega$ and its kinetic energy is $\mathcal{E}_{Ki} = \frac{1}{2}m_i r_i^2 \omega^2$. Each particle in the body has a kinetic energy, and the total kinetic energy, $\mathcal{E}_K$, of the rotating body is the sum of the kinetic energies of the individual particles in the body. That is, $\mathcal{E}_K = \sum \mathcal{E}_{Ki} = \frac{1}{2}(\sum m_i r_i^2)\omega^2$. The quantity in parentheses is the mass moment of inertia I of the body. Therefore:

$$\mathcal{E}_K = \frac{1}{2} I \omega^2 \qquad (14.15)$$

Equation (14.15) defines the rotational kinetic energy of a body in terms of the mass moment of inertia and angular velocity of the body, and it is analogous to the kinetic energy $\mathcal{E}_K = \frac{1}{2}mv^2$ associated with linear motion.

14.8 Angular Work and Power

By definition, the work done by a force is equal to the magnitude of the force times the corresponding displacement. The *angular work done* by a force applied on a rotating body is related to the angular displacement of the body. Consider a body rotating about a fixed axis at O due to an applied force $\underline{F}$. As illustrated in Figure 14.16, let P_1 and P_2 represent the positions of a point in the body at times t_1 and t_2, respectively. In the time interval between t_1 and t_2, the body rotates through an arc of length s

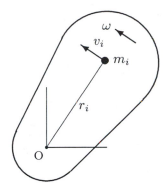

Figure 14.15 *Rotational motion of a body about a fixed axis.*

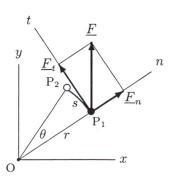

Figure 14.16 *A particle located at P_1 is displaced by an angle θ or arc length s to position P_2.*

or angle θ. The work done by $\underline{F}$ on the body is equal to the magnitude of the component of the force vector in the direction of motion (tangential component, F_t), times the displacement s:

$$W = F_t\, s$$

The arc length is related to the angular displacement through the radius of the circular path of motion as $s = r\theta$. Therefore:

$$W = F_t\, r\, \theta$$

By definition, $F_t r$ is the magnitude M of the torque generated by force $\underline{F}$ about O. Hence:

$$W = M\theta \qquad (14.16)$$

In other words, the work done by a rotation-producing force is equal to the torque generated by the force times the angular displacement of the body. Note that the normal (radial) component of the force vector does no work on a body undergoing rotational motion because there is no motion in the normal direction.

It must be pointed out here that the relationship between angular work done, torque, and angular displacement given in Eq. (14.16) is valid when the torque is constant. The work done by a torque, which is a function of angular displacement, on a body to rotate the body from position 1 to 2 is:

$$W = \int_{\theta_1}^{\theta_2} M\, d\theta \qquad (14.17)$$

Here, θ_1 and θ_2 are the angular displacements of the body at positions 1 and 2, respectively. Equation (14.17) can also be written in terms of the change in angular velocity by noting that $M = I\alpha$ and $\alpha = d\omega/dt$. Using the chain rule of differentiation:

$$M = I\,\alpha = I\,\frac{d\omega}{dt} = I\,\frac{d\omega}{d\theta}\frac{d\theta}{dt} = I\,\omega\,\frac{d\omega}{d\theta}$$

Substituting this into Eq. (14.17):

$$W = \int_{\omega_1}^{\omega_2} I\,\omega\, d\omega = \frac{1}{2}\,I\,\omega_2{}^2 - \frac{1}{2}\,I\,\omega_1{}^2 \qquad (14.18)$$

In Eq. (14.18), ω_1 and ω_2 are the angular velocities of the body at positions 1 and 2, respectively. Equation (14.18), known as the *work–energy theorem in rotational motion*, states that the net angular work done on a rigid body in rotating the body about a fixed axis is equal to the change in the body's rotational kinetic energy.

The rate at which work is done is known as power. The *angular power* describes the rate at which angular work is done. For a constant torque:

$$P = \frac{dW}{dt} = M\,\frac{d\theta}{dt} = M\,\omega \qquad (14.19)$$

That is, the angular power is equal to the product of the applied torque and the angular velocity of the body.

Example 14.4 Consider the knee extension problem analyzed in Example 14.3. As illustrated in Figure 14.17, the person is seated on a table. The upper body is strapped to a back rest and the right thigh is strapped firmly on the table with the lower leg hanging vertically downward. The person is then asked to extend the right lower leg. The angular displacement of the lower leg during knee extension is determined via a goniometer attached to the leg. After a series of computations, it is determined that the lower leg was extended from $\theta = 0°$ to $90°$ in a time period of 0.5 s with an average angular velocity of 3 rad/s by producing an average extensor muscle torque of 90 N-m.

Figure 14.17 *Knee extension.*

Assuming that the mass moment of inertia of the lower leg about the center of rotation of the knee joint is 92 kg-m^2, calculate the average angular kinetic energy produced, angular work done, and angular power generated by the knee extensor muscles to extend the lower leg from $\theta = 0°$ to $90°$.

Solution: The range of motion of the lower leg is $\Delta\theta = 90°$, which is covered in a time period of $\Delta t = 0.5$ s (Figure 14.18). The mass moment of inertia of the lower leg about the knee joint is given as $I_0 = 92$ kg-m^2. The average angular velocity of the lower leg is calculated to be $\overline{\omega} = 3$ rad/s and the average torque produced by the knee extensors is $\overline{M} = 90$ N-m. Therefore, the average angular kinetic energy produced by the knee extensor muscles is:

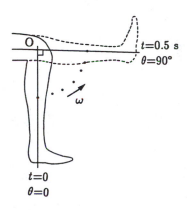

Figure 14.18 *Movement of the lower leg.*

$$\overline{\mathcal{E}_K} = \frac{1}{2}\, I_0\, \overline{\omega}^2 = \frac{1}{2}\,(92)(3)^2 = 414\ \text{J}$$

The average work done by the muscles to extend the lower leg at an angle of $90°$ or $90 \times \pi/180 = 1.57$ rad is:

$$\overline{W} = \overline{M}\, \Delta\theta = (92)(1.57) = 144\ \text{J}$$

The average power generated by the extensors is:

$$\overline{P} = \overline{M}\,\overline{\omega} = (92)(3)^2 = 828\ \text{W}$$

Chapter 15

Impulse and Momentum

15.1 Introduction

In Chapter 12, Newton's second law of motion is presented in the form of "equations of motion." In Chapter 14, the concepts of work and energy are introduced. Based on the same law, "work–energy" and "conservation of energy" methods are devised to facilitate the solutions of specific problems in kinetics. In this chapter, the concepts of linear momentum and impulse will be defined. Newton's second law of motion will be reformulated to introduce other methods for kinetic analyses based on the "impulse–momentum theorem" and the principle of "conservation of linear momentum." These methods will then be applied to analyze the impact and collision of bodies.

15.2 Linear Momentum and Impulse

Consider an object with mass m acted upon by an external force $\underline{F}$. Let $\underline{a}$ be the acceleration of the object under the action of the applied force. The relationship between $\underline{F}$, m, and $\underline{a}$ is described by the equation of motion in the following vector form:

$$\underline{F} = m\,\underline{a} \tag{15.1}$$

By definition, acceleration is the time rate of change of velocity. If the mass of the object is constant, then Eq. (15.1) can be written as:

$$\underline{F} = \frac{d}{dt}(m\,\underline{v}) \tag{15.2}$$

The vector $m\underline{v}$ is called the *linear momentum* (or simply the momentum) of the object, and is denoted by $\underline{p}$:

$$\underline{p} = m\,\underline{v} \tag{15.3}$$

Momentum is a vector quantity. The line of action and direction of the momentum vector is the same as the velocity vector of the object. The magnitude of the momentum is equal to the product of the mass and the speed of the object.

The equation of motion can now be expressed in terms of momentum by substituting Eq. (15.3) into Eq. (15.2):

$$\underline{F} = \frac{d\underline{p}}{dt} \tag{15.4}$$

If there is more than one force acting on the object, then $\underline{F}$ in Eq. (15.4) must be replaced by the vector sum of all forces (the resultant force) acting on the object. Equation (15.4) states that the time rate of change of momentum of an object is equal to the resultant force on the object.

The concept of momentum is particularly useful for analyzing the effects of forces applied in very short time intervals. Such forces are called *impulsive forces*, and the motions associated with

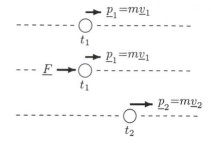

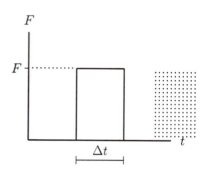

Figure 15.1 *Impulsive force $\underline{F}$ changes the momentum of the object from $\underline{p}_1$ to $\underline{p}_2$.*

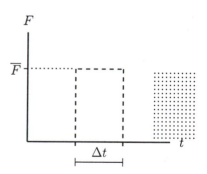

Figure 15.2 *Impulse is the area under the force versus time curve.*

Figure 15.3 *Impulse of a varying force can be determined by considering an average force and an average time.*

them are called *impulsive motions*. Consider an object moving with velocity $\underline{v}_1$ at time t_1 (Figure 15.1). Assume that a force $\underline{F}$ is applied on the object in the time interval between t_1 and t_2, and the velocity of the object is changed to $\underline{v}_2$ at time t_2. Multiplying Eq. (15.4) by dt and integrating it between t_1 and t_2 will yield:

$$\int_{t_1}^{t_2} \underline{F}\, dt = \underline{p}_2 - \underline{p}_1 = m\,\underline{v}_2 - m\,\underline{v}_1 \qquad (15.5)$$

The integral on the left-hand side of Eq. (15.5) represents the *linear impulse* of force $\underline{F}$ on the object in the time interval $\Delta t = t_2 - t_1$. The right-hand side of Eq. (15.5) is equal to the change in momentum $\Delta\underline{p} = \underline{p}_2 - \underline{p}_1$ of the object in the time interval Δt. Therefore:

$$\int_{t_1}^{t_2} \underline{F}\, dt = \Delta\underline{p} \qquad (15.6)$$

This equation is the mathematical representation of the *impulse–momentum theorem*.

Impulse is a vector quantity. It has the same line of action and direction as the impulsive force. In general, the magnitude and direction of the impulsive force may vary in the time interval Δt. If the force is known as a function of time, the impulse can be determined by integrating $\underline{F}$ with respect to time, which will essentially yield the area under the force versus time curve. If the impulsive force is constant (Figure 15.2), then the impulse is simply:

$$\underline{F}\,\Delta t = \Delta\underline{p} \qquad (\underline{F} \text{ constant}) \qquad (15.7)$$

Equation (15.7) can also be used to determine the impulse of a varying force by approximating the force with an average force value (Figure 15.3).

In the equations derived so far, all of the parameters involved (except for the mass m) are vector quantities. These parameters can be represented in terms of their rectangular components along the Cartesian coordinate directions x, y, and z. This approach will yield three independent scalar equations valid along the x, y, and z directions. For example, Eq. (15.5) will yield:

$$\int_{t_1}^{t_2} F_x\, dt = \Delta p_x = p_{x2} - p_{x1} = m\,v_{x2} - m\,v_{x1}$$

$$\int_{t_1}^{t_2} F_y\, dt = \Delta p_y = p_{y2} - p_{y1} = m\,v_{y2} - m\,v_{y1} \qquad (15.8)$$

$$\int_{t_1}^{t_2} F_z\, dt = \Delta p_z = p_{z2} - p_{z1} = m\,v_{z2} - m\,v_{z1}$$

For one-dimensional problems, any one of the above equations is sufficient to analyze the problem. For plane problems, two equations may be needed. While using Eqs. (15.8), it should be kept in mind that force, velocity, momentum, and impulse are vector quantities. For example, consider an object moving in the positive x direction. Assume that between times t_1 and t_2, an impulsive force $\underline{F}_x$ is applied on the object in the direction of motion (Figure 15.4a). The momentum of the object will become:

$$p_{x2} = p_{x1} + \int_{t_1}^{t_2} F_x \, dt$$

That is, the final momentum is equal to the initial momentum plus the impulse. If the force is applied in the direction opposite to that of the motion of the object (Figure 15.4b), then the final momentum of the object is equal to the initial momentum minus the impulse:

$$p_{x2} = p_{x1} - \int_{t_1}^{t_2} F_x \, dt$$

If the motion is in one direction and the impulsive force is applied in a different direction (Figure 15.4c), then it is easier to treat impulse and momentum as vector quantities.

Impulse and momentum have the same dimension (ML/T) and units. Their units in different systems are listed in Table 15.1.

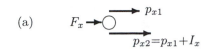
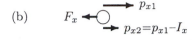
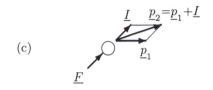

Figure 15.4 *Scalar and vector additions of impulse and momentum.*

Table 15.1 *Units of impulse and momentum.*

SYSTEM	UNITS OF IMPULSE AND MOMENTUM
SI	kg-m/s=N-s
c-g-s	g-cm/s=dyn-s
British	slug-ft/s=lb-s

15.3 Applications of the Impulse–Momentum Method

The concepts of impulse and momentum have numerous applications. These concepts may facilitate the analysis of situations where impact and impulsive forces play a significant role. The impulse–momentum method can be utilized to determine the forces involved in various athletic, sports, and daily activities. This method enables us to investigate the forces exerted by the foot on a football, or by the foot or the forehead on a soccer ball, which is in fact the same force exerted by the ball on the foot or the forehead. Forces involved while walking and running and the forces applied on various joints of the human body, which are transmitted through the feet during landing after a long jump or after a routine in gymnastics, may also be analyzed with the same technique. Other applications of the method include the analyses of forces exerted by a club on a golf ball, a bat on a baseball, and a racket on a tennis ball.

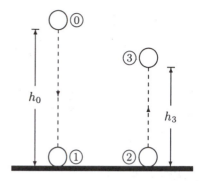

Figure 15.5 *The ball hits the floor and bounces.*

Example 15.1 As illustrated in Figure 15.5, consider a ball of mass $m = 0.25$ kg dropped from a height $h_0 = 1$ m. The ball hits the floor, bounces, and reaches a height $h_3 = 0.75$ m.

Determine the momentum of the ball immediately before and after it collides with the floor. Assuming that the duration of collision (duration of contact) is $\Delta t = 0.01$ s, determine an average force exerted by the floor on the ball during collision.

Solution: Let 0, 1, 2, and 3 correspond to each stage of the action: release, impact, rebound, and reaching the height 0.75 m for the ball. The speed v_1 of the ball at the instant of impact can be determined by utilizing the conservation of energy principle between 0 and 1. Since $v_0 = 0$ and $h_1 = 0$:

$$m\,g\,h_0 = \frac{1}{2}\,m\,v_1{}^2$$

Solving this equation for v_1 will yield:

$$v_1 = \sqrt{2\,g\,h_0} = \sqrt{2(9.8)(1)} = 4.43 \text{ m/s}$$

Similarly, the speed v_2 of the ball immediately after collision can be determined by considering the motion of the ball between 2 and 3. Since $h_2 = 0$ and $v_3 = 0$:

$$\frac{1}{2}\,m\,v_2{}^2 = m\,g\,h_3$$

Solving this equation for v_2 will yield:

$$v_2 = \sqrt{2\,g\,h_3} = \sqrt{2(9.8)(0.75)} = 3.83 \text{ m/s}$$

Momenta of the ball immediately before and after collision are:

$$p_1 = m\,v_1 = (0.25)(4.43) = 1.11 \text{ kg-m/s}\quad(\downarrow)$$
$$p_2 = m\,v_2 = (0.25)(3.83) = 0.96 \text{ kg-m/s}\quad(\uparrow)$$

In vector form:

$$\underline{p}_1 = -1.11\underline{j} \text{ (kg-m/s)}$$
$$\underline{p}_2 = 0.96\underline{j} \text{ (kg-m/s)}$$

The change in momentum of the ball during the course of collision is:

$$\Delta\underline{p} = \underline{p}_2 - \underline{p}_1 = (0.96\underline{j}) - (-1.11\underline{j}) = 2.07\underline{j} \text{ (kg-m/s)}$$

The force exerted by the floor on the ball during the course of collision can be determined by assuming that the impulsive force is approximately constant during the collision. Using the impulse–momentum equation given in Eq. (15.7):

$$\underline{F}\,\Delta t = \Delta \underline{p}$$

Solving this equation for the unknown force $\underline{F}$:

$$\underline{F} = \frac{\Delta \underline{p}}{\Delta t} = \frac{2.07\,\underline{j}}{0.01} = 207\,\underline{j} \ \ (\text{N})$$

Remarks:

- The magnitude of the average impulsive force calculated is much greater than the force of gravity on the ball, which is about 2.5 N. This result suggests that for problems involving collisions, the impulsive force between the colliding bodies (in this case the ball and the floor) is the most dominant, and the effects of all other external forces that may be present can be ignored.

- The average force calculated depends heavily on the duration of contact Δt.

- The fact that $v_2 = 3.83$ m/s is less than $v_1 = 4.43$ m/s suggests that some of the kinetic energy of the ball just before collision is lost as heat during the course of collision. The amount of energy lost can be calculated by considering the difference between the kinetic energies of the ball before and after collision.

Example 15.2 Consider a soccer player kicking a stationary ball (Figure 15.6). If the effect of air resistance is negligible, the ball will undergo a projectile motion (Figure 15.7).

Assuming that the mass of the ball is $m = 0.5$ kg, the horizontal range of motion of the ball is $\ell = 40$ m, the maximum height the ball reaches in the air is $h = 4$ m, and the time at which the foot of the soccer player remains in contact with the ball is $\Delta t = 0.1$ s, determine the momentum of the ball at the instant of takeoff and the impulsive force applied by the player on the ball.

Figure 15.6 *The player applies an impulsive force on the ball.*

Solution: To determine the momentum of the ball at the instant of takeoff, we need to calculate the takeoff speed of the ball and its angle of takeoff first. For this purpose, we can utilize the formulas given in Section 8.10. The angle of takeoff is provided in Eq. (8.39) in terms of the maximum height and horizontal range of motion of the projectile motion in the following form:

$$\theta = \arctan\left(\frac{4\,h}{\ell}\right)$$

Substituting the numerical values of $h = 4$ m and $\ell = 40$ m into the above equation, and carrying out the calculations will yield:

$$\theta = \arctan\left[\frac{4(4)}{40}\right] = 21.8^\circ$$

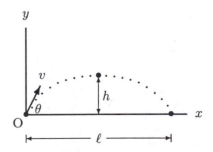

Figure 15.7 *The soccer ball undergoes a projectile motion.*

Using Eq. (8.40), the speed of takeoff can also be determined:

$$v = \frac{\sqrt{2\,g\,h}}{\sin\theta} = \frac{\sqrt{2(9.8)(4)}}{\sin 21.8°} = 23.8 \text{ m/s}$$

The velocity of takeoff $\underline{v}$ has components both in the x and y directions with magnitudes:

$$v_x = v\,\cos\theta = (23.8)(\cos 21.8°) = 22.1 \text{ m/s} \quad (\rightarrow)$$

$$v_y = v\,\sin\theta = (23.8)(\sin 21.8°) = 8.8 \text{ m/s} \quad (\uparrow)$$

Therefore, the takeoff velocity of the ball in vector form is:

$$\underline{v} = 22.1\,\underline{I} + 8.8\,\underline{j} \text{ (m/s)}$$

The momentum of the soccer ball at the instant of takeoff (immediately after impact) can be determined as:

$$\underline{p} = m\,\underline{v} = 0.5(22.1\,\underline{I} + 8.8\,\underline{j}) = 11.0\,\underline{I} + 4.4\,\underline{j} \text{ (kg-m/s)}$$

The ball is stationary and its momentum is zero immediately before the soccer player kicks the ball. Therefore, the change in momentum of the ball during the course of contact with the player's foot is equal to its takeoff momentum. The impulse–momentum equation [Eq. (15.7)] can now be utilized to determine an average impulsive force exerted by the soccer player on the ball:

$$\underline{F} = \frac{\Delta\underline{p}}{\Delta t} = \frac{11.0\,\underline{I} + 4.4\,\underline{j}}{0.1} = 110\,\underline{I} + 44\,\underline{j} \text{ (N)}$$

Therefore, the net or the resultant force applied on the ball by the player is:

$$F = \sqrt{(110)^2 + (44)^2} = 118.5 \text{ N}$$

Note that since action and reaction must have equal magnitudes, F is also the magnitude of the force applied by the ball on the player's foot.

Example 15.3 A *force platform*, as illustrated in Figure 15.8, is a flat, rectangular, force-sensitive device that electronically records forces exerted against its upper surface. This device can be used to measure the impulsive forces involved during walking, running, jumping, and other activities.

Consider the force versus time recording shown in Figure 15.9 for an athlete making vertical jumps on a force platform. The force scale is normalized with the weight of the athlete so that the force reading is zero when the athlete is stationary (standing still or crouching). The reason for normalizing the force measurement is to be able to disregard the effect of gravitational

Figure 15.8 *A force platform.*

acceleration on the impulse calculations. In this case, a positive force means a force exerted on the platform due to factors other than the weight of the person. The force versus time graph has three distinct regions. An initial "takeoff push" during which the athlete exerts a positive force on the platform, an "airborne" region during which the athlete is not in contact with the platform, and a "landing" period in which the athlete again exerts impulsive forces on the platform. These regions are approximated with rectangular areas A_1 and A_2, and a triangular area A_3. Boundaries of these regions are shown with dashed lines in Figure 15.9. The approximate force applied and the duration of takeoff push are about $F_1 = 600$ N and $\Delta t_1 = 0.3$ s. The force reading is about $F_2 = -800$ N while the athlete is airborne, and the athlete remains in the air for about $\Delta t_2 = 0.4$ s. This suggests that the weight of the athlete is about 800 N. Therefore, the athlete has a mass of about $m = (800 \text{ N})/(9.8 \text{ m/s}^2) = 82$ kg. The maximum impulsive force the athlete exerts on the platform during landing is about $F_3 = 1000$ N, which reduces to zero (in the normalized scale) in a time interval of about $\Delta t_3 = 0.4$ s.

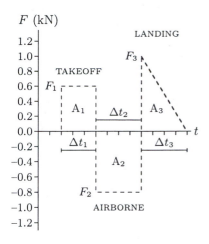

Figure 15.9 *Normalized force versus time plot for the athlete.* (1 kN = 1000 N.)

Determine an approximate takeoff velocity of the center of gravity of the athlete, calculate the height of jump, and determine the impulse and momentum of the athlete during landing by using the approximated areas under the force versus time curve.

Solution: Under the suggested assumptions, the analysis of the problem is quite straightforward. All we need to do is to utilize the impulse–momentum equation. The rectangular area designated by A_1 under the force versus time curve is equal to the impulse during takeoff:

$$A_1 = F_1 \, \Delta t_1 = (600)(0.3) = 180 \text{ N-s}$$

The velocity of the athlete before takeoff is zero, and therefore, the impulse is converted into momentum at the instant of takeoff. If v_T is the speed of takeoff, then the impulse–momentum equation [Eq. (15.7)] takes the following simplified form:

$$F_1 \, \Delta t_1 = \Delta p = m \, v_T$$

Solving this equation for the takeoff speed will yield:

$$v_T = \frac{F_1 \, \Delta t_1}{m} = \frac{180}{82} = 2.2 \text{ m/s}$$

To determine the height of the jump, we can utilize the conservation of energy principle. At the instant of takeoff, the athlete has a kinetic energy of $m v_t^2 / 2$ and zero potential energy. As the athlete ascends, the kinetic energy is converted to potential energy. At the instant when the athlete reaches the maximum height h, the kinetic energy of the athlete becomes zero and the potential energy is mgh. Therefore:

$$\frac{1}{2} m \, v_t^2 = m \, g \, h$$

Solving this equation for the maximum elevation of the athlete's center of gravity will yield:

$$h = \frac{v_2{}^2}{2\,g} = \frac{(2.2)^2}{2\,(9.8)} = 0.25 \text{ m}$$

The impulse applied by the athlete on the force platform during landing is equal to the triangular area, A_3, under the force versus time curve:

$$A_3 = \frac{F_3\,\Delta t_3}{2} = \frac{(1000)(0.4)}{2} = 200 \text{ N-s}$$

During the course of landing the downward velocity of the athlete is reduced to zero and the impulse calculated during landing is essentially equal to the momentum, p_L, of the athlete at the instant of landing:

$$p_L = A_3 = 200 \text{ N-s}$$

Note that we can also calculate the landing speed of the athlete as:

$$v_L = \frac{p_L}{m} = \frac{200}{82} = 2.4 \text{ m/s}$$

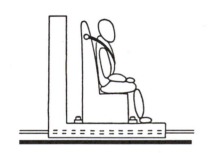

Figure 15.10 *A crash test.*

Example 15.4 A laboratory crash test is set up to measure the endurance of seat belts for automobile passengers (Figure 15.10). The initial horizontal speed of the test vehicle carrying a 70 kg dummy is set to 100 kilometers per hour. The speed of the vehicle and the dummy is brought to zero in a time interval of 0.1 s.

Assuming that the frictional effects are negligibly small, determine an average horizontal force applied by the dummy on the seat belt.

Solution: In the absence of frictional forces, the dummy would have continued moving along the positive x direction (toward the right) if the seat belt were not constraining its motion. The seat belt is applying a force on the dummy in the negative x direction, which brings the dummy to rest in a time interval of $\Delta t = 0.1$ s. At the instant when the brakes are applied, the speed of the vehicle and the dummy is $v_1 = 100$ km/h or 27.8 m/s. Therefore, the momentum of the dummy at the same instant is:

$$p_1 = m\,v_1 = (70)(27.8) = 1946 \text{ kg-m/s}$$

In vector form, $\underline{p}_1 = 1946\,\underline{I}$ kg-m/s. In a time interval of $\Delta t = 0.1$ s, the speed of the dummy is brought to zero. Therefore, the momentum of the dummy at the end of 0.1 s is zero:

$$\underline{p}_2 = 0$$

According to the impulse–momentum theorem, the change in momentum of the dummy must be equal to the impulse applied by the seat belt on the dummy:

$$\underline{F}\, \Delta t = \underline{p}_2 - \underline{p}_1$$

Or:

$$\underline{F} = -\frac{\underline{p}_1}{\Delta t} = -\frac{1946\,\underline{I}}{0.1} = -19460\,\underline{I} \quad (N)$$

Therefore, the seat belt applied a force of $F = 19{,}460$ N on the dummy in the negative x direction. Since action and reaction must have equal magnitudes, F is also the magnitude of the force applied by the dummy on the seat belt. If the objective was to design a proper seat belt, this result suggests that the seat belt material must be chosen to withstand a force of more than about 19.5 kN.

15.4 Conservation of Linear Momentum

The equation of motion [Eq. (15.4)] relates the resultant of forces applied on an object and the time rate of change of momentum of the object. When the resultant force on an object is zero (i.e., when the object is in equilibrium), then the time rate of change of momentum is also zero, and the momentum of the object is constant. This condition is known as the *principle of conservation of linear momentum*. Conservation of momentum is particularly useful for impact and collision analyses.

Consider two objects that interact with each other. Assume that these objects are isolated from their surroundings so that there are no external forces present except for the forces they exert onto each other. In other words, the effects of external forces are negligibly small as compared to the forces they exert onto each other, which is particularly true in the case of impact and collision. Suppose that at some time t the two objects have momenta $\underline{p}_1$ and $\underline{p}_2$, respectively. Let $\underline{F}_{12}$ be the force on object 1 applied by object 2, and $\underline{F}_{21}$ the force on object 2 applied by object 1. Using Eq. (15.4):

$$\underline{F}_{12} = \frac{d\underline{p}_1}{dt} \qquad \underline{F}_{21} = \frac{d\underline{p}_2}{dt}$$

According to Newton's third law, the action and reaction must be equal in magnitude and opposite in direction ($\underline{F}_{12} = -\underline{F}_{21}$), or:

$$\underline{F}_{12} + \underline{F}_{21} = \frac{d\underline{p}_1}{dt} + \frac{d\underline{p}_2}{dt} = \frac{d}{dt}(\underline{p}_1 + \underline{p}_2) = 0$$

This condition of equilibrium is equivalent to:

$$\underline{p}_1 + \underline{p}_2 = \text{constant} \qquad (15.9)$$

Equation (15.9) represents the principle of conservation of linear momentum for two interacting bodies that form an isolated system. It states that whenever two objects collide their total momentum will remain constant, regardless of the nature of the forces involved between the two.

15.5   Impact and Collisions

When two bodies collide, they deform to a certain extent because of the forces involved. We know that the amount of deformation depends on the objects' material properties, the extent and duration of applied forces, and other conditions such as temperature. In general, an object may undergo an *elastic* deformation, a *plastic* deformation, or both. The elastic deformation is recoverable upon release of the force causing the deformation, whereas plastic deformations are permanent.

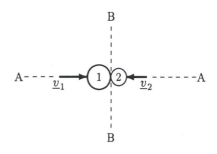

Figure 15.11 *Direct central impact.*

There are several types of collisions, which may be distinguished either with respect to the orientations of the impact velocities of the two objects, or according to the nature of deformations they undergo. As illustrated in Figure 15.11, two objects may collide "head-on," which is known as the *direct central impact*. In this case, the velocities of the objects are collinear with the line of impact. The *line of impact* (AA) is a fictitious line passing through the mass centers of the colliding objects, and it is perpendicular to the *plane of contact* (BB), which is tangent to the contacting surfaces. Figure 15.12 illustrates the *oblique central impact* of two objects. In this case, the impact velocities of the objects are not collinear with the line of impact.

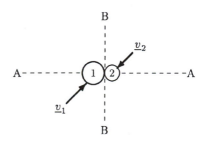

Figure 15.12 *Oblique central impact.*

Collisions can also be distinguished based on the nature of the deformations occurring during the course of a collision. An *elastic* or *perfectly elastic collision* is defined as a collision in which both the total momentum and total kinetic energy of the system (i.e., the two objects) are conserved. For an *inelastic* or *plastic collision*, on the other hand, only the total momentum of the system is conserved. During an inelastic collision, some of the kinetic energies of the colliding objects are dissipated as heat. The extreme case of inelastic collision is called the *perfectly inelastic collision*. This is a collision in which the objects stick together after the collision and move with a common velocity.

Whether a collision is an elastic or an inelastic one, the total momentum of the system is conserved. Therefore, the analyses of impact and collision problems are based on the conservation of momentum principle. Consider the collision of two objects with masses m_1 and m_2. Let $\underline{v}_1$ and $\underline{v}_2$ refer to the velocities of objects 1 and 2, respectively. Also let subscripts "i" and "f" denote the instants immediately before and after the collision. In the time interval between t_i and t_f, the principle of conservation of momentum (Eq. 15.9) can be expressed as:

$$m_1 \underline{v}_{1i} + m_2 \underline{v}_{2i} = m_1 \underline{v}_{1f} + m_2 \underline{v}_{2f} \qquad (15.10)$$

Characteristics of different types of collisions will be provided by considering one-dimensional cases first. These concepts will then be expanded to analyze two-dimensional collisions by utilizing the vectorial properties of the parameters involved.

15.6 One-Dimensional Collisions

For a collision along a straight line (i.e., direct central impact), Eq. (15.10) can be simplified in the following scalar form:

$$m_1 v_{1i} + m_2 v_{2i} = m_1 v_{1f} + m_2 v_{2f} \qquad (15.11)$$

This equation is valid both for elastic and inelastic collisions.

15.6.1 Perfectly inelastic collision

A perfectly inelastic collision is one in which the objects stick together and move with a common velocity v_f after the collision (Figure 15.13). To determine v_f, it is sufficient to consider the conservation of momentum principle during the collision. Substituting $v_{1f} = v_{2f} = v_f$ into Eq. (15.11), and solving it for v_f will yield:

$$v_f = \frac{m_1 v_{1i} + m_2 v_{2i}}{m_1 + m_2} \qquad (15.12)$$

Before collision:

After collision:

Figure 15.13 *Perfectly inelastic collision.*

Example 15.5 Figure 15.14 illustrates a *ballistic pendulum*, which may be a block of wood suspended by light wires. This simple device can be used to measure the velocity of a bullet. A bullet of mass m_1 fired at a stationary block of mass m_2 will penetrate the block, and the bullet–block system with mass $m_1 + m_2$ will swing to a height h.

Assuming that the bullet remains in the block, determine an expression for the initial speed v_{1i} of the bullet immediately before impact in terms of m_1, m_2, and h.

Solution: This is a typical example of perfectly inelastic collision. The velocity of the block before impact is zero. Immediately after the collision, the block gains a kinetic energy. Designating the speed immediately after impact by v_f, then the bullet–block system has a kinetic energy of $(m_1 + m_2)v_f{}^2/2$, which is converted to a potential energy of $(m_1 + m_2)gh$ as the block swings to a height h. Applying the conservation of energy principle:

$$\frac{1}{2}(m_1 + m_2)v_f{}^2 = (m_1 + m_2)g h$$

Solving this equation for the common speed v_f of the bullet and the block right after impact:

$$v_f = \sqrt{2 g h} \qquad (i)$$

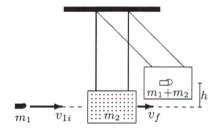

Figure 15.14 *Ballistic pendulum.*

During the course of impact the momentum is conserved. Since the velocity of the block before impact is zero, Eq. (15.12) takes the following simpler form:

$$v_f = \frac{m_1\, v_{1i}}{m_1 + m_2}$$

Solving this equation for v_{1i} and using Eq. (i) will yield:

$$v_{1i} = \left(\frac{m_1 + m_2}{m_1}\right)\sqrt{2\,g\,h}$$

Before collision:

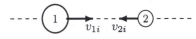

After collision:

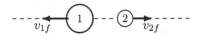

Figure 15.15 *Perfectly elastic collision.*

15.6.2 Perfectly elastic collision

A perfectly elastic collision is one in which the total kinetic energy is conserved as well as the total momentum of the objects involved (Figure 15.15). The condition of conservation of total kinetic energy during the course of collision can be written as:

$$\frac{1}{2}\,m_1\,v_{1i}{}^2 + \frac{1}{2}\,m_2\,v_{2i}{}^2 = \frac{1}{2}\,m_1\,v_{1f}{}^2 + \frac{1}{2}\,m_2\,v_{2f}{}^2 \qquad (15.13)$$

In this case, Eqs. (15.11) and (15.13) must be solved simultaneously for the unknowns v_{1f} and v_{2f}. This will yield:

$$v_{1f} = \left(\frac{m_1 - m_2}{m_1 + m_2}\right)v_{1i} + \left(\frac{2\,m_2}{m_1 + m_2}\right)v_{2i} \qquad (15.14)$$

$$v_{2f} = \left(\frac{2\,m_1}{m_1 + m_2}\right)v_{1i} + \left(\frac{m_2 - m_1}{m_1 + m_2}\right)v_{2i} \qquad (15.15)$$

At this point, it should be remembered that velocity is a vector quantity and that appropriate signs for all velocities involved must be included in Eqs. (15.12), (15.14), and (15.15). For this purpose, a positive direction of motion must be chosen. Velocities acting in the same direction must be considered to be positive, and those acting in the opposite direction must be taken to be negative.

Before collision:

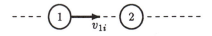

After collision:

Figure 15.16 *Perfectly elastic collision of two identical billiard balls.*

Example 15.6 As illustrated in Figure 15.16, consider a perfectly elastic collision of two billiard balls with equal masses m. Before the collision, ball 1 has a speed of v_{1i} and ball 2 is stationary.

Assuming a direct central impact, determine the velocities of the billiard balls immediately after the collision.

Solution: The fact that the masses of the balls are equal (i.e., $m_1 = m_2 = m$) and that one of the balls was stationary before the collision ($v_{2i} = 0$) simplifies the problem considerably. Since the masses are equal, $m_1 - m_2 = m_2 - m_1 = 0$ and $2m_1/(m_1 + m_2)$ $= 1$. Therefore, the first term on the right-hand side of Eq. (15.14) and the second term on the right-hand side of Eq. (15.15) are zero. The second term on the right-hand side of Eq. (15.14) is zero as well because $v_{2i} = 0$. Hence, choosing the right to be the positive direction of motion, Eqs. (15.14) and (15.15) yield:

$$v_{1f} = 0$$
$$v_{2f} = v_{1i} \quad (\rightarrow)$$

Note that the kinetic energy and momentum of ball 1 immediately before the collision is totally transferred into kinetic energy and momentum for ball 2 during the collision.

15.6.3 Elasto-plastic collision

In reality, a material can undergo both elastic and plastic deformations. To facilitate the analyses of the impact characteristics of objects made up of such materials, a concept called the *coefficient of restitution* is developed, which is defined as the ratio of the relative velocity of separation and the relative velocity of approach. If e represents the coefficient of restitution between two materials, then:

$$e = \frac{\text{relative velocity of separation}}{\text{relative velocity of approach}} \quad (15.16)$$

The coefficient of restitution can take values between 0 and 1, such that $e = 0$ for perfectly inelastic impact and $e = 1$ for perfectly elastic impact. The coefficient of restitution is a positive number, and while calculating e the absolute values of the relative velocities of separation and approach must be considered. Consider the two objects shown in Figure 15.17. Before the collision, object 1 is moving toward the right with velocity v_{1i} and object 2 is moving toward the left with velocity v_{2i}. Therefore, the relative velocity of approach is $v_{1i} + v_{2i}$. Assume that after the collision, the objects move in opposite directions with velocities v_{1f} and v_{2f}. Then the relative velocity of separation is $v_{1f} + v_{2f}$. If the objects were moving in the same direction before the collision, then the relative velocity of approach is $v_{1i} - v_{2i}$. If they move in the same direction after the collision, then the relative velocity of separation is $v_{1f} - v_{2f}$. If one of the objects is stationary before and after the collision, then the coefficient of restitution can be determined by considering the ratio of the final and initial velocities of the object. In fact, this is the easiest to measure the coefficient of restitution between two materials.

The coefficient of restitution depends on the material properties of the objects involved in a collision. Temperature is also a

Velocities of approach:

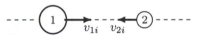

Velocities of separation:

Figure 15.17 *The coefficient of restitution between two colliding objects is defined as the ratio of relative velocities of separation and approach.*

factor that can influence the coefficient of restitution, because temperature can alter the mechanical properties of materials. For example, a ball will bounce better after being heated. If the ball is air-filled, such as a tennis ball, heat can also increase the internal pressure of the ball. A highly inflated ball will bounce better than a flat ball. When a ball is deformed during impact, some of the energy is dissipated as heat. The rise in temperature is important in games like tennis and squash where the ball is impacted at a relatively high rate. Another factor that may affect the coefficient of restitution between two materials is the relative velocity of approach. The higher the velocity of approach is, the lower the coefficient of restitution will be.

Example 15.7 Consider the ball analyzed in Example 15.1, which has a mass $m = 0.25$ kg. As illustrated in Figure 15.18, the ball is dropped from a height $h_0 = 1$ m. After hitting the floor, the ball bounces and reaches a height $h_3 = 0.75$ m. Using the conservation of energy principle between 0 and 1, and between 2 and 3, we calculated that the ball has speeds $v_1 = 4.43$ m/s and $v_2 = 3.83$ m/s immediately before and after impact.

What is the coefficient of restitution between the ball and the floor, and how much energy is lost during the course of impact?

Solution: By definition, the coefficient of restitution is equal to the ratio of the velocities of separation and approach. In this case, the floor is stationary before and after the impact. Therefore, the speeds v_1 and v_2 of the ball before and after impact are also the magnitudes of the relative velocities of approach and separation, respectively. Therefore, the coefficient of restitution between the ball and the floor is:

$$e = \frac{v_2}{v_1} \qquad (i)$$

Substituting the numerical values and carrying out the calculations:

$$e = \frac{3.83}{4.43} = 0.86$$

The amount of energy lost (converted into heat, sound, or internal potential energy) during the course of impact can be determined by calculating the difference in the kinetic energies of the ball before and after the impact:

$$\varepsilon_{K2} - \varepsilon_{K1} = \frac{1}{2} m \left(v_2^2 - v_1^2\right) = -0.62 \text{ J}$$

The coefficient of restitution [represented by Eq. (*i*) for the ball hitting a stationary surface and bouncing back] can alternatively be written in terms of the initial height h_0 from which the ball

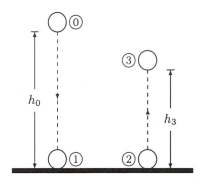

Figure 15.18 *The ball hits the floor and bounces.*

is dropped and the final height h_3 to which the ball bounced. In Example 15.1, using the conservation of energy principle, we determined that $v_1 = \sqrt{2gh_0}$ and $v_2 = \sqrt{2gh_3}$. Substituting these into Eq. (i) will yield:

$$e = \sqrt{\frac{h_3}{h_0}} \qquad\qquad (ii)$$

The U.S. National Collegiate Athletic Association (NCAA) rule requiring a basketball dropped from a height $h_0 = 1.8\,\text{m}$ to bounce a height of about $h_3 = 1.0$ m is equivalent to requiring a coefficient of restitution of about $e = \sqrt{1.0/1.8} = 0.75$.

15.7 Two-Dimensional Collisions

As discussed in Section 15.4, if two interacting objects are isolated from their surroundings (if the effects of external forces are negligible as compared to the impulsive forces present), then the total momentum of the system is conserved. This principle of conservation of momentum is applied to analyze direct central impact, or one-dimensional collision, of some simple systems. By considering the vectorial properties of the parameters involved, the concepts developed for one-dimensional collisions can be expanded to analyze the oblique central impact or two-dimensional collision of two objects.

For a two-dimensional collision problem, the conservation of linear momentum principle must be applied in two coordinate directions. In addition, the nature of the collision must be known. For example, if the collision is perfectly elastic, then the total kinetic energy of the system is also conserved. If the collision is perfectly inelastic, then the velocities of the objects after the collision are equal.

The following example will illustrate some of the concepts involved in two-dimensional collision problems.

Example 15.8 Figure 15.19 illustrates an instant during a pool game. What the pool player wishes to do is to hit the stationary target ball (ball 2) by the cue ball (ball 1) so as to move the target ball toward and into the corner pocket. Consider that the cue ball is given a velocity of $v_{1i} = 5$ m/s toward the target ball, and that a line connecting the center of mass and the geometric center of the corner pocket make an angle $\theta = 45°$, as shown in Figure 15.20. The rectangular coordinates x and y are chosen in such a way that they respectively coincide with the line of impact (perpendicular to the contacting surfaces), and the plane of contact (tangential to the contacting surfaces). Assume that

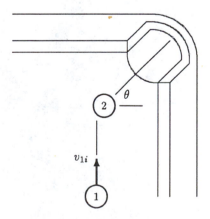

Figure 15.19 *Oblique central impact of two pool balls.*

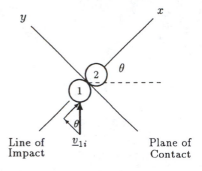

Figure 15.20 *Before collision.*

the cue ball hits the target ball at a point along the line of impact, and that the balls have equal mass.

Neglecting the effects of friction, rotation, and gravity, determine the velocities of the cue and target balls immediately after collision if the coefficient of restitution between them is $e = 0.8$.

Solution: This problem can be analyzed individually along the x and y directions first, and then by combining the results by utilizing the vectorial properties of velocity. Before impact, the target ball is stationary, and the cue ball has velocity components both in the x and y directions. Therefore:

$$(v_{1i})_x = v_{1i}\,\sin\theta = (5)(\sin 45°) = 3.54 \text{ m/s}$$

$$(v_{1i})_y = v_{1i}\,\cos\theta = (5)(\cos 45°) = 3.54 \text{ m/s}$$

$$(v_{2i})_x = 0$$

$$(v_{2i})_y = 0$$

Since the frictional effects are negligible, the motion along the y direction is equivalent to two balls moving along parallel lines without a chance of collision. The motion along the x direction, on the other hand, is equivalent to the direct central impact of the two balls.

(a) Motion along the y direction

In the y direction, there is no collision. Therefore, the momentum of ball 1 in the y direction is conserved:

$$m\,(v_{1i})_y = m\,(v_{1f})_y$$

$$(v_{1f})_y = (v_{1i})_y = 3.54 \text{ m/s} \qquad (i)$$

Utilizing the conservation of momentum principle for ball 2 in the y direction:

$$m\,(v_{2i})_y = m\,(v_{2f})_y$$

$$(v_{2f})_y = (v_{2i})_y = 0 \qquad (ii)$$

(b) Motion along the x direction:

In the x direction, the balls undergo a direct central impact, and therefore, the total momentum of the system is conserved. Assume that the positive direction of motion is along the positive x axis (i.e., toward the corner pocket), and that after collision, both balls move along the positive x direction. In other words, both $(v_{1f})_x$ and $(v_{2f})_x$ are positive. Since the balls have equal masses, the conservation of linear momentum along the x direction can be written as:

$$m\,(v_{1i})_x + m\,(v_{2i})_x = m\,(v_{1f})_x + m\,(v_{2f})_x$$

Eliminating the masses, and since $(v_{2i})_x = 0$:

$$(v_{1f})_x + (v_{2f})_x = (v_{1i})_x \qquad (iii)$$

The coefficient of restitution for the collision is given as $e = 0.8$. By definition, e is equal to the ratio of the relative velocities of separation and approach. Before impact, ball 2 is stationary. Hence the relative velocity of approach in the x direction is equal to the initial speed of ball 1 in the x direction. Because of the assumed directions of motion after the collision, the relative velocity of separation in the x direction is equal to the difference of the velocity components of the balls along the x direction. Assuming that $(v_{2f})_x$ is larger than $(v_{1f})_x$:

$$e = \frac{(v_{2f})_x - (v_{1f})_x}{(v_{1i})_x}$$

This equation can also be written as:

$$(v_{2f})_x - (v_{1f})_x = e\,(v_{1i})_x \qquad (iv)$$

We have two equations, Eqs. (iii) and (iv), for two unknowns, $(v_{2f})_x$ and $(v_{1f})_x$. Solving these equations simultaneously will yield:

$$(v_{1f})_x = \frac{(e-1)}{2}\,(v_{1i})_x \qquad (v)$$

$$(v_{2f})_x = \frac{(e+1)}{2}\,(v_{1i})_x \qquad (vi)$$

Substituting $(v_{1i})_x = 3.54$ m/s and $e = 0.8$ into these equations will yield:

$$(v_{1f})_x = -0.35$$

$$(v_{2f})_x = 3.19$$

Hence, the velocities of the balls after the collision are:

$$\underline{v}_{1f} = -0.35\,\underline{I} + 3.54\,\underline{j} \quad (\text{m/s})$$

$$\underline{v}_{2f} = 3.19\,\underline{I} \quad (\text{m/s})$$

Immediately after the collision, the target ball will move with a speed of 3.19 m/s toward the corner pocket (i.e., along the positive x direction). As illustrated in Figure 15.21, the cue ball will move with a speed of $v_{1f} = \sqrt{(0.35)^2 + (3.54)^2} = 3.56$ m/s along a direction that makes an angle $\arctan(3.54/0.35) = 84°$ with the negative x direction, or at an angle $\beta = 84° - 45° = 39°$ with the horizontal.

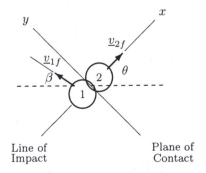

Line of Impact

Plane of Contact

Figure 15.21 *After the collision.*

While utilizing the equations for the coefficient of restitution and conservation of momentum, we assumed that the velocity components of both balls would be in the positive x direction. As a result of our computations, we determined a negative value for $(v_{1f})_x$, which means that it is acting along the negative x direction.

If the collision were a perfectly elastic one (i.e., $e = 1$), then Eqs. (v) and (vi) would yield $(v_{1f})_x = 0$ and $(v_{1f})_y = (v_{1i})_x$. Therefore,

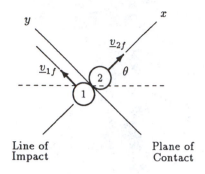

Figure 15.22 *Perfectly elastic collision of two pool balls.*

the velocity vectors for the balls after the collision would be:

$$\underline{v}_{1f} = 3.54\,\underline{j}\quad(\text{m/s})$$
$$\underline{v}_{2f} = 3.54\,\underline{I}\quad(\text{m/s})$$

Immediately after the collision, the target ball would move with a speed of 3.54 m/s along the positive x direction (toward the corner pocket) and the cue ball would move with the same speed along the positive y direction, as illustrated in Figure 15.22. The balls would move at right angles to each other after the collision.

15.8   Angular Impulse and Momentum

The rotational analog of linear momentum is called *angular momentum*. Angular momentum is defined as the product of the mass moment of inertia and the angular velocity of the body undergoing rotational motion and is commonly denoted with L:

$$L = I\,\omega \tag{15.17}$$

The *impulse–momentum theorem for rotational motion* relates applied torque and change in angular momentum. If a torque with magnitude M is applied to a rotating body in the time interval between t_1 and t_2 so that the angular momentum of the body is changed from L_1 to L_2, then the impulse–momentum theorem for rotational motion states that:

$$\int_{t_1}^{t_2} M\,dt = L_2 - L_1 \tag{15.18}$$

The left-hand side of Eq. (15.18) is the *angular impulse*, and M is the *impulsive torque*. If the torque is constant, then the integral in Eq. (15.18) can be evaluated to yield:

$$M\,\Delta t = \Delta L = I\,\Delta\omega \tag{15.19}$$

That is, a constant impulsive torque M applied on a body in the time interval $\Delta t = t_2 - t_1$ will change the angular velocity of the body from ω_1 to ω_2. Consequently, the angular momentum of the body will change from L_1 to L_2. The angular velocity and momentum of the body will increase if the torque is applied in the direction of motion.

Note that rotational kinetic energy, angular work done, and power have the same dimensions and units as their linear counterparts. On the other hand, the dimension of angular momentum is $[M][L^2]/[T]$ and has the unit of kg-m^2/s in SI.

15.9   Summary of Basic Equations

Table 15.2 provides a list of the basic equations necessary for rotational motion analyses about a fixed axis (circular motion) along with the equations for one-dimensional translational motion analyses. Note that the linear and angular quantities are analogous to each other in such pairs as x and θ, v and ω, a and α, m and I, F and M, and p and L.

Table 15.2 *Equations of translational and rotational motion.*

TRANSLATIONAL MOTION	ROTATIONAL MOTION (CIRCULAR)
VELOCITY	
$v = \dfrac{dx}{dt}$	$\omega = \dfrac{d\theta}{dt}$
ACCELERATION	
$a = \dfrac{dv}{dt}$	$\alpha = \dfrac{d\omega}{dt}$
KINEMATIC RELATIONS FOR CONSTANT ACCELERATION	
$x = x_0 + v_0 t + \frac{1}{2} a_0 t^2$	$\theta = \theta_0 + \omega_0 t + \frac{1}{2} \alpha_0 t^2$
$v = v_0 + a_0 t$	$\omega = \omega_0 + \alpha_0 t$
$v^2 = v_0^2 + 2a_0(x - x_0)$	$\omega^2 = \omega_0^2 + 2\alpha_0(x - x_0)$
EQUATION OF MOTION	
$F = ma$	$M = I\alpha$
WORK DONE	
$W = \displaystyle\int_{x_1}^{x_2} F\,dx$	$W = \displaystyle\int_{\theta_1}^{\theta_2} M\,d\theta$
KINETIC ENERGY	
$\mathcal{E}_K = \frac{1}{2} m v^2$	$\mathcal{E}_K = \frac{1}{2} I \omega^2$
WORK-ENERGY	
$W = \frac{1}{2} m (v_2^2 - v_1^2)$	$W = \frac{1}{2} I (\omega_2^2 - \omega_1^2)$
POWER	
$P = Fv$	$P = M\omega$
MOMENTUM	
$p = mv$	$L = I\omega$
IMPULSE-MOMENTUM	
$\displaystyle\int_{t_1}^{t_2} F\,dt = p_2 - p_1$	$\displaystyle\int_{t_1}^{t_2} M\,dt = L_2 - L_1$

15.10 Kinetics of Rigid Bodies in Plane Motion

In most situations (e.g., when the effects due to air resistance are neglected), the size and shape of an object do not affect its

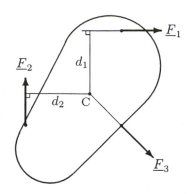

Figure 15.23 *A system of three forces acting on the body.*

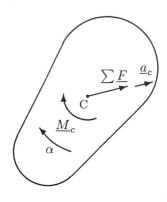

Figure 15.24 *The multi-force system can be reduced to a one-force and one-moment system.*

translational motion characteristics. The size and shape (i.e., inertial effects) must be taken into consideration if the object is undergoing a rotational motion. Now that we have defined the basic concepts behind the rotational motion of rigid bodies, we can integrate our knowledge about translational and rotational motions to investigate their general motion characteristics.

Consider the rigid body illustrated in Figure 15.23. Let m be the total mass of the body, and C be the location of its mass center. There are three coplanar forces acting on the body. Force F_3 is not producing any torque about C because its line of action is passing through C (i.e., its moment arm is zero). Forces $\underline{F}_1$ and $\underline{F}_2$ are producing clockwise moments about C with magnitudes $M_1 = d_1 F_1$ and $M_2 = d_2 F_2$, respectively. As illustrated in Figure 15.24, the three-force system can be reduced to a one-force and one-moment system such that $\sum \underline{F} = \underline{F}_1 + \underline{F}_2 + \underline{F}_3$ is the net or the resultant force acting on the body and $\underline{M}_c = \underline{M}_1 + \underline{M}_2$ is the resultant of the couple-moments as measured about C. $\sum \underline{F}$ causes the body to translate and $\underline{M}_c$ causes it to rotate about C.

Recall that the translational motion of a body depends on its mass and the net force applied on it. Newton's second law of motion states that:

$$\sum \underline{F} = m \, \underline{a}_c \qquad (15.20)$$

$\underline{a}_c$ is the acceleration of the mass center of the body, and Eq. (15.20) accounts for its translational motion. The rotational motion of the body depends on its mass moment of inertia and the net torque applied on it:

$$\sum \underline{M}_c = I_c \, \underline{\alpha} \qquad (15.21)$$

In Eq. (15.21), I_c is the mass moment of inertia of the body about the axis perpendicular to the plane of rotation and passing through the mass center of the body, and α is the angular acceleration. Note that Eq. (15.20) is valid for any point within the body, but Eq. (15.21) is correct only about the mass center at C. For two-dimensional motion analyses in the xy plane, Eqs. (15.20) and (15.21) will yield three scalar equations:

$$\sum F_x = m \, a_{cx} \qquad (15.22)$$

$$\sum F_y = m \, a_{cy} \qquad (15.23)$$

$$\sum M_c = I_c \, \alpha \qquad (15.24)$$

These three equations are the governing equations of motion for studying the two-dimensional general motion characteristics of bodies.

Note that if $M_c = 0$, then the body is in pure translation, the body is in pure rotation when $F_x = 0$ and $F_y = 0$, and the body is said to be in equilibrium when $F_x = 0$, $F_y = 0$, and $M_c = 0$.

Appendix A

Plane Geometry

A.1 Angles

An *angle* is formed by two intersecting straight lines. The angles identified as θ in Figure A.1 that are formed by two intersecting straight lines aa and bb are equal and called *opposite angles*. If the two lines are perpendicular to one another, then the angles formed are called *right angles*. A right angle is equal to 90°. Graphically, right angles are usually denoted by small square boxes. An angle is called *acute* if it is smaller than 90°, and is called *obtuse* if it is greater than 90°.

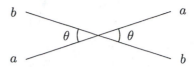

Figure A.1 *Opposite angles.*

The angles identified as θ in Figure A.2 are equal and called *alternate angles*. These angles are formed by a straight line cc intersecting two parallel straight lines aa and bb.

The angles identified as θ in Figure A.3 are equal. In this case, straight line cc is perpendicular to bb, and dd is perpendicular to aa. The geometry illustrated in Figure A.3 is utilized extensively in physics and mechanics, for example, while analyzing motions on inclined surfaces. For such analyses, aa represents the horizontal, bb represent the inclined surface that makes an angle θ with the horizontal, cc is perpendicular to the inclined surface, and dd represents the vertical.

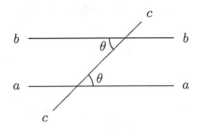

Figure A.2 *Alternate angles.*

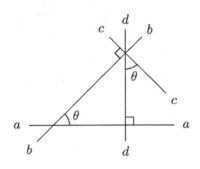

Figure A.3 *Lines bb and cc, and aa and dd are perpendicular.*

A.2 Triangles

A *triangle* is a geometric shape with three sides and three angles (Figure A.4). For any triangle, the sum of the three angles is equal to 180°:

$$\alpha + \beta + \theta = 180°$$

A.3 Law of Sines

For any triangle, such as the one in Figure A.4, the angles and sides of the triangle are related through the *law of sines*, which states that:

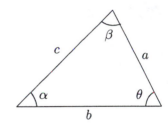

Figure A.4 *A triangle.*

$$\frac{\sin \alpha}{a} = \frac{\sin \beta}{b} = \frac{\sin \theta}{c} \qquad (A.1)$$

The definition of sine (abbreviated as sin) is given in Section A.6.

A.4 The Right-Triangle

A *right-triangle* is formed if one of the angles of a triangle is equal to 90°. For the right-triangle shown in Figure A.5, angle θ is equal to 90°, and the sum of the remaining angles is also equal to 90°:

$$\theta = \alpha + \beta = 90°$$

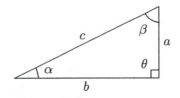

Figure A.5 *The right-triangle.*

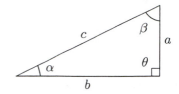

Figure A.6 *The right-triangle.*

In Figure A.6, side c of the triangle opposite to the right angle (angle θ) is called the *hypotenuse*, and it is the longest side of the right-triangle. With respect to angle α, b is the length of the *adjacent* side and a is the length of the *opposite* side.

A.5 Pythagorean Theorem

The *Pythagorean theorem* states that the square of the length of the hypotenuse of a right-triangle is equal to the sum of the squares of the lengths of the other sides of the right-triangle:

$$c^2 = a^2 + b^2 \tag{A.2}$$

Considering the square-root of both sides:

$$c = \sqrt{a^2 + b^2} \tag{A.3}$$

A.6 Sine, Cosine, and Tangent

The *sine* of an acute angle (angles smaller than the right angle) in a right-triangle is equal to the ratio of the lengths of the opposite side and the hypotenuse. For the right-triangle in Figure A.6:

$$\sin \alpha = \frac{a}{c} \qquad \sin \beta = \frac{b}{c} \tag{A.4}$$

The *cosine* of an acute angle in a right-triangle is the ratio of the lengths of the adjacent side and the hypotenuse:

$$\cos \alpha = \frac{b}{c} \qquad \cos \beta = \frac{a}{c} \tag{A.5}$$

Note that the sine of angle α in Figure A.6 is equal to the cosine of angle β, and the cosine of angle α is equal to the sine of angle β. Also note that Eqs. (A.4) and (A.5) for angle α can alternatively be written as:

$$a = c \sin \alpha$$
$$b = c \cos \alpha \tag{A.6}$$

Furthermore, we can take the squares of a and b in Eq. (A.6) and substitute them in Eq. (A.2) to obtain:

$$c^2 = a^2 + b^2$$
$$c^2 = (c \cos \alpha)^2 + (c \sin \alpha)^2$$
$$c^2 = c^2(\cos^2 \alpha + \sin^2 \alpha)$$

Dividing both sides of the above equation by c^2 will yield an important trigonometric identity:

$$\cos^2 \alpha + \sin^2 \alpha = 1 \tag{A.7}$$

The *tangent* of an acute angle in a right-triangle is the ratio of the lengths of the opposite side and the adjacent side:

$$\tan \alpha = \frac{a}{b} \qquad \tan \beta = \frac{b}{a} \qquad (A.8)$$

The sine, cosine, and tangent of a few selected angles are listed in Table A.1. Note that the tangent of an angle is also equal to the ratio of the sine and the cosine of that angle:

$$\tan \alpha = \frac{\sin \alpha}{\cos \alpha} \qquad \tan \beta = \frac{\sin \beta}{\cos \beta} \qquad (A.9)$$

It should also be noted that a right-triangle can uniquely be defined if the lengths of two sides of the triangle are known, or one of the acute angles along with one of the sides of the triangle are known. The remaining sides and angles can be determined using trigonometric identities.

Table A.1 *Sine, cosine, and tangent of selected angles.*

Angle	sin	cos	tan
0°	0.000	1.000	0.000
30°	0.500	0.866	0.577
45°	0.707	0.707	1.000
60°	0.866	0.500	1.732
90°	1.000	0.000	

A.7 Inverse Sine, Cosine, and Tangent

Sometimes the sine, cosine, or the tangent of an angle is known and the task is to determine the angle itself. For this purpose, *inverse* sine, cosine, and tangent are defined such that:

If $\sin \alpha = A$ then $\alpha = \sin^{-1}(A)$

If $\cos \alpha = B$ then $\alpha = \cos^{-1}(B)$

If $\tan \alpha = C$ then $\alpha = \tan^{-1}(C)$

Inverse sine, cosine, and tangent are alternatively referred as arcsine, arccosine, and arctangent that are abbreviated as arcsin, arccos, and arctan, respectively.

Example A.1 For the right-triangle shown in Figure A.7, if $a = 4$ and $b = 3$, determine angles α and β, and the length c of the hypotenuse.

Solution: From Eq. (A.3):

$$c = \sqrt{a^2 + b^2}$$

Substitute the numerical values of a and b, and carry out the calculations:

$$c = \sqrt{4^2 + 3^2} = \sqrt{16 + 9} = \sqrt{25} = 5$$

The cosine of angle α is:

$$\cos \alpha = \frac{b}{c} = \frac{3}{5} = 0.6$$

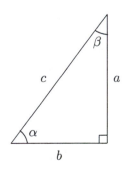

Figure A.7 *Example A.1.*

Take the inverse cosine of both sides:

$$\alpha = \cos^{-1}(0.6) = 53.13°$$

The cosine of angle β is:

$$\cos \beta = \frac{a}{c} = \frac{4}{5} = 0.8$$

Take the inverse cosine of both sides:

$$\beta = \cos^{-1}(0.8) = 36.87°$$

Check whether the results are correct:

$$\alpha + \beta \stackrel{?}{=} 53.13° + 36.87° \stackrel{\checkmark}{=} 90°$$

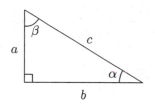

Figure A.8 *Example A.2.*

Example A.2 For the right-triangle shown in Figure A.8, if $b = 2$ and $\alpha = 30°$, determine angle β, and sides a and c.

Solution: The sum of the acute angles of a right-triangle must be equal to 90°. Therefore:

$$\beta = 90° - \alpha = 90° - 30° = 60°$$

Consider the cosine of angle α:

$$\cos \alpha = \frac{b}{c}$$

Multiply both sides by c and divide by $\cos \alpha$:

$$c = \frac{b}{\cos \alpha}$$

Substitute the numerical values of b and α:

$$c = \frac{2}{\cos 30°} = \frac{2}{0.866} = 2.31$$

Consider the tangent of angle α:

$$\tan \alpha = \frac{a}{b}$$

Multiply both sides by b:

$$a = b \, \tan \alpha$$

Substitute the numerical values of b and α:

$$a = (2)(\tan 30°) = (2)(0.577) = 1.15$$

Check if the results are correct:

$$a^2 + b^2 = c^2$$

$$(1.15)^2 + (2)^2 \stackrel{?}{=} (2.31)^2$$

$$5.3 \stackrel{\checkmark}{=} 5.3$$

Appendix B

Vector Algebra

B.1 Definitions

Most of the concepts in mechanics are either scalar or vector. A *scalar* quantity has a magnitude only. Concepts such as mass, energy, power, mechanical work, and temperature are scalar quantities. For example, it is sufficient to say that an object has 80 kilograms (kg) of mass. A *vector* quantity, on the other hand, has both a magnitude and a direction associated with it. Force, moment, velocity, and acceleration are examples of vector quantities. To describe a force fully, one must state how much force is applied and in which direction it is applied. The magnitude of a vector is also a scalar quantity, which is always a positive number.

It should be noted that both scalars and vectors are special forms of a more general category of all quantities in mechanics called *tensors*. Scalars are also known as "zero-order tensors," whereas vectors are "first-order tensors." Concepts such as stress and strain, on the other hand, are "second-order tensors." In this definition, the order corresponds to the power n in 3^n. For scalars, n is zero, and therefore, $3^0 = 1$, or only one quantity (magnitude) is necessary to define a scalar quantity. For vectors, n is one, and therefore, $3^1 = 3$. That is, three quantities (components in three directions) are necessary to define a vector quantity in three-dimensional space. On the other hand, n is two for second-order tensors, and since $3^2 = 9$, nine quantities (three components in three planes) are needed to define concepts such as stress and strain.

B.2 Notation

There are various notations used to refer to vector quantities. In this text, we shall use letters with underbars. For example, $\underline{A}$ will refer to a vector quantity, whereas A without the underbar will be used to refer to a scalar quantity. In graphical solutions, vectors are commonly represented by arrows, as illustrated in Figure B.1. The orientation of the arrow indicates the *line of action* of the vector. The arrowhead denotes the *direction* of the vector by defining its *sense* along its line of action. If the vector represents, for example, an applied force, then the base (tail) of the arrow corresponds to the *point of application* of the force vector. If there is a need to show more than one vector in a single drawing, the length of each arrow must be proportional to the magnitude of the vector it is representing. The *magnitude* of a vector quantity is always a positive number corresponding to the numerical measure of that quantity. There are two ways of referring to the magnitude of a vector quantity: either by dropping the underbar (A) or by enclosing the vector quantity with a set of vertical lines known as the absolute sign ($|\underline{A}|$).

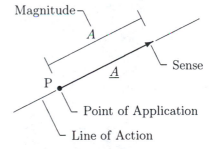

Figure B.1 *Graphical representation of vector $\underline{A}$.*

B.3 Multiplication of a Vector by a Scalar

Let $\underline{A}$ be a vector quantity with magnitude A, and m be a scalar quantity. The product $m\underline{A}$ is equal to a new vector, $\underline{B} = m\underline{A}$, such that it has the same direction as vector $\underline{A}$ but a magnitude equal to m times A. For example, if $m = 2$, then the magnitude of the product vector $\underline{B}$ is twice as large as the magnitude of vector $\underline{A}$.

B.4 Negative Vector

Let $\underline{A}$ be a vector quantity with magnitude A. $-\underline{A}$ is called a *negative vector* and it differs from vector $\underline{A}$ in that $-\underline{A}$ has a direction opposite to that of vector $\underline{A}$ (Figure B.2). Since magnitudes of vector quantities are positive scalar quantities, both $-\underline{A}$ and $\underline{A}$ have the same magnitude equal to A. Therefore, for vector quantities, a negative sign implies change in direction and has nothing to do with the magnitude of the vector.

B.5 Addition of Vectors: Graphical Methods

There are two ways of adding two or more vectors graphically: parallelogram and triangle or tail-to-tip methods. Consider the two vectors $\underline{A}$ and $\underline{B}$ shown in Figure B.3. Vector $\underline{A}$ is pointing toward point Q and vector $\underline{B}$ is pointing toward point R. The *parallelogram* method of adding two vectors involves the construction of a parallelogram by drawing a line at the tip of one of the vectors parallel to the other vector, and repeating the same thing for the second vector. If S corresponds to the point of intersection of these parallel lines, then an arrow drawn from point P toward S represents a vector that is equal to the vector sum of $\underline{A}$ and $\underline{B}$ (Figure B.4). The third vector thus obtained is called the *resultant* or the *net* vector. In Figure B.4, the resultant vector is identified as $\underline{C}$ which can be mathematically expressed as:

$$\underline{A} + \underline{B} = \underline{C} \qquad (B.1)$$

The *triangle* or *tail-to-tip* method of adding two vectors graphically is illustrated in Figure B.5. In this case, without changing its orientation, one of the vectors to be added is translated to the tip of the other vector in such a way that the tip of one of the vectors coincides with the tail of the other. An arrow drawn from the tail of the first vector toward the tip of the second vector represents the resultant vector.

Note that while performing vector addition, the order of appearance of vectors is arbitrary. That is, the addition of vectors is a *commutative* operation. The sum of $\underline{A}$ and $\underline{B}$, and the sum of $\underline{B}$ and $\underline{A}$ result in the same vector $\underline{C}$:

$$\underline{A} + \underline{B} = \underline{B} + \underline{A} = \underline{C} \qquad (B.2)$$

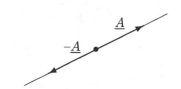

Figure B.2 *Negative vector.*

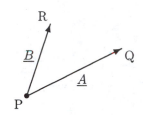

Figure B.3 *Vectors $\underline{A}$ and $\underline{B}$.*

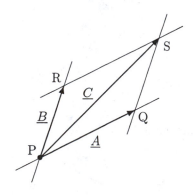

Figure B.4 *Parallelogram.*

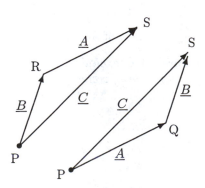

Figure B.5 *Tail-to-tip method.*

B.6   Subtraction of Vectors

The subtraction of one vector from another can easily be done by noting that the negative of any vector is a vector of the same magnitude pointing in the opposite direction. For example, as illustrated in Figure B.6, to subtract vector B from vector A, vector $-B$ can be added to A to determine the resultant vector.

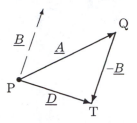

Figure B.6 $D = A - B$.

Note that subtraction of vector A from B follows a similar approach. If D and E are two vectors such that

$$A - B = D$$
$$B - A = E$$

then since $A - B = -(B - A)$, vectors D and E have an equal magnitude ($D = E$) but opposite directions ($D = -E$).

B.7   Addition of More Than Two Vectors

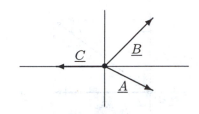

Figure B.7 *Three vectors.*

As illustrated in Figure B.7, let A, B, and C be three vectors such that all three vectors lie on a two-dimensional surface (a *coplanar* system of vectors). There are a number of different ways to add three vectors together. Since the addition of vectors is a commutative operation, the order of appearance of vectors to be added does not influence the resultant vector.

The three vectors shown in Figure B.7 are graphically added together in Figure B.8 by using the tail-to-tip method. In the case illustrated, vector C is added to vector B which is added to vector A to obtain the resultant vector D. That is:

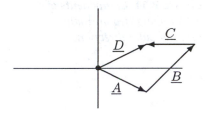

Figure B.8 *Vector sum of A, B, and C.*

$$A + B + C = D$$

The parallelogram method of adding the same vectors is illustrated in Figure B.9. In this case, two (B and C) of the three vectors are added together first. The resultant (vector E) of this operation is then added to vector A and the overall resultant (vector D) is determined. The sequence of additions illustrated in Figure B.9 can be mathematically expressed as:

$$A + (B + C) = A + E = D$$

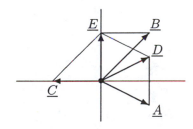

Figure B.9 *Parallelogram method of adding three vectors.*

B.8   Projection of Vectors

As illustrated in Figure B.10, let s represent a line and A be a vector whose line of action makes an angle θ with s. Assume that line s and vector A lie on a plane surface. To determine the projection or the component of vector A on s, drop a straight line from the tip (point Q) of the vector that cuts the line defined by s at right angles. If R denotes the point of intersection of these two lines and A_s is the length of the line segment between points P

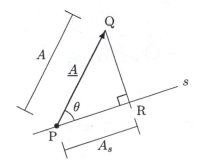

Figure B.10 *Projection of a vector on a given direction.*

and R, then A_s is the *projection* or *scalar component* of vector $\underline{A}$ on s. Note that points P, Q, and R define a right-triangle, and therefore:

$$A_s = A \cos\theta \qquad (B.3)$$

B.9 Resolution of Vectors

The resolution of a vector into its components is the reverse action of adding two vectors. In Figure B.11, x and y indicate the horizontal and vertical, respectively. $\underline{A}$ is a vector acting on the xy-plane. To determine the component of $\underline{A}$ along the x direction, drop a vertical line from the tip of the vector that cuts the horizontal line defined by x at right angles. Similarly, to determine the component of $\underline{A}$ along the y direction, draw a horizontal line passing through the tip of the vector that cuts the vertical line defined by y at right angles. These operations result in two similar right-triangles forming a rectangle (a parallelogram). The lengths of the sides of this rectangle represent the *scalar components* of vector $\underline{A}$ along the x and y directions, which can be determined by utilizing the properties of right-triangles:

$$A_x = A \cos\theta$$
$$A_y = A \sin\theta \qquad (B.4)$$

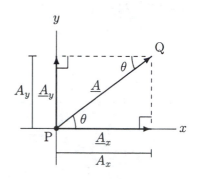

Figure B.11 *Components of a vector along two mutually perpendicular directions.*

In Eq. (B.4), θ is the angle that vector $\underline{A}$ makes with the horizontal. Vector $\underline{A}$ can also be represented as the sum of its *vector components* $\underline{A}_x$ and $\underline{A}_y$ along the x and y directions, respectively:

$$\underline{A} = \underline{A}_x + \underline{A}_y \qquad (B.5)$$

B.10 Unit Vectors

Usually it is convenient to express a vector $\underline{A}$ as the product of its magnitude A and a vector $\underline{a}$ of unit magnitude that has the same direction as vector $\underline{A}$ (Figure B.12). $\underline{a}$ is called the *unit vector*. For a given vector, its unit vector can be obtained by dividing that vector with its magnitude:

$$\underline{a} = \frac{\underline{A}}{A} \qquad (B.6)$$

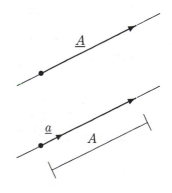

Figure B.12 $\underline{A} = A\underline{a}$.

The original vector $\underline{A}$ can now be expressed as:

$$\underline{A} = A\,\underline{a} \qquad (B.7)$$

Since the magnitude of $\underline{a}$ is equal to the magnitude of $\underline{A}$ (which is A) divided by A, the magnitude of vector $\underline{a}$ is always equal to one. Unit vectors are used as a convenient way of describing directions.

B.11 Rectangular Coordinates

To be able to define the position of an object in space and to be able to analyze changes in position over time, measurements must be made relative to a reference frame or a coordinate system. There are a number of widely used coordinate systems. Among these, the *Cartesian* or *rectangular coordinate system* is the one most commonly used. As illustrated in Figure B.13, the Cartesian coordinate system consists of three mutually perpendicular axes.

The concept of the unit vector has an important application in the construction of coordinate systems. The unit vectors along the Cartesian coordinate axes are so frequently used that they have widely accepted symbols. Symbols i, j, and k are commonly used to refer to *unit coordinate vectors* indicating positive x, y, and z directions, respectively.

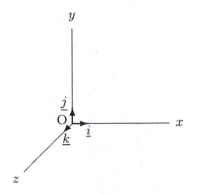

Figure B.13 *Cartesian or rectangular coordinates.*

B.12 Addition of Vectors: Trigonometric Method

We have seen the addition and subtraction of vectors, and the resolution of vectors into their components by means of graphical methods that may be time-consuming and not accurate. Faster and more precise results can be obtained through the use of trigonometric identities.

To apply the *trigonometric method* of addition and subtraction of vectors, one must first resolve each vector into its components. For example, consider the two vectors shown in Figure B.14. Vectors $\underline{A}$ and $\underline{B}$ have magnitudes equal to A and B, and they make angles α and β with the horizontal (x axis). The scalar components of $\underline{A}$ and $\underline{B}$ along the x and y directions can be determined by utilizing the properties of right-triangles. For vector $\underline{A}$:

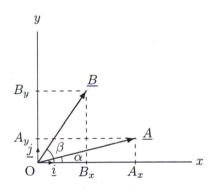

Figure B.14 *Vectors $\underline{A}$ and $\underline{B}$.*

$$A_x = A \cos \alpha$$
$$A_y = A \sin \alpha \qquad \text{(B.8)}$$

A_x and A_y are the scalar components of $\underline{A}$ along the x and y directions, respectively. Making use of the unit vectors $\underline{i}$ and $\underline{j}$ that identify positive x and y directions, the vector components of $\underline{A}$ in rectangular coordinates can be determined easily from its scalar components:

$$\underline{A}_x = A_x \underline{i}$$
$$\underline{A}_y = A_y \underline{j} \qquad \text{(B.9)}$$

Now, vector $\underline{A}$ can alternately be expressed as:

$$\begin{aligned} \underline{A} &= \underline{A}_x + \underline{A}_y \\ &= A_x \underline{i} + A_y \underline{j} \qquad \text{(B.10)} \\ &= A \cos \alpha \, \underline{i} + A \sin \alpha \, \underline{j} \end{aligned}$$

Note that A_x, A_y, and A correspond to the sides of a right-triangle with A being the hypotenuse. Therefore:

$$A = \sqrt{(A_x)^2 + (A_y)^2} \qquad (B.11)$$

Similarly, vector $\underline{B}$ can be expressed as:

$$\begin{aligned}
\underline{B} &= \underline{B}_x + \underline{B}_y \\
&= B_x\,\underline{i} + B_y\,\underline{j} \qquad (B.12) \\
&= B\,\cos\beta\,\underline{i} + B\,\sin\beta\,\underline{j}
\end{aligned}$$

Once the vectors are expressed in terms of their components, the next step is to add (or subtract) the x components of all vectors together. This will yield the x component of the resultant vector. Similarly, adding the y components of all vectors will give the y component of the resultant vector. For example, the addition of vectors $\underline{A}$ and $\underline{B}$ can be performed as:

$$\begin{aligned}
\underline{A} + \underline{B} &= (A_x\,\underline{i} + A_y\,\underline{j}) + (B_x\,\underline{i} + B_y\,\underline{j}) \\
&= (A_x + B_x)\underline{i} + (A_y + B_y)\underline{j} \qquad (B.13) \\
&= (A\cos\alpha + B\cos\beta)\underline{i} + (A\sin\alpha + B\sin\beta)\underline{j}
\end{aligned}$$

If $\underline{C}$ refers to the vector sum of $\underline{A}$ and $\underline{B}$, then:

$$\begin{aligned}
\underline{C} &= \underline{A} + \underline{B} \\
&= \underline{C}_x + \underline{C}_y \qquad (B.14) \\
&= C_x\,\underline{i} + C_y\,\underline{j}
\end{aligned}$$

Comparing Eqs. (B.13) and (B.14) one can conclude that:

$$\begin{aligned}
C_x &= A_x + B_x = A\cos\alpha + B\cos\beta \\
C_y &= A_y + B_y = A\sin\alpha + B\sin\beta
\end{aligned} \qquad (B.15)$$

If A, B, α, and β are known, then C_x and C_y can be determined from Eqs. (B.15). Note that C_x, C_y, and C also form a right-triangle, with C being the hypotenuse (Figure B.15). Therefore:

$$C = \sqrt{(C_x)^2 + (C_y)^2} \qquad (B.16)$$

If γ represents the that angle vector $\underline{C}$ makes with the x axis, then:

$$\gamma = \tan^{-1}\left(\frac{C_y}{C_x}\right) \qquad (B.17)$$

Note that $\tan^{-1}$ is called the *inverse tangent* or *arctangent* (abbreviated as arctan).

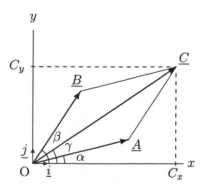

Figure B.15 *Vector sum of $\underline{A}$ and $\underline{B}$ is equal to $\underline{C}$.*

Subtraction of one vector from another is as straightforward as adding the two. For example:

$$\underline{A} - \underline{B} = (A_x\,\underline{i} + A_y\,\underline{j}) - (B_x\,\underline{i} + B_y\,\underline{j})$$
$$= (A_x - B_x)\underline{i} + (A_y - B_y)\underline{j} \qquad (B.18)$$
$$= (A\cos\alpha - B\cos\beta)\underline{i} + (A\sin\alpha - B\sin\beta)\underline{j}$$

Example B.1 Vector $\underline{A}$ in Figure B.16 is such that its magnitude is $A = 5$ units and it makes an angle $\alpha = 36.87°$ with the horizontal.

(a) Determine the scalar components of $\underline{A}$ along the horizontal and vertical.
(b) Express $\underline{A}$ in terms of its components.

Solution:

(a) Using Eqs. (B.8):

$$A_x = A\cos\alpha = 5\cos(36.87°) = 4$$
$$A_y = A\sin\alpha = 5\sin(36.87°) = 3$$

(b) From Eq. (B.10):

$$\underline{A} = A_x\,\underline{i} + A_y\,\underline{j} = 4\,\underline{i} + 3\,\underline{j}$$

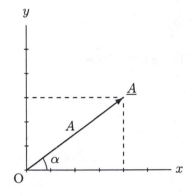

Figure B.16 *Example B.1.*

Example B.2 The component of vector $\underline{B}$ in Figure B.17 along the positive x direction is measured as 8 units, and its component in the negative y direction is 12 units.

(a) Express vector $\underline{B}$ in terms of its components.
(b) Determine the magnitude of vector $\underline{B}$ and angle β that it makes with the positive x direction.

Solution:

(a) Vector $\underline{B}$ can be expressed as:

$$\underline{B} = 8\,\underline{i} - 12\,\underline{j}$$

(b) Note that $B_x = 8$ and $B_y = 12$. Hence:

$$B = \sqrt{(B_x)^2 + (B_y)^2} = \sqrt{(8)^2 + (12)^2} = \sqrt{208} = 14.42$$

$$\beta = \tan^{-1}\left(\frac{B_y}{B_x}\right) = \tan^{-1}\left(\frac{12}{8}\right) = 56.31°$$

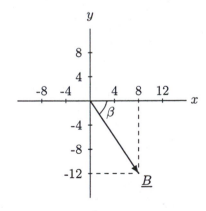

Figure B.17 *Example B.2.*

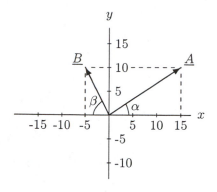

Figure B.18 *Example B.3.*

Example B.3 Consider vectors $\underline{A}$ and $\underline{B}$ shown in Figure B.18. The scalar components of these vectors are measured as:

$$A_x = 15 \qquad B_x = 5$$
$$A_y = 10 \qquad B_y = 10$$

(a) Express $\underline{A}$ and $\underline{B}$ in terms of their components.
(b) Determine the magnitudes of $\underline{A}$ and $\underline{B}$, and angles α and β.
(c) Determine $\underline{A} + \underline{B}$.
(d) Determine $\underline{A} - \underline{B}$.
(e) Determine $\underline{B} - \underline{A}$.
(f) Determine $-\underline{A} - \underline{B}$.

Solution:

(a)

$$\underline{A} = 15\,\underline{i} + 10\,\underline{j}$$
$$\underline{B} = -5\,\underline{i} + 10\,\underline{j}$$

(b) Magnitudes of $\underline{A}$ and $\underline{B}$:

$$A = \sqrt{(A_x)^2 + (A_y)^2} = \sqrt{15^2 + 10^2} = 18.03$$

$$B = \sqrt{(B_x)^2 + (B_y)^2} = \sqrt{5^2 + 10^2} = 11.18$$

Angles α and β:

$$\alpha = \tan^{-1}\left(\frac{10}{15}\right) = 33.69°$$

$$\beta = \tan^{-1}\left(\frac{10}{5}\right) = 63.43°$$

(c) Let $\underline{C} = \underline{A} + \underline{B}$. Then:

$$\underline{C} = (15\underline{i} + 10\underline{j}) + (-5\underline{i} + 10\underline{j})$$
$$= (15 - 5)\underline{i} + (10 + 10)\underline{j}$$
$$= 10\underline{i} + 20\underline{j}$$

(d) Let $\underline{D} = \underline{A} - \underline{B}$. Then:

$$\underline{D} = (15\underline{i} + 10\underline{j}) - (-5\underline{i} + 10\underline{j})$$
$$= (15 + 5)\underline{i} + (10 - 10)\underline{j}$$
$$= 20\underline{i}$$

(e) Let $\underline{E} = \underline{B} - \underline{A}$. Then:

$$\underline{E} = (-5\underline{i} + 10\underline{j}) - (15\underline{i} + 10\underline{j})$$
$$= (-5 - 15)\underline{i} + (10 - 10)\underline{j}$$
$$= -20\underline{i}$$

(f) Let $\underline{F} = -\underline{A} - \underline{B}$. Then:

$$\begin{aligned}
\underline{F} &= -(15\underline{i} + 10\underline{j}) - (-5\underline{i} + 10\underline{j}) \\
&= (-15 + 5)\underline{i} + (-10 - 10)\underline{j} \\
&= -10\underline{i} - 20\underline{j}
\end{aligned}$$

The resultant vectors $\underline{C}$, $\underline{D}$, $\underline{E}$, and $\underline{F}$ are illustrated in Figure B.19. A careful examination of these results shows that vectors $\underline{C}$ and $\underline{F}$ form a pair of negative vectors with equal magnitude and opposite directions, and so are vectors $\underline{D}$ and $\underline{E}$. Also note that vectors $\underline{D}$ and $\underline{E}$ have no components along the y direction.

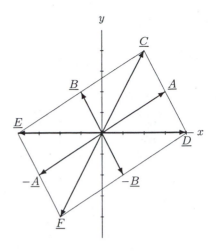

Figure B.19 *Addition and subtraction of vectors.*

B.13   Three-Dimensional Components of Vectors

The trigonometric method of adding and subtracting vectors with components in the x, y, and z directions is an extension of the principles introduced for two-dimensional systems. With respect to the Cartesian coordinate system, to define a vector quantity such as $\underline{A}$ uniquely, either all three components A_x, A_y, and A_z of vector $\underline{A}$, or the magnitude A of the vector along with the angles the vector makes with the x, y, and z directions, must be provided (Figure B.20). We can alternatively express $\underline{A}$ as:

$$\begin{aligned}
\underline{A} &= \underline{A}_x + \underline{A}_y + \underline{A}_z \\
&= A_x\,\underline{i} + A_y\,\underline{j} + A_z\,\underline{k}
\end{aligned} \tag{B.19}$$

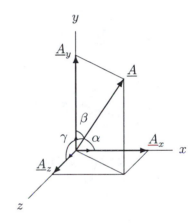

Figure B.20 *Components of a vector in three directions.*

If α, β, and γ refer to the angles that vector $\underline{A}$ makes with the x, y, and z directions, respectively, then:

$$\begin{aligned}
A_x &= A \cos \alpha \\
A_y &= A \cos \beta \\
A_z &= A \cos \gamma
\end{aligned} \tag{B.20}$$

Assume that there is a second vector, $\underline{B}$, that is to be added to vector $\underline{A}$. If $\underline{C}$ represents the resultant vector, then:

$$\begin{aligned}
\underline{C} &= \underline{A} + \underline{B} \\
&= (A_x\,\underline{i} + A_y\,\underline{j} + A_z\,\underline{k}) + (B_x\,\underline{i} + B_y\,\underline{j} + B_z\,\underline{k}) \\
&= (A_x + B_x)\underline{i} + (A_y + B_y)\underline{j} + (A_z + B_z)\underline{k}
\end{aligned} \tag{B.21}$$

Note that the components of the resultant vector $\underline{C}$ in the x, y, and z directions are:

$$\begin{aligned}
C_x &= A_x + B_x \\
C_y &= A_y + B_y \\
C_z &= A_z + B_z
\end{aligned} \tag{B.22}$$

As is true for coplanar systems, the addition of two or more vector quantities having components in all three directions requires addition of the x components of all vectors together. The same procedure must be repeated for the y and z components. Subtraction of two or more three-dimensional vector quantities follows a similar approach.

B.14   Dot (Scalar) Product of Vectors

Two of the most commonly encountered vector quantities in mechanics are the force and displacement vectors. In mechanics, *work done* by a force is defined as the product of the force component in the direction of displacement times the magnitude of displacement. Although force and displacement are vector quantities, their product – work – is a scalar quantity. An operation that represents such situations concisely is called the *dot* or *scalar product*.

The dot product of any two vectors is defined as a scalar quantity equal to the product of magnitudes of the two vectors multiplied by the cosine of the smaller angle between the two. For example, consider vectors $\underline{A}$ and $\underline{B}$ in Figure B.21. The dot product of these vectors is:

$$\underline{A} \cdot \underline{B} = A\,B\,\cos\theta \tag{B.23}$$

In Eq. (B.23), A and B are the magnitudes of the two vectors, and θ is the smaller angle between them. The operation given by Eq. B.23 is representative of first projecting vector $\underline{A}$ onto the line of action of vector $\underline{B}$ (or vice versa) and then multiplying the magnitudes of the projected component and the other vector.

Note that the dot product may result in positive or negative quantities depending on whether the smaller angle between the vectors is less or greater than $90°$. Since $\cos 90° = 0$, the dot product of two vectors is equal to zero if the vectors are perpendicular to one another. Unit vectors $\underline{i}$ and $\underline{j}$, which define the positive x and y directions, respectively, are also vector quantities with magnitudes equal to unity. The concept of the dot product is applicable to unit vectors as well:

$$\begin{aligned}
\underline{i} \cdot \underline{i} = \underline{j} \cdot \underline{j} = \underline{k} \cdot \underline{k} = 1\,1\,\cos 0° = 1 \\
\underline{i} \cdot \underline{j} = \underline{j} \cdot \underline{k} = \underline{k} \cdot \underline{i} = 1\,1\,\cos 90° = 0
\end{aligned} \tag{B.24}$$

To take the dot product of the vectors shown in Figure B.21, we can first represent each vector in terms of its components along the x and y directions, and then apply the dot product to the unit vectors:

$$\begin{aligned}
\underline{A} \cdot \underline{B} &= (A_x\,\underline{i} + A_y\,\underline{j}) \cdot (B_x\,\underline{i} + B_y\,\underline{j}) \\
&= A_x B_x(\underline{i} \cdot \underline{i}) + A_x B_y(\underline{i} \cdot \underline{j}) \\
&\quad + A_y B_x(\underline{j} \cdot \underline{i}) + A_y B_y(\underline{j} \cdot \underline{j}) \\
&= A_x B_x + A_y B_y
\end{aligned} \tag{B.25}$$

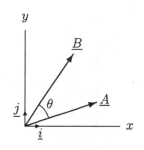

Figure B.21 *Dot (scalar) product of two vectors.*

The following are some of the properties of the dot product.

- The dot product of two vectors is a commutative operation:

$$\underline{A} \cdot \underline{B} = \underline{B} \cdot \underline{A}$$

- The dot product is a distributive operation:

$$\underline{A} \cdot (\underline{B} + \underline{C}) = \underline{A} \cdot \underline{B} + \underline{A} \cdot \underline{C}$$

- A vector multiplied by itself as a dot product is equal to the square of the magnitude of the vector:

$$\underline{A} \cdot \underline{A} = A^2$$

- The dot product of two three-dimensional vector quantities $\underline{A}$ and $\underline{B}$ expressed in terms of their Cartesian components is:

$$\underline{A} \cdot \underline{B} = A_x B_x + A_y B_y + A_z B_z$$

- The scalar component of a vector along a given direction is equal to the dot product of the vector times the unit vector along that direction. For example, the x and y components of a vector $\underline{A}$ are:

$$A_x = \underline{A} \cdot \underline{i} \qquad A_y = \underline{A} \cdot \underline{j}$$

B.15   Cross (Vector) Product of Vectors

There are other interactions between vector quantities that result in other vector quantities. An example of such interactions is the moment or torque generated by an applied force. The mathematical tool developed to define such interactions is called the *cross* or *vector product*.

Consider the vectors $\underline{A}$ and $\underline{B}$ in Figure B.22. The cross product of these vectors is equal to a third vector, say $\underline{C}$. The commonly used mathematical notation for the cross product is:

$$\underline{A} \times \underline{B} = \underline{C} \qquad (B.26)$$

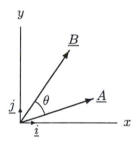

Figure B.22 *Two vectors $\underline{A}$ and $\underline{B}$.*

The vector $\underline{C}$ has a magnitude equal to the product of the magnitudes of vectors $\underline{A}$ and $\underline{B}$ times the sine of the smaller angle (angle θ in Figure B.22) between the two:

$$C = A B \sin \theta \qquad (B.27)$$

Vector $\underline{C}$ has a direction perpendicular to the plane defined by vectors $\underline{A}$ and $\underline{B}$. For example, if both $\underline{A}$ and $\underline{B}$ are in the xy plane, then $\underline{C}$ acts in the z direction.

The sense of vector $\underline{C}$ can be determined by the *right-hand-rule*. To apply this rule, first the fingers of the right hand are pointed in the direction of vector $\underline{A}$ (the first vector), and then they are curled toward vector $\underline{B}$ (the second vector) in such a way as to cover the smaller angle between $\underline{A}$ and $\underline{B}$. The extended thumb points in the direction of the product vector $\underline{C}$.

Note again that the concept of the cross product is applicable to the Cartesian unit vectors $\underline{i}$, $\underline{j}$, and $\underline{k}$ that are mutually perpendicular. Since $\sin 0° = 0$ and $\sin 90° = 1$, one can write:

$$\begin{aligned}
\underline{i} \times \underline{i} &= \underline{j} \times \underline{j} = \underline{k} \times \underline{k} = 0 \\
\underline{i} \times \underline{j} &= \underline{k} \qquad \underline{j} \times \underline{i} = -\underline{k} \\
\underline{j} \times \underline{k} &= \underline{i} \qquad \underline{k} \times \underline{j} = -\underline{i} \\
\underline{k} \times \underline{i} &= \underline{j} \qquad \underline{i} \times \underline{k} = -\underline{j}
\end{aligned} \qquad (B.28)$$

Using the above relations between the unit vectors, one can determine mathematical relations to evaluate the vector products that would include both the magnitude and the direction of the product vector. For example, if $\underline{A}$ and $\underline{B}$ are two vectors in the xy-plane (Figure B.23), then:

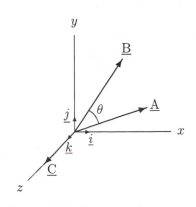

Figure B.23 $\underline{C} = \underline{A} \times \underline{B}$.

$$\begin{aligned}
\underline{C} &= \underline{A} \times \underline{B} \\
&= (A_x \underline{i} + A_y \underline{j}) \times (B_x \underline{i} + B_y \underline{j}) \\
&= A_x B_x (\underline{i} \times \underline{i}) + A_x B_y (\underline{i} \times \underline{j}) \\
&\quad + A_y B_x (\underline{j} \times \underline{i}) + A_y B_y (\underline{j} \times \underline{j}) \\
&= A_x B_x (0) + A_x B_y (\underline{k}) + A_y B_x (-\underline{k}) + A_y B_y (0) \\
&= (A_x B_y - A_y B_x) \underline{k}
\end{aligned} \qquad (B.29)$$

The following are some of the properties of the cross product.

• The cross product is not a commutative operation:

$$\underline{A} \times \underline{B} \neq \underline{B} \times \underline{A}$$

But

$$\underline{A} \times \underline{B} = -\underline{B} \times \underline{A}$$

• The cross product is a distributive operation:

$$\underline{A} \times (\underline{B} + \underline{C}) = \underline{A} \times \underline{B} + \underline{A} \times \underline{C}$$

• The vector product of two three-dimensional vector quantities $\underline{A}$ and $\underline{B}$ expressed in terms of their Cartesian components is:

$$\underline{A} \times \underline{B} = (A_y B_z - A_z B_y)\underline{i} + (A_z B_x - A_x B_z)\underline{j} + (A_x B_y - A_y B_x)\underline{k}$$

Example B.4 Vectors $\underline{A}$ and $\underline{B}$ shown in Figure B.24 are given in terms of their Cartesian components:

$$\underline{A} = 15\,\underline{i} + 10\,\underline{j}$$
$$\underline{B} = -5\,\underline{i} + 10\,\underline{j}$$

(a) Evaluate the scalar product $c = \underline{A} \cdot \underline{B}$.
(b) Evaluate the vector product $\underline{D} = \underline{A} \times \underline{B}$.
(c) Evaluate the vector product $\underline{E} = \underline{B} \times \underline{A}$.

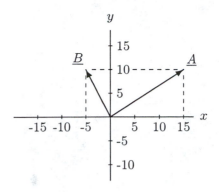

Solution:

(a)

$$c = \underline{A} \cdot \underline{B}$$
$$= A_x B_x + A_y B_y$$
$$= (15)(-5) + (10)(10)$$
$$= 25$$

Figure B.24 *Example B.4.*

(b)

$$\underline{D} = \underline{A} \times \underline{B}$$
$$= (A_x B_y - A_y B_x)\underline{k}$$
$$= \left[(15)(10) - (10)(-5)\right]\underline{k}$$
$$= 200\,\underline{k}$$

(c)

$$\underline{E} = \underline{B} \times \underline{A} = -\underline{D}$$
$$= (B_x A_y - B_y A_x)\underline{k}$$
$$= \left[(-5)(10) - (10)(15)\right]\underline{k}$$
$$= -200\,\underline{k}$$

Note that the product vectors $\underline{D}$ and $\underline{E}$ have an equal magnitude of 200 units. Vector $\underline{D}$ has a counterclockwise direction, whereas vector $\underline{E}$ is clockwise. Both vectors are acting in the direction perpendicular to the surface of the page, $\underline{D}$ out of the page and $\underline{E}$ into the page.

B.16 Exercise Problems

Problem B.1 Consider the system of four coplanar vectors $\underline{A}$, $\underline{B}$, $\underline{C}$, and $\underline{D}$ illustrated in Figure B.25. The magnitudes of these vectors are such that $A = 4$, $B = 7$, $C = 3$, and $D = 5$. Vector $\underline{A}$ acts in the positive x direction, vector $\underline{B}$ makes an angle 30° with the positive x axis, vector $\underline{C}$ acts in the positive y direction, and vector $\underline{D}$ makes an angle 45° with the negative x axis.

(a) Determine the scalar components of vectors $\underline{A}$, $\underline{B}$, $\underline{C}$, and $\underline{D}$ along the x and y directions.

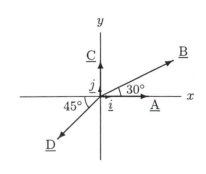

Figure B.25 *Problem B.1.*

(b) Express vectors $\underline{A}$, $\underline{B}$, $\underline{C}$, and $\underline{D}$ in terms of their components along the x and y directions.
(c) Determine vector $\underline{E} = \underline{A} + \underline{D}$.
(d) Determine vector $\underline{F} = \underline{C} + \underline{B}$.
(e) Determine vector $\underline{G} = \underline{A} + \underline{C} - \underline{B}$.
(f) Determine vector $\underline{H} = \underline{A} + \underline{B} + \underline{C} + \underline{D}$.
(g) Calculate the magnitudes of $\underline{E}$, $\underline{F}$, $\underline{G}$, and $\underline{H}$.

Answers:

(a) $A_x = 4$, $A_y = 0$, $B_x = 6.06$, $B_y = 3.50$, $C_x = 0$, $C_y = 3$, $D_x = 3.54$, $D_y = 3.54$
(b) $\underline{A} = 4\underline{i}$, $\underline{B} = 6.06\underline{i} + 3.50\underline{j}$, $\underline{C} = 3\underline{j}$, $\underline{D} = -3.54\underline{i} - 3.54\underline{j}$
(c) $\underline{E} = 0.46\underline{i} - 3.54\underline{j}$
(d) $\underline{F} = 6.06\underline{i} + 6.50\underline{j}$
(e) $\underline{G} = -2.06\underline{i} - 0.50\underline{j}$
(f) $\underline{H} = 6.52\underline{i} + 2.96\underline{j}$
(g) $E = 3.57$, $F = 8.89$, $G = 2.12$, $H = 7.16$

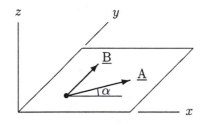

Figure B.26 *Problem B.2.*

Problem B.2 Consider vectors $\underline{A}$ and $\underline{B}$ illustrated in Figure B.26. These vectors act on the xy-plane and have magnitudes $A = 10$ and $B = 7$ units. Vector $\underline{A}$ makes an angle $\alpha = 15°$ with the positive x axis and vector $\underline{B}$ acts along the positive y direction.

(a) Express $\underline{A}$ and $\underline{B}$ in terms of their components.
(b) Evaluate the scalar product $m = \underline{A} \cdot \underline{B}$.
(c) Evaluate the vector product $\underline{C} = \underline{A} \times \underline{B}$.

Answers:

(a) $\underline{A} = 9.66\underline{i} + 2.59\underline{j}$, $\underline{B} = 7\underline{j}$
(b) $m = 18.13$
(c) $\underline{C} = 67.62\underline{k}$

Appendix C

Calculus

C.1 Functions

In general, as a result of a series of experiments, a relationship between two quantities may be established. This relationship may be represented in the form of a table, a graph, or a mathematical equation called *function*. Once a function relating the two quantities is established, changes in one as a result of changes in the other quantity may be predicted without resorting to additional experiments.

Most physical laws are expressed in the form of mathematical equations. For example, velocity is defined as the time rate of change of position. If we can measure the change of position of an object over time and express that change in terms of a function, then we can also determine the velocity of that object simply by subjecting the function to certain mathematical operations without actually measuring the velocity of the object.

Consider two quantities X and Y. Assume that these quantities are related, such that a change in quantity X causes quantity Y to vary. Assume that a series of 10 experiments are conducted, where by varying quantity X, corresponding Y values are measured. The variations in Y with respect to variations in X can be represented by various schemes. For example, the data collected can be presented in tabular form (Table C.1). However, information can be extracted more easily from a diagram than from a set of numbers. A diagram can be constructed by assigning the horizontal coordinate (abscissa) to the input X and the vertical coordinate (ordinate) to the output Y (Figure C.1). Each pair of X and Y values recorded in Table C.1 will then correspond to a point on this diagram. A curve can be obtained by connecting these points, which will represent the graph of the data obtained.

Another way of representing the relationship between X and Y may be by means of a function. Functions are usually denoted by single letters, such as f or g. For example, $Y = f(X)$ implies that Y is a function of X. X in $Y = f(X)$ is the "input" or "cause" of an operation or process, while Y is the "output" or "effect." Usually, the input of a function is the *independent variable* and the output is the *dependent variable*. There are various ways of determining functions relating two or more quantities. One method is to compare the graph obtained as a result of an experiment or observation (such as the one in Figure C.1) with the graphs of known functions. For example, the straight line in Figure C.1 indicates that quantity Y is a "linear" function of X, and that the relationship between X and Y can be represented by the equation $Y = 1 + 2X$. Therefore, the function relating X and Y is $f(X) = 1 + 2X$. It is clear from this discussion that to be able to establish the most suitable function, one has to be familiar with the characteristics of commonly encountered functions.

Table C.1 *Experimental data.*

Experiment	X	Y
1	0.5	2.0
2	1.0	2.9
3	1.5	4.1
4	2.0	5.2
5	2.5	6.0
6	3.0	6.9
7	3.5	8.0
8	4.0	8.9
9	4.5	10.0
10	5.0	11.0

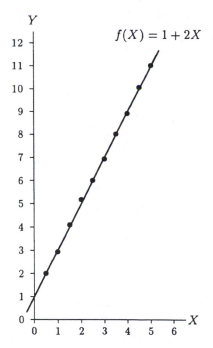

Figure C.1 *Y is a function of X, such that* $Y = f(X) = 1 + 2X$.

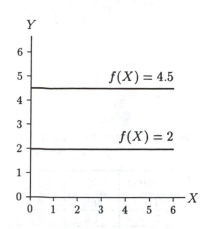

Figure C.2 *Constant functions.*

There are a few basic functions in calculus that may be sufficient to describe a large number of physical phenomena. These basic functions include constant, power, trigonometric, logarithmic, and exponential functions. Basic functions can be combined together in various ways to construct more complex functions.

C.1.1 Constant functions

Constant functions can be represented in the following general form:

$$f(X) = c \qquad \text{(C.1)}$$

Here, c is a symbol with a constant numerical value. For example, $Y = 2$ and $Y = 4.5$ are constant functions. Graphs of constant function are horizontal lines, as illustrated in Figure C.2.

C.1.2 Power functions

Power functions can be represented in the following general form:

$$f(X) = X^r \qquad \text{(C.2)}$$

In Eq. (C.2), r can be any real or integer number including zero and the ratio of numbers. Note that a constant function can be considered a power function for which $r = 0$, because $X^0 = 1$. The following are examples of power functions:

$$f(X) = X^1 = X$$
$$f(X) = X^3 = X \cdot X \cdot X$$
$$f(X) = X^{-1} = \frac{1}{X}$$
$$f(X) = X^{-2} = \frac{1}{X^2}$$
$$f(X) = X^{0.5} = X^{\frac{1}{2}} = \sqrt{X}$$

Graphs of some power functions are illustrated in Figure C.3. For a given function, its graph can be obtained by assigning a value to X, substituting that value to the function, solving the function for the corresponding Y value, repeating this for a number of X values, plotting each pair of X and Y values on the graph paper, and connecting them with a curve.

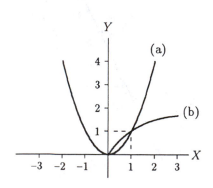

Figure C.3 *Examples of power functions. (a)* $f(X) = X^2$ *and (b)* $f(X) = \sqrt{X}$.

C.1.3 Linear functions

Linear functions can be represented in the following general form:

$$f(X) = a + bX \qquad \text{(C.3)}$$

In Eq. (C.3), a and b are some constant coefficients. The graph of a linear function is a straight line with coefficient a representing the point at which the straight line intersects the Y axis and coefficient b is the slope of the line. Examples of linear functions include:

$$f(X) = 1 + 2X$$
$$f(X) = 0.5 - 5X$$
$$f(X) = X$$

The graphs of some linear functions are illustrated in Figure C.4. Note that you need only two points to draw the graph of a linear function.

C.1.4 Quadratic functions

Quadratic functions can be represented in the following general form:

$$f(X) = a + bX + cX^2 \qquad (C.4)$$

In Eq. (C.4), a, b, and c are real or integer, positive or negative numbers. Coefficients a and b can be zero. Examples of quadratic functions include:

$$f(X) = 2 + X - 0.5X^2$$
$$f(X) = 5 + X^2$$
$$f(X) = -3X + 4X^2$$
$$f(X) = X^2$$

The distinguishing characteristic of these functions is that the highest power of X appearing in these equations is two. However, for example, $f(X) = 1 + X^{0.5} - 3X^2$ is not a quadratic function, because of the term that carries $X^{0.5}$. Since $X^0 = 1$ and $X^1 = X$, quadratic functions can also be represented as:

$$f(X) = aX^0 + bX^1 + cX^2$$

Graphs of quadratic functions are parabolas. The graphs of selected quadratic functions are illustrated in Figure C.5.

C.1.5 Polynomial functions

A *polynomial function* is one for which

$$f(X) = A_0 + A_1 X + A_2 X^2 + A_3 X^3 + \cdots + A_n X^n \qquad (C.5)$$

Coefficients $A_0, A_1, \ldots A_n$ in Eq. (C.5) are real or integer, positive or negative, zero or nonzero numbers, and n is a positive integer number corresponding to the highest power of X. Power n defines the "order" or the polynomial. For example,

$$f(X) = 1 - X - 2X^2 + 5X^3$$

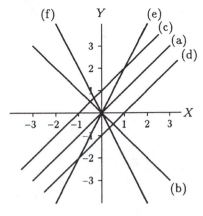

Figure C.4 *Linear functions.*
(a) $Y = X$, (b) $Y = -X$,
(c) $Y = 1 + X$, (d) $Y = -1 + X$,
(e) $Y = 2X$, and (f) $Y = -2X$.

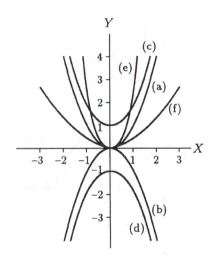

Figure C.5 *Quadratic functions.*
(a) $Y = X^2$, (b) $Y = -X^2$,
(c) $Y = 1 + X^2$, (d) $Y = -1 - X^2$,
(e) $Y = 3X^2$ and (f) $Y = X^2/3$.

is a polynomial of order 3 with coefficients $A_0 = 1$, $A_1 = -1$, $A_2 = -2$, and $A_3 = 5$. The function

$$f(X) = 2 + 3X^2$$

is also a polynomial of order 2 with coefficients $A_0 = 2$, $A_1 = 0$, and $A_2 = 3$. It is also a quadratic function with coefficients $a = 2$, $b = 0$, and $c = 3$.

Note that constant, linear, and quadratic functions are special forms of polynomial functions. A constant function is also a zero-order polynomial, a linear function is a first-order polynomial, and a quadratic function is a second-order polynomial.

C.1.6 Trigonometric functions

If a quantity Y depends on another quantity X through a trigonometric relationship, such as $Y = \sin(X)$ or $Y = \cos(X)$, then Y is said to be a *trigonometric function* of X. $f(X) = \sin(X)$ and $f(X) = \cos(X)$ are called the *sine* and *cosine functions*, respectively. The graphs of these functions are illustrated in Figures C.6 and C.7.

X in $Y = \sin(X)$ and $Y = \cos(X)$ can be measured either in *degrees* or in *radians*. An angle of $180°$ is called π radians with $\pi = 3.1416$. Degrees can be converted into radians using:

$$\text{radian} = \frac{\pi}{180} \times \text{degree}$$

For example, $0° = 0$ rad, $45° = \pi/4$ rad, $90° = \pi/2$ rad, $180° = \pi$ rad, $270° = 3\pi/2$ rad, $360° = 2\pi$ rad, and $720° = 4\pi$ rad.

Trigonometric functions are cyclic or periodic in the sense that their graphs repeat a pattern. The graphs of $Y = \sin(X)$ and $Y = \cos(X)$ in Figures C.6 and C.7 repeat after every 2π radians or $360°$. This means that the *period* of $Y = \sin(X)$ and $Y = \cos(X)$ is 2π radians. Furthermore, Y in $Y = \sin(X)$ and $Y = \cos(X)$ assume values between -1 and $+1$. Therefore, the *amplitude* of $Y = \sin(X)$ and $Y = \cos(X)$ is 1.

$Y = \sin(X)$ and $Y = \cos(X)$ are the simplest forms of trigonometric functions. Sine functions can be expressed in a more general form as:

$$f(X) = a \sin(bX) \tag{C.6}$$

Here, a and b are some constants. The sine function defined by Eq. (C.6) has an amplitude a and a period $2\pi/b$. For example, as illustrated in Figure C.8, the amplitude and period of the function $f(X) = 3\sin(\frac{X}{2})$ are 3 and 4π, respectively. The sine function in Eq. (C.6) can further be generalized as:

$$f(X) = a \sin(bX + c) \tag{C.7}$$

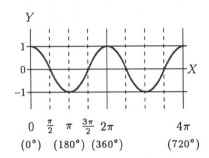

Figure C.6 $Y = f(X) = \cos X$.

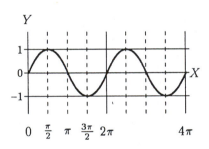

Figure C.7 $Y = f(X) = \sin X$.

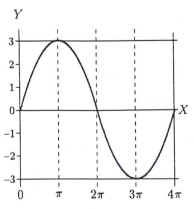

Figure C.8 $Y = 3\sin\left(\frac{X}{2}\right)$.

Here, a is the amplitude, $2\pi/b$ is the period, and the graph of this function is shifted by c to the right or left as compared to the graph of the function $f(X) = a\sin(bX)$. For example, the graph of $f(X) = 3\sin(\frac{X}{2} - \frac{\pi}{2})$ is the one shown in Figure C.9.

Note that the graph of a sine function that is shifted by $\pi/2$ is essentially the graph of a negative cosine function. In other words, $\sin(\frac{X}{2} - \frac{\pi}{2}) = -\cos(\frac{X}{2})$. There are a number of other trigonometric identities and formulas that are useful in handling trigonometric functions. Some of these formulas are provided in section C.4.

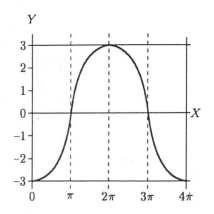

Figure C.9 $Y = 3\sin(\frac{X}{2} - \frac{\pi}{2})$.

C.1.7   Exponential and logarithmic functions

Functions such as 3^X and $(\frac{1}{2})^X$ are called *exponential functions*. The general form of exponential functions is b^X, where b is called the *base*. The most popular base in calculus is e, an irrational number between 2.71 and 2.72, and the function e^X or exp X is often referred to as the exponential function.

The exponential function exp X has an inverse, called the *natural logarithmic function* denoted by ln X. It can also be written as $\log_e X$ and called the logarithm with base e. The properties of exp X and ln X are such that:

$$
\begin{aligned}
\text{if} \quad \ln X = Y \quad &\text{then} \quad \exp Y = X \\
\ln(\exp X) = X \quad &\text{and} \quad \exp(\ln X) = X \\
\ln 1 = 0 \quad &\text{and} \quad \exp 0 = 1
\end{aligned}
$$

Graphs of exp X and ln X are shown in Figure C.10. Additional properties of exponential and logarithmic functions include:

$$\exp X \; \exp Y = \exp(X + Y)$$

$$\frac{\exp X}{\exp Y} = \exp(X - Y)$$

$$\exp(-X) = \frac{1}{\exp X}$$

$$(\exp X)^Y = \exp(XY)$$

$$\ln(XY) = \ln X + \ln Y$$

$$\ln\left(\frac{X}{Y}\right) = \ln X - \ln Y$$

$$\ln\left(\frac{1}{X}\right) = -\ln X$$

$$\ln(X^Y) = Y \ln X$$

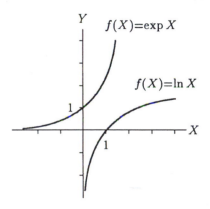

Figure C.10 *Exponential and logarithmic functions.*

Note that it is not allowed to take the "ln" of a negative number, but it is possible for ln X to come out negative. In other words, ln X is defined for X greater than zero. On the other hand, exp X is defined for all X. However, exp X is always positive.

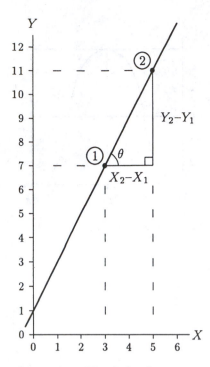

Figure C.11 *The derivative represents slope.*

C.2 The Derivative

The derivative is one of the fundamental mathematical operations used extensively to determine slopes of curves, maximums and minimums, and it has many other applications. The process of finding the derivative of a function is called *differentiation*. The branch of calculus dealing with the derivative is called *differential calculus*.

The derivative of a function represents the slope or slopes of the graph of that function. For example, consider the linear equation $Y = 1 + 2X$ whose graph is shown in Figure C.11. The slope of the line representing $Y = 1 + 2X$ can be determined by considering any two points 1 and 2 along the line, such as those points with coordinates $X_1 = 3$ and $Y_1 = 7$, and $X_2 = 5$ and $Y_2 = 11$. The slope of a line is defined by the tangent of the angle that the line makes with the horizontal. For the line shown in Figure C.11:

$$\text{slope} = \tan\theta = \frac{Y_2 - Y_1}{X_2 - X_1} = \frac{11 - 7}{5 - 3} = 2$$

Therefore, the derivative of the function $Y = 1 + 2X$ must be equal to 2. We shall demonstrate that this is the same 2 in front of the X in $Y = 1 + 2X$.

The slope of a straight line is constant. The curve that is not a straight line has many slopes, and it is usually difficult to predict the varying slopes of curves. Differentiation makes it easier to determine the slopes of curves.

In general, the derivative of a function is another function. For a function $Y = f(X)$ there are many symbols used to denote the derivative. For example:

$$f' \qquad f'(X) \qquad \frac{df}{dX} \qquad Y' \qquad \frac{dY}{dX}$$

C.2.1 Derivatives of basic functions

The graph of a constant function $f(X) = c$ is a horizontal line whose slope is zero. Therefore, the derivative of constant functions is always zero:

$$\frac{d}{dX}(c) = 0 \qquad\qquad (C.8)$$

Graph of the linear function $f(X) = X$ is a straight line with a slope equal to 1. Therefore:

$$\frac{d}{dX}(X) = 1 \qquad\qquad (C.9)$$

It is easy to find the derivatives of functions such as $f(X) = c$

and $f(X) = X$ using slopes. However, the graphs of power functions such as X^3, $\sqrt{X}$, and X^{-2} have varying slopes, and it is not easy to predict their derivatives. To differentiate a power function X^r, multiply the function by its power and reduce the power by 1:

$$\frac{d}{dX}(X^r) = rX^{r-1} \tag{C.10}$$

The following examples illustrate the use of Eq. (C.10).

Function, $f(X)$	Derivative, f'
X^4	$4X^3$
X^{-2}	$-2X^{-3}$
$X^{2.3}$	$2.3X^{1.3}$
$\sqrt{X} = X^{\frac{1}{2}}$	$\frac{1}{2}X^{-\frac{1}{2}} = \frac{1}{2\sqrt{X}}$

If the power r is zero, then $X^0 = 1$ regardless of what X is. Therefore, 1 is X^0 or a power function with $r = 0$, and the *power rule* as given in Eq. (C.10) can be applied to differentiate it:

$$\frac{d}{dX}(1) = \frac{d}{dX}(X^0) = 0X^{-1} = 0$$

Note that anything multiplied by zero is equal to zero. The function $f(X) = X$ is also a power function with $r = 1$. Therefore:

$$\frac{d}{dX}(X) = \frac{d}{dX}(X^1) = 1X^0 = 1$$

The definition of the derivative can be utilized for the differentiation of other basic functions. We shall adopt the following definitions without presenting any proofs:

$$\frac{d}{dX}(\sin X) = \cos X \tag{C.11}$$

$$\frac{d}{dX}(\cos X) = -\sin X \tag{C.12}$$

$$\frac{d}{dX}(\exp X) = \exp X \tag{C.13}$$

$$\frac{d}{dX}(\ln X) = \frac{1}{X} \tag{C.14}$$

C.2.2 The constant multiple rule

To differentiate a function in the form of a product of a constant c and another function $f(X)$, take the derivative of the function and then multiply it with the constant:

$$\frac{d}{dX}[cf(X)] = c\frac{df}{dX} = cf' \tag{C.15}$$

The following examples illustrate the use of Eq. (C.15).

Function, $f(X)$	Derivative, f'
$2X^2$	$4X$
$-1.25X^{-3}$	$3.75X^{-4}$
$0.5\cos X$	$-0.5\sin X$
$-3\exp X$	$-3\exp X$

C.2.3 The sum rule

The derivative of a function in the form of the sum of two functions is equal to the sum of the derivatives of the functions. If $f_1(X)$ and $f_2(X)$ represent two functions, then the derivative of the function $f(X) = f_1(X) + f_2(X)$ is:

$$\frac{d}{dX}[f_1(X) + f_2(X)] = f_1' + f_2' \tag{C.16}$$

For example:

$$\frac{d}{dX}(3X - 2\sin X) = \frac{d}{dX}(3X) + \frac{d}{dX}(-2\sin X) = 3 - 2\cos X$$

The sum rule can also be applied to take the derivative of linear and quadratic functions. If a, b, and c are constants, then:

$$\frac{d}{dX}(a + bX) = b \tag{C.17}$$

$$\frac{d}{dX}(a + bX + cX^2) = b + 2cX \tag{C.18}$$

Recall that linear and quadratic functions are special forms of polynomial functions. If A_0, A_1, A_2, ..., and A_n are some constants, then the derivative of a polynomial is:

$$\frac{d}{dX}(A_0 + A_1X + A_2X^2 + \cdots + A_nX^n)$$

$$= A_1 + 2A_2X + \cdots + nA_nX^{n-1} \tag{C.19}$$

Note that differentiation reduces the order of the polynomial by one. For example, the derivative of a quadratic function (second-order polynomial) is a linear function (first-order polynomial). Similarly, the derivative of a third-order polynomial is a quadratic function. For example:

$$\frac{d}{dX}(3 - 2X + 5X^2 - X^3) = -2 + 10X - 3X^2$$

C.2.4 The product rule

The derivative of a function in the form of a product of two functions is equal to the first function multiplied by the derivative of the second function plus the second function multiplied by the derivative of the first function. If $f_1(X)$ and $f_2(X)$ are two functions, then the derivative of the function $f(X) = f_1(X)f_2(X)$ is:

$$\frac{d}{dX}[f_1(X)f_2(X)] = f_1 f_2' + f_1' f_2 \qquad (C.20)$$

For example:

$$\begin{aligned}
\frac{d}{dX}(X^2 \cos X) &= (X^2)\frac{d}{dX}(\cos X) + (\cos X)\frac{d}{dX}(X^2) \\
&= (X^2)(-\sin X) + (\cos X)(2X) \\
&= 2X \cos X - X^2 \sin X
\end{aligned}$$

Note that the application of the product rule can be expanded to include functions in the form of the product of more than two functions. For example:

$$\frac{d}{dX}[f_1(X)f_2(X)f_3(X)] = f_1 f_2 f_3' + f_1 f_2' f_3 + f_1' f_2 f_3$$

C.2.5 The quotient rule

The derivative of a function in the form of a ratio of two functions is equal to the derivative of the function in the numerator times the function in the denominator, minus the function in the numerator times the derivative of the function in the denominator, all divided by the square of the function in the denominator. If $f_1(X)$ and $f_2(X)$ represent two functions, then the derivative of the function $f(X) = f_1(X)/f_2(X)$ is:

$$\frac{d}{dX}\left[\frac{f_1(X)}{f_2(X)}\right] = \frac{f_1' f_2 - f_1 f_2'}{f_2^2} \qquad (C.21)$$

For example:

$$\begin{aligned}
\frac{d}{dX}\left(\frac{2+3X^2}{X^2}\right) &= \frac{(X^2)\frac{d}{dX}(2+3X^2) - (2+3X^2)\frac{d}{dX}(X^2)}{(X^2)^2} \\
&= \frac{(X^2)(6X) - (2+3X^2)(2X)}{X^4} \\
&= \frac{6X^3 - 4X - 6X^3}{X^4} \\
&= -\frac{4}{X^3}
\end{aligned}$$

Alternatively:

$$\frac{d}{dX}\left(\frac{2+3X^2}{X^2}\right) = \frac{d}{dX}\left(\frac{2}{X^2}+3\right)$$

$$= \frac{(X^2)\frac{d}{dX}(2) - (2)\frac{d}{dX}(X^2)}{(X^2)^2} + \frac{d}{dX}(3)$$

$$= \frac{(X^2)(0) - (2)(2X)}{X^4} + (0)$$

$$= -\frac{4}{X^3}$$

Note that trigonometric functions $\tan X$, $\cot X$, $\sec X$, and $\csc X$ are various quotients of $\sin X$ and $\cos X$. Therefore, the quotient rule can be applied to determine their derivatives. For example:

$$\frac{d}{dX}(\tan X) = \frac{d}{dX}\left(\frac{\sin X}{\cos X}\right)$$

$$= \frac{(\cos X)\frac{d}{dX}(\sin X) - (\sin(X)\frac{d}{dX}(\cos X)}{(\cos X)^2}$$

$$= \frac{(\cos X)(\cos X) - (\sin X)(-\sin X)}{\cos^2 X}$$

$$= \frac{\cos^2 X + \sin^2 X}{\cos^2 X}$$

Since $\cos^2 X + \sin^2 X = 1$:

$$\frac{d}{dX}(\tan X) = \frac{1}{\cos^2 X} = \sec^2 X \qquad (C.22)$$

Similarly:

$$\frac{d}{dX}(\cot X) = -\frac{1}{\sin^2 X} = -\csc^2 X \qquad (C.23)$$

$$\frac{d}{dX}(\sec X) = \frac{\sin X}{\cos^2 X} = \sec X \tan X \qquad (C.24)$$

$$\frac{d}{dX}(\csc X) = -\frac{\cos X}{\sin^2 X} = -\csc X \cot X \qquad (C.25)$$

C.2.6 The chain rule

Sometimes functions appear in forms other than those analyzed in previous sections. None of the above rules can be applied to differentiate functions such as $\cos(X^3)$, $\exp(2-5X)$, and $\ln(4X)$. For example, consider the function $f(X) = \exp(2-5X)$. We have already seen the derivatives of $\exp X$ and $2-5X$, but the derivative of $\exp(2-5X)$ follows a different rule. To take the derivative of $f(X) = \exp(2-5X)$, define the terms in the parentheses as $Z = 2-5X$, so that the original function can be reduced to a form $f(Z) = \exp Z$. The *chain rule* states that:

$$\frac{df}{dX} = \frac{df}{dZ}\frac{dZ}{dX} \qquad (C.26)$$

Now, we can take the derivative of $f(X) = \exp(2 - 5X)$.

$$\frac{d}{dX}[\exp(2 - 5X)] = \frac{d}{dZ}(\exp Z)\frac{d}{dX}(Z)$$
$$= (\exp Z)(-5)$$
$$= -5\exp(2 - 5X)$$

The chain rule has a vast number of applications. For example, to take the derivative of $Y = \sin(X^2)$, let $Z = X^2$ and $Y = \sin Z$. The derivative of Y with respect to Z is $\cos X$, and the derivative of Z with respect to X is $12X$. Therefore:

$$\frac{d}{dX}[\sin(X^2)] = \frac{d}{dZ}(\sin Z)\frac{d}{dX}(Z)$$
$$= (\cos Z)(2X)$$
$$= 2X\cos(X^2)$$

Consider the function $Y = (3 + X^2)^2$. There are two ways to take the derivative of this function. One way is by expanding the parentheses, writing the function as $Y = 9 + 6X^2 + X^4$, and then taking the derivative:

$$\frac{d}{dX}[(3 + X^2)^2] = \frac{d}{dX}(9 + 6X^2 + X^4)$$
$$= 0 + 12X + 4X^3$$
$$= 12X + 4X^3$$

The second way is to let $Z = 3 + X^2$ so that $Y = Z^2$, taking individual derivatives, and then applying the chain rule:

$$\frac{d}{dX}[(3 + X^2)^2] = \frac{d}{dZ}(Z^2)\frac{d}{dX}(3 + X^2)$$
$$= (2Z)(2X)$$
$$= 12X + 4X^3$$

For any Z that is a function of X, applications of the chain rule can be summarized in the following form:

$$\frac{d}{dX}(Z^n) = nZ^{n-1}\frac{dZ}{dX} \qquad (C.27)$$

$$\frac{d}{dX}(\sin Z) = \cos Z\frac{dZ}{dX} \qquad (C.28)$$

$$\frac{d}{dX}(\cos Z) = -\sin Z\frac{dZ}{dX} \qquad (C.29)$$

$$\frac{d}{dX}(\exp Z) = \exp Z\frac{dZ}{dX} \qquad (C.30)$$

$$\frac{d}{dX}(\ln Z) = \frac{1}{Z}\frac{dZ}{dX} \qquad (C.31)$$

C.2.7   Implicit differentiation

Sometimes a quantity may be an *implicit function* of another quantity. For example, in

$$Y^2 - X^2 = 4$$

Y is an implicit function of X. The same equation can be rewritten in an *explicit* form as:

$$Y = (4 + X^2)^{\frac{1}{2}}$$

The derivative of Y with respect to X can now be determined by applying the chain rule:

$$\frac{dY}{dX} = \frac{1}{2}(4 + X^2)^{-\frac{1}{2}}(2X) = \frac{X}{(4 + X^2)^{\frac{1}{2}}}$$

Note that the derivative of Y with respect to X could also be determined directly from the implicit expression, by taking the derivative of both sides of the equation with respect to X:

$$2Y\frac{dY}{dX} - 2X = 0$$

$$\frac{dY}{dX} = \frac{X}{Y} = \frac{X}{(4 + X^2)^{\frac{1}{2}}}$$

C.2.8   Higher derivatives

The derivative of $f'(X)$ of a function $f(X)$ is also a function. The derivative of $f'(X)$ is yet another function, called the *second derivative* of $f(X)$ and denoted by $f''(X)$. Some of the notations used for the second derivative are:

$$f'' \qquad f''(X) \qquad \frac{d^2 f}{dX^2} \qquad Y'' \qquad \frac{d^2 Y}{dX^2}$$

Similarly, $f'''(X)$ refers to the first derivative of $f''(X)$, the second derivative of $f'(X)$, and the third derivative of $f(X)$.

Consider the following examples:

$$Y = 1 + 3X - X^2 + 2X^3$$
$$Y' = 3 - 2X + 6X^2$$
$$Y'' = -2 + 12X$$
$$Y''' = 12$$

$$Y = X\sin(2X)$$
$$Y' = \sin(2X) + 2X\cos(2X)$$
$$Y'' = 2\cos(2X) + 2\cos(2X) - 4X\sin(2X)$$
$$= 4\cos(2X) - 4X\sin(2X)$$
$$Y''' = -8\sin(2X) - 4\sin(2X) - 8X\cos(2X)$$
$$= -12\sin(2X) - 8X\cos(2X)$$

C.3 The Integral

In Section C.2 we concentrated on finding the derivative $f'(X)$ of a given function $f(X)$. We have seen that the differentiation of basic functions is relatively simple and straightforward, and the differentiation of relatively complex functions is possible using the derivatives of basic functions along with the rules for the derivatives of combinations (sums, products, quotients) of functions.

Next we want to determine the functions whose derivatives are known. The reversed operation of differentiation is called *integration*. As compared to differentiation, integration is more difficult. There are no standard product, quotient, or chain rules for integration. In the absence of sufficient rules to integrate combinations of functions, it is a common practice to use integral tables.

Integration has many applications, such as area and volume calculations, work and energy computations.

The integral of a function $Y = f(X)$ with respect to X can be expressed in two ways:

$$\int f(X)\, dX \qquad (C.32)$$

$$\int_a^b f(X)\, dX \qquad (C.33)$$

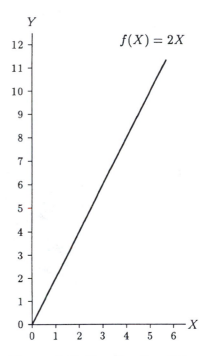

Figure C.12 *Function $Y = 2X$.*

Equation (C.32) is called the *indefinite* integral, and Eq. (C.33) is called the *definite* integral. The integral symbol $\int$ is an elongated "S" which stands for summation, the function $f(X)$ to be integrated is called the *integrand*, dX is an increment in X, and a and b in Eq. (C.33) are called the *lower* and *upper limits* of integration, respectively.

Consider the function $Y = f(X) = 2X$ whose graph is shown in Figure C.12. Also consider $F_1(X) = X^2$. The derivative of $F_1(X)$ is essentially equal to $f(X)$, and therefore, $F_1(X)$ can be the integral of $f(X) = 2X$. Now consider another function $F_2(X) = X^2 + c_0$ where c_0 is a constant. The derivative of $F_2(X)$ is again equal to $f(X) = 2X$ because the derivative of a constant is zero. Note that F_1 is in fact a special form of F_2 for which $c_0 = 0$. Therefore, the indefinite integral of $f(X) = 2X$ is:

$$\int (2X)\, dX = X^2 + c_0$$

Here, c_0 is called the *constant of integration*. Note that the indefinite integral of a function is another function that is not unique. There are different solutions for different values of c_0. These different solutions have parallel graphs (Figure C.13), all of which have slopes expressed by the function $2X$.

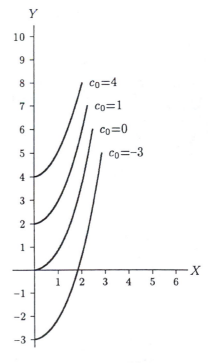

Figure C.13 *Slope of functions $Y = X^2 + c_0$ for different c_0 values is equal to $2X$.*

The definite integral of a given function is unique. To evaluate the definite integral of a function $f(X)$ between a and b, first take the integral of the given function. For definite integrals, the constant of integration cancels out during the course of integration. Therefore, it can be assumed that the constant of integration is zero. If $F(X)$ represents the integral of $f(X)$, then evaluate the values of $F(X)$ at $X = a$ and $X = b$ by subsequently substituting the numerical values of a and b wherever you have X in $F(X)$. In other words, evaluate $F(a)$ and $F(b)$. The definite integral of $f(X)$ between a and b is equal to $F(b)$ minus $F(a)$:

$$\int_a^b f(X)\,dX = F(b) - F(a)$$

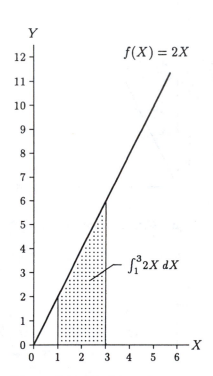

$f(X) = 2X$

$\int_1^3 2X\,dX$

Figure C.14 *The integral represents area.*

For example, consider the function $Y = f(X) = 2X$. Assume that the integral of $f(X)$ between $X = 0$ and $X = 3$ is to be evaluated. That is, $a = 0$ and $b = 3$. Recall that the integral of $f(X) = 2X$ is $F(X) = X^2$ (with $c_0 = 0$). The steps to be followed while evaluating the definite integral are as follows:

Step 1: $$\int_0^3 (2X)\,dX = [X^2]_0^3$$

Step 2: $$= [(3^2) - (0^2)]$$

Step 3: $$= (9 - 0) = 9$$

That is, determine $F(X)$ by taking the integral of the given function (Step 1), evaluate $F(b)$ and $F(a)$ by substituting the upper and lower limits of integration (Step 2), and evaluate the difference $F(b) - F(a)$ (Step 3).

The physical meaning of the definite integral $\int_a^b f(X)\,dX$ is such that it represents the area bounded by the given function $f(X)$, the X axis, and the vertical lines passing through $X = a$ and $X = b$. For example, the integral of function $f(X) = 2X$ between $X = 1$ and $X = 3$ is equal to the shaded area in Figure C.14, which is 8.

C.3.1 Properties of indefinite integrals

Integrals of basic functions:

$$\int dX = X + c_0$$

$$\int c\,dX = cX + c_0 \qquad \text{(constant } c) \qquad \text{(C.34)}$$

$$\int X\,dX = \frac{X^2}{2} + c_0$$

$$\int X^2\,dX = \frac{X^3}{3} + c_0$$

$$\int X^r \, dX = \frac{X^{r+1}}{r+1} + c_0 \qquad (r \neq -1) \qquad \text{(C.35)}$$

$$\int \sin X \, dX = -\cos X + c_0 \qquad \text{(C.36)}$$

$$\int \cos X \, dX = \sin X + c_0 \qquad \text{(C.37)}$$

$$\int \exp X \, dX = \exp X + c_0 \qquad \text{(C.38)}$$

$$\int \frac{1}{X} \, dX = \ln X + c_0 \qquad (X > 0) \qquad \text{(C.39)}$$

Constant multiple and sum rules:

$$\int [cf(X)] \, dX = c \int f(X) \, dX \qquad \text{(C.40)}$$

$$\int [f_1(X) + f_2(X)] \, dX = \int f_1(X) \, dX + \int f_2(X) \, dX \qquad \text{(C.41)}$$

Examples:

$$\int 2\cos X \, dX = 2 \int \cos X \, dX = 2\sin X + c_0$$

$$\int \left(5 - \frac{1}{2} X^3\right) dX = 5 \int dX - \frac{1}{2} \int X^3 \, dX$$

$$= 5(X) - \frac{1}{2} \left(\frac{X^4}{4}\right) + c_0$$

$$= 5X - \frac{1}{8} X^4 + c_0$$

$$\int \left(\frac{1}{X^2} + \sin X\right) dX = \int X^{-2} \, dX + \int \sin X \, dX$$

$$= \left(\frac{X^{-1}}{-1}\right) + (-\cos X) + c_0$$

$$= -\frac{1}{X} - \cos X + c_0$$

C.3.2 Properties of definite integrals

Let $F(X)$ represent the integral of a function $f(X)$ with respect to X. That is:

$$F(X) = \int f(X) \, dX$$

Also let $f_1(X)$ and $f_2(X)$ be two other funtions. Then:

$$\int_a^b f(X) \, dX = F(b) - F(a) \qquad \text{(C.42)}$$

$$\int_a^b [cf(X)] \, dX = c \int_a^b f(X) \, dX = c[F(b) - F(a)] \qquad \text{(C.43)}$$

$$\int_a^b [f_1(X) + f_2(X)]\, dX = \int_a^b f_1(X)\, dX + \int_a^b f_2(X)\, dX \quad \text{(C.44)}$$

$$\int_a^b f(X)\, dX + \int_b^c f(X)\, dX = \int_a^c f(X)\, dX \quad \text{(C.45)}$$

$$\int_a^b f(X)\, dX = -\int_b^a f(X)\, dX \quad \text{(C.46)}$$

Definite integrals of basic functions:

$$\int_a^b dX = [X]_a^b = b - a$$

$$\int_a^b X\, dX = \left[\frac{X^2}{2}\right]_a^b = \left[\frac{b^2}{2} - \frac{a^2}{2}\right] = \frac{1}{2}(b^2 - a^2)$$

$$\int_a^b X^2\, dX = \left[\frac{X^3}{3}\right]_a^b = \left[\frac{b^3}{3} - \frac{a^3}{3}\right] = \frac{1}{3}(b^3 - a^3)$$

$$\int_a^b \sin X\, dX = [-\cos X]_a^b = -\cos b + \cos a$$

$$\int_a^b \cos X\, dX = [\sin X]_a^b = \sin b - \sin a$$

$$\int_a^b \exp X\, dX = [\exp X]_a^b = \exp b - \exp a$$

$$\int_a^b \frac{1}{X}\, dX = [\ln X]_a^b = \ln b = \ln a$$

Examples:

$$\int_{45°}^{90°} \cos X\, dX = [\sin X]_{45°}^{90°} = \sin 90° - \sin 45° = 0.3$$

$$\int_1^2 (4X + 9X^2)\, dX = \int_1^2 (4X)\, dX + \int_1^2 (9X^2)\, dX$$

$$= 4\left[\frac{X^2}{2}\right]_1^2 + 9\left[\frac{X^3}{3}\right]_1^2$$

$$= 2[X^2]_1^2 + 3[X^3]_1^2$$

$$= 2(2^2 - 1^2) + 3(2^3 - 1^3)$$

$$= 2(4 - 1) + 3(8 - 1)$$

$$= 2(3) + 3(7) = 27$$

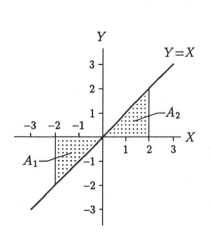

Figure C.15 *Caution while evaluating definite integrals.*

Caution. Consider the linear function $f(X) = X$. The graph of this function is shown in Figure C.15. Assume that the definite integral of this function between $X = -2$ and $X = 2$ will be evaluated. Since the definite integral represents the area enclosed by the function, the X axis, and vertical lines passing through the lower and upper limits, the integration of $f(X) = X$ between $X = -2$ and $X = 2$ should yield the shaded area $(A_1 + A_2)$ in Figure C.15. From the geometry, we can readily determine that $(A_1 = A_2) = 2$, or the total shaded area is equal to 4 units. Now, we evaluate the integral:

$$\int_{-2}^{2} X\, dX = \left[\frac{X^2}{2} \right]_{-2}^{2} = \frac{1}{2}[(2)^2 - (-2)^2] = \frac{1}{2}[4 - 4] = 0$$

This unexpected result occurred because in the interval between the lower and upper limits of integration, the function crosses the X axis, and also that a negative value is calculated for the area designated as A_1 in Figure C.15 during the process of integration. To avoid such a mistake, the proper evaluation of the definite integral should follow the steps provided below:

$$\int_{-2}^{2} X\, dX = \left| \int_{-2}^{0} X\, dX \right| + \left| \int_{0}^{2} X\, dX \right|$$

$$= \left| \left[\frac{X^2}{2} \right]_{-2}^{0} \right| + \left| \left[\frac{X^2}{2} \right]_{0}^{2} \right|$$

$$= \frac{1}{2}\left| (0)^2 - (-2)^2 \right| + \frac{1}{2}\left| (2)^2 - (0)^2 \right|$$

$$= \frac{1}{2}|-4| + \frac{1}{2}|4| = \frac{4}{2} + \frac{4}{2} = 2 + 2 = 4$$

Therefore, before evaluating the definite integral of a function $f(X)$ with respect to X, one must first draw a simple graph of the function to be integrated to check whether the graph of the function crosses the X axis in the interval between the lower and upper limits of integration. For example, if a function $f(X)$ will be integrated with respect to X between a and b, and the curve of the function crosses the X axis at c and d such that $a < c < d < b$, then the integral should first be expressed as:

$$\int_{a}^{b} f(X)\, dX = \left| \int_{a}^{c} f(X)\, dX \right| = \left| \int_{c}^{d} f(X)\, dX \right| = \left| \int_{d}^{b} f(X)\, dX \right|$$

C.3.3 Methods of integration

There is no standard rule for integrating functions in the form of the product and quotient of other functions whose integrals are known. For such functions there are integral tables and a number of methods of integration, including:

- Integration by Substitution

- Integration by Parts

- Integration by Trigonometric Substitution

- Integration by Partial Fraction Decomposition

- Numerical Integration

For example, assume that the following integral will be evaluated:

$$\int 2X \sin(X^2)\, dX$$

The function $2X \sin(X^2)$ can be integrated by observing that $2X$ is the derivative of X^2. If we let $Z = X^2$, then $dZ = 2X\, dX$. By substituting Z and dZ into the given integral, we can write:

Substitution: $$\int 2X \sin(X^2)\, dX = \int \sin Z\, dZ$$

Integration: $$= -\cos Z + c_0$$

Back substitution: $$= -\cos(X^2) + c_0$$

The method used here is called "integration by substitution." Detailed descriptions of these methods are beyond the scope of this text. For further information, the interested reader should refer to calculus textbooks.

C.4 Trigonometric Identities

Negative angle formulas:

$$\sin(-X) = -\sin(X)$$
$$\cos(-X) = \cos(X)$$

Addition formulas:

$$\sin(X + Y) = \sin(X)\cos(Y) + \cos(X)\sin(Y)$$
$$\sin(X - Y) = \sin(X)\cos(Y) - \cos(X)\sin(Y)$$
$$\cos(X + Y) = \cos(X)\cos(Y) - \sin(X)\sin(Y)$$
$$\cos(X - Y) = \cos(X)\cos(Y) + \sin(X)\sin(Y)$$

Product formulas:

$$2\sin(X)\cos(Y) = \sin(X + Y) + \sin(X - Y)$$
$$2\cos(X)\sin(Y) = \sin(X + Y) - \sin(X - Y)$$
$$2\cos(X)\cos(Y) = \cos(X + Y) + \cos(X - Y)$$
$$2\sin(X)\sin(Y) = \cos(X - Y) - \cos(X + Y)$$

Factoring formulas:

$$\sin(X) + \sin(Y) = 2\cos\left(\frac{X - Y}{2}\right)\sin\left(\frac{X + Y}{2}\right)$$

$$\sin(X) - \sin(Y) = 2\cos\left(\frac{X+Y}{2}\right)\sin\left(\frac{X-Y}{2}\right)$$

$$\cos(X) + \cos(Y) = 2\cos\left(\frac{X+Y}{2}\right)\cos\left(\frac{X-Y}{2}\right)$$

$$\cos(X) - \cos(Y) = 2\sin\left(\frac{X+Y}{2}\right)\sin\left(\frac{X-Y}{2}\right)$$

Double-angle formulas:

$$\sin(2X) = 2\sin(X)\cos(X)$$
$$\cos(2X) = \cos^2(X) - \sin^2(X)$$

Half-angle formulas:

$$2\sin^2\left(\frac{X}{2}\right) = 1 - \cos(X)$$

$$2\cos^2\left(\frac{X}{2}\right) = 1 + \cos(X)$$

Pythagorean identity:

$$\sin^2(X) + \cos^2(X) = 1$$

Reduction formulas:

$$\cos\left(\frac{\pi}{2} - X\right) = \sin(X) \qquad \cos(\pi - X) = -\cos(X)$$

$$\sin\left(\frac{\pi}{2} - X\right) = \cos(X) \qquad \sin(\pi - X) = \sin(X)$$

Sine and cosine are the basic trigonometric functions. Other trigonometric functions, namely, the tangent (tan), cotangent (cot), secant (sec), and cosecant (csc), can be derived from sine and cosine using the following definitions:

$$\tan(X) = \frac{\sin(X)}{\cos(X)} \qquad \cot(X) = \frac{\cos(X)}{\sin(X)} = \frac{1}{\tan(X)}$$

$$\sec(X) = \frac{1}{\cos(X)} \qquad \csc(X) = \frac{1}{\sin(X)}$$

C.5 The Quadratic Formula

An algebraic equation such as $3x = 7$ is a *linear equation* because the unknown x appears to the first power. This equation can be solved to determine that $x = 7/3$. An equation such as $x^2 + 3x = 5$, on the other hand, is a *quadratic equation*. In this case, the highest power of x is two. Note that $x^2 + 3x^{0.5} = 5$ is not a quadratic equation, because of the term that carries $x^{0.5}$.

Quadratic equations can be written in the following general form:

$$ax^2 + bx + c = 0 \qquad \text{(C.47)}$$

In Eq. (C.47), a, b, and c are some known numbers, and x is the unknown parameter. The general solution of Eq. (C.47) for x is:

$$x = \frac{-b \pm \sqrt{b^2 - 4ac}}{2a} \qquad \text{(C.48)}$$

The $\pm$ sign indicates that there are two solutions for x.

Caution. Note that in Eq. (C.48), b^2 must be greater than or equal to $4ac$, so that $(b^2 - 4ac)$ has a positive value. Otherwise, $(b^2 - 4ac)$ will have a negative value. The square root of a negative number is an "imaginary" number as opposed to a "real" number.

For example, assume that the following quadratic equation will be solved for x:

$$x^2 + 2x = 8$$

First, rewrite this equation in the form given in Eq. (C.47) by subtracting 8 from both sides:

$$x^2 + 2x - 8 = 0$$

Compare this with Eq. (C.47) to observe that:

$$a = 1 \qquad b = 2 \qquad c = -8$$

Substitute a, b, and c into Eq. (C.48):

$$x = \frac{-(2) \pm \sqrt{(2^2) - 4(1)(-8)}}{2(1)} = \frac{-2 \pm \sqrt{36}}{2} = \frac{-2 \pm 6}{2}$$

Consider the plus sign:

$$x = \frac{-2 + 6}{2} = \frac{4}{2} = 2$$

Consider the minus sign:

$$x = \frac{-2 - 6}{2} = \frac{-8}{2} = -4$$

Note that $x = 2$ and $x = -4$ are the "roots" of the quadratic equation. The original equation can now be expressed as:

$$(x - 2)(x + 4) = 0$$

C.6 Exercise Problems

Problem C.1 Show that slopes of the graph of the function $Y = 1 - X^2$ at $X = -2$, $X = 0$, and $X = 2$ are $4, 0$, and -4, respectively.

Problem C.2 Evaluate the derivatives of the following functions with respect to X.

Functions	Answers
$Y = X^2 \sin X$	$2X \sin X + X^2 \cos X$
$Y = \dfrac{X}{\cos X}$	$\dfrac{\cos X + X \sin X}{\cos^2 X}$
$Y = \dfrac{\cos X}{X}$	$\dfrac{-X \sin X - \cos X}{X^2}$
$Y = \sqrt{X - X^3}$	$\dfrac{1 - 3X^2}{2\sqrt{X - X^3}}$

Problem C.3 Show that the integral of the function $Y = \cos X$ with respect to X between $X = 0°$ and $X = 180°$ is 2. (Examine the graph of the function before evaluating the integral.)

Problem C.4 Evaluate the following integrals.

Integrals	Answers
$\int (X - \sin X) \, dX$	$\dfrac{X^2}{2} + \cos X + c_0$
$\int \left(1 - \dfrac{1}{X^2}\right) dX$	$X + \dfrac{1}{X}$
$\int_1^2 \left(1 - \dfrac{1}{X^2}\right) dX$	0.5
$\int_{0°}^{45°} (\cos X + \sin X) \, dX$	1

Problem C.5 Show that the roots of the quadratic equation $x^2 - 7.5x = 4$ are $x = 8$ and $x = -0.5$.

Index